2012 A Family Brief

2012 A Family Brief

"*The Science is all in*"

Rogue Saber

To order additional copies of this book, contact:
Xlibris Corporation
1-888-795-4274
www.Xlibris.com
Orders@Xlibris.com
64966

I would like to state that there are really no chapters in this book to separate it's information. It was written as a briefing document for my own personal family. But, after giving it to them, it was deemed that it should be published for all of man-kind to receive as they too are all part of my human family.—(Sabre)

30% of this document contains copied information from *www. projectcamelot.org*, and it is untouched.

50% of this document was copied from *www.projectavalon.net*. This is a site also run by *www.projectcamelot.org*, and sourced from a book called (The Nexus Report—"An Unknown Form Of Energy Comes Our Way") written by an individual under the pseudonym "Astralwalker".

Finally, there is my own 20% that is just a summation of the surrounding material put into my own words. There are web sites sited all over this document, and I believe all credit has been given where credit is due. It is not my intention to take any credit for the works that were copied to be presented in this medium. It is my only concern that this information be brought out into the open for the public to read and decide for themselves what it is that is really going on around us all on this little blue planet we all call "Home".

Other works contained within this briefing are sited from *www.exopolitics. com*, *www.disclosureproject.com*, *www.cseti.com*, *www.americanantigravity. com*, *www.nasa.gov*, *www.eaglesdisobey.net*, and Dr. Richard Sauder from his book, ***"Underwater and Underground Bases"***.

Please, I would like to make sure that all of the appropriate permissions and credits are given to the right people in association with the right works, expressing thanks and gratitude for all who have worked so hard to find the truth.—(Sabre) would be the explanation of the facts of the case. On the other hand, wave theory of light had many applications and was indeed becoming the accepted theory. But after giving it some thought, he realized that the two applications, all of these would be impossible to reconcile in the present state of the question.

Forward

In 1994 I met an individual through a mutual friend with whom I had started to build an interesting relationship with. This young man, whose name escapes me now, was about 3 to 5 years younger than me, but we shared a common interest that brought us both together, that interest was in Extraterrestrials and UFOs.

This is the interesting part of our friendship because his father was, at the time, a Colonel in the AIR-FORCE. So for a few weeks I would go to his house and he would go into his father's office area and bring out this little blue book that had a complete disclosure of the ET presence. Along with mostly hand drawn pictures, this book also had a photograph of an alien who looked exactly like a human in it. The alien was sitting at a table wearing a normal business suit with other men. He looked to have blond hair and blue eyes. His hair was full like that of a man in his late 20's. He was very handsome, and quite frankly now that I think about it, he looked a lot like James Dean, the famous Hollywood actor. The caption under the photo said that the aliens name was Val Valiant Thor that he hailed from the planet Mercury and he flew here in a ship named the Victory One.

This book disclosed the fact that an army was being built underground in a Colorado laboratory combining cows and men to create a giant zombie warrior of some kind. Some of the details I have forgotten, but the one thing I will never forget is the drawings that were contained in that little light blue book that we used to read and talk about every single day in 1994. Amazingly, I don't know how they ended up on the web, but they did. Now either this young man has grown up and posted these pictures on the web himself, or someone else had this book as well, and they posted them. None the less that's where I copied them, from off of the web, ***these are those very pictures.***

ROOM LIGHT: PINK-PURPLE
BRIGHT IN SOME AREAS

HUNDREDS OF THESE IN
VARIOUS STAGES OF GROWTH.

WISPY HAIR, "ALMOST NOSE"
MOUTH LOOKS "SEALED"

WOMB LOOKS GREY
VEINS (?) LOOK DARK GREY
CREATURE WHITE-PALE
EYES - DARK LIDS (?)
CAN'T FIND GENDER
2 TOES - 3 FINGERS

LIQUID - AMBER COLOR
NOT COMPLETELY CLEAR

LOOKS LIKE GLASS TUBE,
BUT ABOUT 5 FT TALL

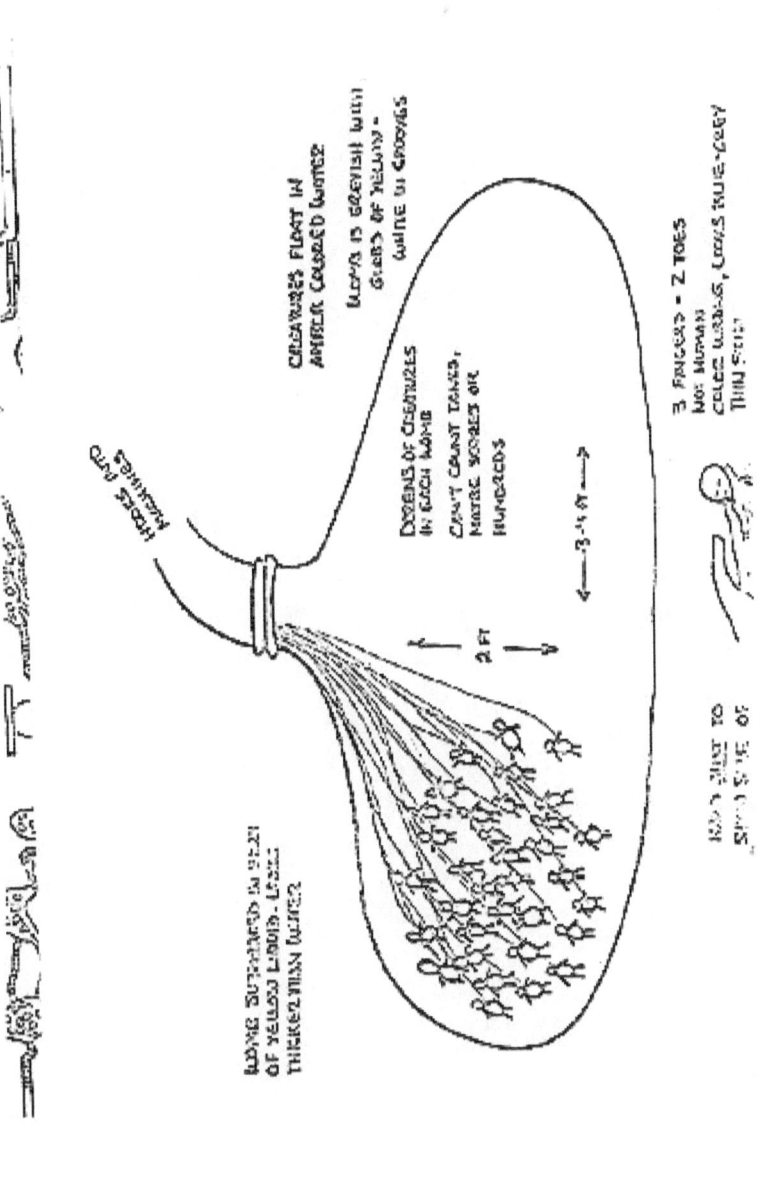

CREATURES FLOAT IN AMBER COLORED WATER

BODY IS GREYISH WITH GOBBS OF YELLOW — ORANGE IN GROOVES

DOZENS OF CREATURES IN EACH TUBE

CAN'T COUNT TENTACLES, MAYBE SCORES OR HUNDREDS

←— ~3-4 FT —→

←→ 2 FT

LONG SUSPENDED IN VEIN OF YELLOW LIQUID. LESS TRANSLUCENT WATER.

HOLES INTO MACHINES

3 FINGERED - 2 TOES NO THUMBS COLOR VARIES, LOOKS BLUE-GREY THIN SKIN

HOW FAST TO SOLVE ....

Introduction

"The quest for true knowledge begins for me as far back as I can remember, and the tastes that I have been given have only fueled my interests, driving my thirst for it into unquenchable limits."—(Sabre)

This book began as an attempt to build a briefing document for my personal family members concerning the 2012 "End of time" scenario. My hope was to let my family know what the Military was up to and then work with them to devise a plan of similar action with the hopes of doing exactly as the Military was planning to do . . . Survive.

In 2008 I was already well aware of the Mayan Calender's end date for the current time, and start date of the "New Era", to be ushered in with the great alignment of galactic center, which didn't sound too menacing. But, what I was now hearing from people around me were dooms-day scenarios about a pole flip that would sink Continents, induce earthquakes, cause volcanic eruptions, and include the collapse of the great oceanic Atlantic conveyor that was to be followed up by a semi-global ice-age brought on by the failure of the oceans movements of waters.

Wow. That was quite a drastically different start on the way this "New Era" was supposed to begin than I was led to believe it would begin. So, I began to research this material further for myself to find out just what exactly we did know scientifically and what its astrological implications were in association with this whole "Galactic Alignment" reference.

My first step was to look at the tectonic plates of this planet to locate fault lines and identify the most stable Continent available that would provide the greatest probability for survival. To my surprise North America had been probed and was listed as the best location on the globe to be as far as a stable land mass was concerned.

The next thing I considered were the locations of all of the known volcanoes on the planet and followed that up by finding out which ones were dormant or currently active. This information led me back to the tectonic plates to understand how the earth's mantle is formed by hot molten liquid magma. This in turn led me back to the "Ring of Fire", which is a formation of volcanoes that create a circle. Starting from Hawaii, they are found all along the Eastern shores of Russia up to the Polar Ice Caps edge stringing themselves across the Great Divide to Alaska and down the Western shores of Canada to North America, to Mexico and finally into South America where, if you head back out to sea, you would find the start and end point of this circle, once again Hawaii.

Now this is curious, as I had continued to investigate, I found that in some instances these dormant volcanoes had been re-engineered by 1st world governments and were now being used as underground bases. The magma had been rerouted by modern tunneling techniques and would no longer be permitted to escape the depths of the underworld through their normal routes. This meant that it was rechanneled to join existing underground lava flows to escape through other exits on the surface of the planet. That means that the next time a volcano blows it's top it has the potential to do at least twice the damage or more than it did before in the area of it's geographical location.

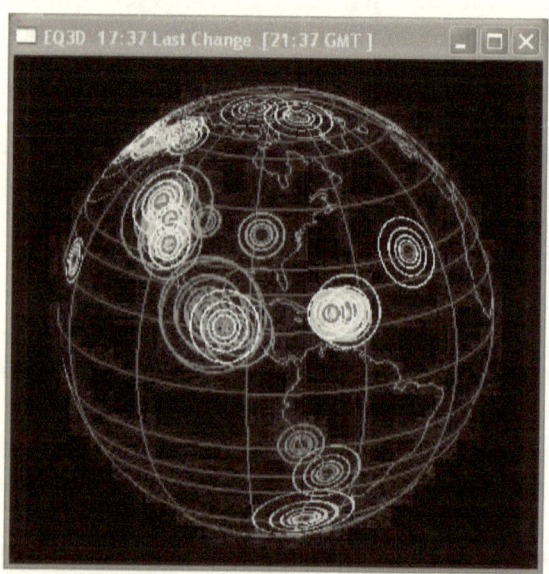

Now this led me into thinking about Mount Shasta in California, and it's ties to Telos City, which is rumored to be a huge base underneath Mount Shasta. These are just my personal thoughts, but after looking at this satellite imaging of both volcanic, (Red & Pink) and earthquake activity, (White), I have come to the logical conclusion that even though it looks like Shasta is active it is not, that's Mount Rainier, and Mount St. Helen. That has got to be why Saint Helen erupted back in the 80's. They must have rerouted other volcanoes into her tunnel system and when they did, of course there was going to be a little release from the excess pressure. But as I said, this is just speculation on my part.

Ok, back to the task at hand, trying to survive extinction level catastrophes'. From here the next logical thing to look at was, if some of these volcanoes blew their tops, and they did it in culmination or even in succession to one another, sulfur-dioxide and just volcanic ash fall out could potentially kill everyone and everything in the Northern Hemisphere. At this point the 1,033 foot Anak-Krackatoa is being surveyed and is currently known as the most dangerous volcano on the planet. The last time this volcano erupted was 1883 and the sound was recorded as being 13,000 times louder than the H-bomb that was dropped on Hiroshima blowing dirt and ash 17 miles into the sky. Now this volcano has been guaranteed by geologists to erupt again. Imagine if only 10 of these volcanoes were to erupt, but they all did it in the same week continuing to erupt on and off for a week each.

How do you prepare for something like that? You dig! You dig deep, and you dig hard, and that is exactly what your governments have been doing. They have been spending trillions doing it and in most cases have sold their entire countries to enable them to do so.

I have spent a great deal of time and energy wading through all of the subrogation discerning credible information from disinformation and misinformation and believe me there is a pile of it out there to be gone through before one can finally get to the bottom of it all.

One of my objectives was to get myself on the cutting edge of what it was that was really going on around this planet before I could attempt to inform anyone else about what I thought it was. And believe me, I am not going to even remotely try to insinuate that I have it all figured out, because there is just too much going on in small projects that are working in concert with larger ones, but ultimately they are all tied back into the theaters main project, which is, to preserve the destiny of Man-Kind.

I urge you to read this material for yourself knowing that this information was sited from the most credible sources on the planet available to date. I have done all the wading that needed to be done for you so that we can now "SEE" the path that we have been being led down for God only knows how long. Follow these sources!

This is a path that leads all the way back to the beginning of man-kind in a place called Sumer, a place where real science was taught to mankind by those who have taken credit for creating us. Once you read what is contained with-in this book you should be able to see right through the tapestry of nonsense that has misguided a large number of people on this planet for a long, long time.

This is where my quest for truth really begins and it is by no means finished. I look very much forward to the coming of the Galactic Alignment in 2012 and the abilities and freedom it will bring to us all as a race of sentient beings on this little blue planet we call home.

I want to thank Dr. Stephen Greer, *www.disclosureproject.com*, *www. cseti.com* for all that he has started and continues to pursue and I also want to thank Bill Ryan and Kerry Cassidy, *www.projectcamelot.org*, *www.projectavalon.net* for their contribution of persistence to the cause, also Micheal Salla Ph.D for continuing to "lay it all on the line" at *www. exopolitics.org*, But I would like to give a special thanks to Dr.s Burisch and McDowell, *www.eaglesdisobey.net* for everything that they continue to do especially for those less fortunate. To my cousin Bob who encouraged me publish all of this, and to my wife, who is, and has continued to be, all that I've ever asked God for. To those of you continuing work in the Black Budget Projects, remember who you speak for, lest when they come for you, who will speak up for you. Keep coming forward to report information on current operations; you are man-kinds front line. I pray for all of you and your continued success in your respective fields, may GOD bless you all.—(Sabre)

2012
A FAMILY BRIEF

"The science is all in."

(What follows is a paper written by Dr. Michael E. Salla, describing how the existence of Extraterrestrials was dealt with and how the back engineering projects were covered up and kept from Congress.)—Sabre

POLITICAL MANAGEMENT OF THE EXTRATERRESTRIAL PRESENCE: THE CHALLENGE TO DEMOCRACY AND LIBERTY IN AMERICA
Research Study # 5
Published July 4, 2003, www.exopolitics.org
© Michael E. Salla, PhD

ABSTRACT

Political management of the extraterrestrial (ET) presence on Earth has evolved during the course of the successive US Presidential administrations that had to deal with the policy issues that arose once irrefutable proof of such a presence was brought to the attention of policy makers This evolution of responses by Presidential administrations can be broken into five historic phases which demonstrate a gradual erosion of Presidential executive oversight, and the growing autonomy/independence of clandestine organizations embedded in military-intelligence and national security branches of government. This erosion of Presidential/executive oversight in the form of ultimate control of the ET presence being gradually taken away from elected Presidents and/or their congressionally approved political appointments appears to have all the characteristics of a 'political coup' Presidents, in the case of Republicans, have been reduced to rubber stamps for those controlling political management of ET issues; or, in the case of Democratic Presidents, to political irrelevance.

In this paper, I will identify the political management approach taken in the various phases of how US administrations/clandestine organizations have responded to the ET presence. I will analyze the gradual erosion of executive oversight of these clandestine organizations, and the latter's increasing influence over the executive branch of government. I will also outline the increased role of Corporate America and the Council of Foreign Relations in politically managing the ET presence. I further examine the most current political management style as evidenced by the US led military intervention into Iraq and what this suggests for how the ET presence will be managed in the future. I conclude by identifying how political management of the ET presence has been conducted in a way that represents a threat to the principles of democracy and liberty in the US.

Dr. Michael E. Salla has held academic appointments in the School of International Service, American University, Washington DC (1996-2001), and the Department of Political Science, Australian National University, Canberra, Australia (1994-96) He taught as an adjunct faculty member at George Washington University, Washington DC., in 2002. He is currently researching methods of Transformational Peace as a Researcher in Residence in the Center for Global Peace (2001-2003) and directing the Center's Peace Ambassador Program which uses transformational peace techniques for individual self-empowerment. He has a PhD in Government from the University of Queensland, Australia, and an MA in Philosophy from the University of Melbourne, Australia he is the author of *The Hero Journey Toward a Second American Century (Greenwood Press, 2002) and co-editor/author of three other books, and authored more than seventy articles, chapters, and book reviews on peace, ethnic conflict and conflict resolution. He has conducted research and fieldwork in the ethnic conflicts in east Timor, Kosovo, Macedonia, and Sri Lanka. He has organized a number of international workshops involving mid to high level participants from these conflicts. He has an academic website at* ***http://www.american. edu/salla/***

Political Management of the Extraterrestrial Presence—
The Challenge to Democracy and Liberty in America [1]

Introduction

Political management of the extraterrestrial (ET) presence on Earth has evolved during the course of the successive US Presidential administrations that had to deal with the policy issues that arose once irrefutable proof of such a presence was brought to the attention of policy makers. This evolution of responses by Presidential administration can be broken into five historic phases which demonstrate a gradual erosion of Presidential/executive oversight, and the growing autonomy/independence of clandestine organizations embedded in military-intelligence and national security branches of government. This erosion of Presidential/executive oversight in the form of ultimate control of the ET presence being gradually taken away from elected Presidents and/or their congressionally approved political appointments appears to have all the characteristics of a 'political coup'. [2] Presidents, in the case of Republicans, have been reduced to rubber stamps for those controlling political management of ET issues; or, in the case of democratic Presidents, to political irrelevance.

The first political management phase was crisis management during the Roosevelt administration when the ET presence became enmeshed within the foreign policy crisis of the Second World War, and was effectively controlled by a complex of scientific-military institutions set up under executive oversight to conduct the war effort. The second phase was an effort by the Truman administration in the post-war era to establish a framework for politically managing the ET presence through a series of ad hoc committees responsible for setting policy and coordinating response by the scientific-military-intelligence communities.

The third phase was a comprehensive effort by the Eisenhower administration to manage the ET presence through improved policy coordination between different clandestine organizations embedded within military/intelligence and national security branches of government, and introducing more prominent roles for Corporate America and foreign policy elite's in the political management of the ET presence.

The fourth phase was the effective loss of direct Presidential/Executive oversight during the latter part of the Eisenhower administration and the Kennedy administration, and the independence of clandestine organizations created to deal with the ET presence. There is evidence to suggest that the recent military campaign in Iraq marks a disturbing fifth phase in the political management of the ET presence where clandestine organizations effectively take control of a foreign government for the exclusive purpose of managing the ET presence.

For the purpose of this study, 'political management' will be defined as a coordinated series of policies for dealing with a set of issues that have important public policy implications In the case of political management of the ET presence, this refers to the need for developing a coordinated and strategic approach to the ET presence that satisfactorily deals with all its public policy dimensions. More importantly, political management of the ET presence involves coordinating the various agendas, reverse engineering programs, covert military operations, intelligence gathering operations, and policy studies undertaken by a variety of clandestine organizations embedded within the military, intelligence and national security branches of government.

Chief among the clandestine organizations to be identified as taking the lead in politically managing the ET presence in the US is Majestic 12' (MJ-12—aka 'PI-40' and 'Special Studies Group') is embedded within the Covert Operations Committee of the National Security Council. [3] The prominence of Nelson Rockefeller and Dr. Henry Kissinger in influencing this clandestine organization, and the supporting roles given to Corporate America and elite policy study groups such as the Council of Foreign Relations, gives important insight into how the political management of the UFO presence has been historically conducted. This casts considerable light on motivations for the recent military intervention in Iraq, and the likelihood that this marks an important watershed in the political management of the ET presence.

In this paper, I will identify the political management approach taken in the various phases of how US administrations/clandestine organizations have responded to the ET presence. I will analyze the gradual erosion of executive oversight of these clandestine organizations, and the latter's increasing influence over the executive branch of government. I will also outline the increased role of Corporate America and the Council of Foreign Relations in politically managing the ET presence. I further examine the most current political management style as evidenced by the US led military intervention into Iraq and what this suggests for how the ET presence will be managed in the future. I conclude by identifying how political management of the ET presence has been conducted in a way that represents a threat to the principles of democracy and liberty in the US.

Phase One—Crisis Management of the Extraterrestrial Presence

Evidence for the emergence of ET piloted craft over US skies has been claimed to exist from as early as the 19th Century. [4] The emergence of the ET presence as a phenomenon that required political management can be dated to the first instance where US policy makers had to grapple with irrefutable evidence of an ET presence and its tremendous policy implications. There is evidence from 'whistle blower' testimonies from clandestine government organizations that an ET craft crashed off the coast of California in 1941, and its secret retrieval was what initially set off efforts by the Roosevelt administration to politically manage the ET presence. [5] Furthermore, a famous incident in 1942 occurred where there was a naval bombardment in response to what at first perceived to be a Japanese air raid, but which closer evidence suggested was an intelligently piloted UFO. [6] US participation in the Second World War from 1942 meant that these astonishing events involving ET piloted spacecraft craft had to be politically managed in the context of a global military conflict requiring coordinated policy responses that involved the nation al survival of the US.

The approach taken by the Roosevelt administration was based on maintaining tight secrecy given the assumption that the 'enemy'—Japan and Nazi Germany—would take any advantage of the ET presence it could discover through its intelligence assets to bring about defeat of US forces on the battlefield. The Roosevelt administration delegated control of the ET presence to the US Department of War (renamed the Department of Defense) who were immediately aware of the military significance of such a development. Any technology and knowledge acquired from the ET presence would be used to develop weapons technology that could bring victory on the battlefield.

There is considerable evidence from witness testimonies that the US Navy led this clandestine military effort with a top secret project seeking to develop 'stealth technologies' for Navy ships from as early as 1943. [7] Dubbed the 'Philadelphia Experiment' this project established the primacy of the US Navy in reverse engineering ET technology, and subordination of the whole ET issue within military institutions. An important characteristic of this phase of the political management of the

ET presence was the leading role played by military funded scientific laboratories that would play the critical role of reverse engineering ET technology This 'military-scientific complex' was critical to the war effort and for responding to the ET presence.

As Commander-in-Chief in a war time situation, President Roosevelt's political management of the ET presence was synonymous with the political management of the Manhattan Project which produced the first atomic bombs, and other secret weapons technologies that were part of the war effort. Both secrecy and a clear chain of command was required, and no effort would be spared to fund scientific laboratories working directly under the military to utilize this ET 'presence' for battlefield success. There is no evidence that the Roosevelt administration developed any special organizational structures for dealing with the ET presence, other than simply subordinating the whole ET issue to the Department of War that was conducting the war effort. As Commander in Chief, Roosevelt and his most senior advisors would be extensively informed and played the key role in exercising the necessary executive oversight for military projects utilizing technology and intelligence gained from ET sources. Most importantly, the Second World War meant there would be no congressional oversight of the ET presence since the latter's existence and military significance required the utmost secrecy due to dire national security implications it had for the War effort.

Phase Two—The Truman Administration and the Decision to Maintain Secrecy

When Harry Truman became President in 1945, a 'successful' outcome of the Second World War was already clear. This meant that a more organized institutional structure could be developed for politically managing the ET presence. The national security threat to the US was now over as f as the general public were concerned, which meant that there would soon be pressure for Congressional oversight and public disclosure of the clandestine programs conducted by military and intelligence branches of government. Since the US military exercised complete operational and logistical control over all aspects of the ET presence during the emergency conditions of the war, there needed to be a process for deciding how to politically manage the extensive policy implications of such a presence. Undoubtedly, the first policy issue to be confronted was the extent to which the ET presence should be disclosed to Congress and the general public. More importantly, there needed to be an institution created for ensuring policy coordination between the different military and intelligence units that were working on different operational aspects of the ET presence, and, critically, a means of ensuring that the President and his principal advisors would be sufficiently well informed to maintain executive oversight of the entire military-scientific-intelligence community that interfaced with the ET presence.

In 1947, Truman gave executive approval in the form of a memo to then Secretary of Defense, James Forrestal, for the creation of a clandestine committee to be formed that would play these three crucial roles of politically managing public disclosure of the ET presence, policy coordination of the various projects associated with the ET presence, and executive oversight of clandestine organizations dealing with the ET presence. [8] Titled Majestic 12 (MJ-12), this group initially comprised 12 senior individuals from the military, intelligence and civilian sectors who formed an ad hoc committee. MJ-12 was therefore a clandestine political entity created to politically manage all aspects of the ET presence in order to provide the best policy advice to the President. Significantly, MJ-12 was embedded within the National Security Council which was formed at the same time to coordinate policy recommendations from different government, military and intelligence departments into a coherent set of policy recommendations from which Presidents could choose. The way in which policy advice would be gained was through a series of ad hoc committees that would be formed to investigate specific aspects of the ET presence to make policy issues concerning the ET presence on the basis of policy advice gained from top secret ad hoc committees.

A factor which increasingly impacted on the political management of the ET presence was the rise in public sightings of UFO craft in the post-war period making it more difficult to maintain a public policy of secrecy. In 1947, there was an extraordinary increase in the number of public sightings of UFO craft which led to a groundswell of support of an official response, and public disclosure of the ET presence. [10] The most famous of these was the 1947 Roswell incident which has spawned numerous books and testimonies from various individuals and officials. [11] This led to the decision of the Air Force to launch an official public examination of the UFO presence. Project Blue Book began in 1952 and was the official public successor to earlier Air Force investigations of UFOs from 1947-48 (Project Sign), and 1948-52 (Project Grudge). [12]

In the midst of the clamor generated by the public for news on the UFO presence, the Truman administration was evidently advised by the MJ-12 committee to maintain strict secrecy of the ET presence, while continuing to exercise executive oversight of clandestine projects concerning ET activity and their technology. Evidence of the nature of the executive control of the ET presence and the decision by MJ-12 to maintain public secrecy can be found in the circumstances surrounding the dismissal and death of James Forrestal as Truman's Secretary of Defense in 1949. Secretary Forrestal had, according to whistleblower testimonies, developed a clear difference of opinion on how the ET presence should be politically managed, and was said to have favored public disclosure. [13] Forrestal, who was a member of the MJ-12 committee, was thwarted by President Truman, his principal advisors and others on the MJ-12 committee, who decided that the whole ET presence had to be politically managed in a way that maintained strict secrecy, thereby denying the general public and Congress the truth about the ET presence. Forrestal was dismissed due to what was officially claimed to be a 'nervous breakdown' and later 'committed suicide' from the 6th floor of the Bethesda Naval Hospital. [14] According to several military 'whistleblowers', Forrestal was murdered [15].

In conclusion, the political management of the ET presence by the Truman administration was one of firm executive oversight where he would be advised by his appointed committees such as MJ-12 in how to deal with the ET presence. MJ-12 would provide policy recommendations for coordination and oversight of clandestine organizations embedded in military and intelligence departments, and the military funded scientific laboratories that pursued reverse-engineering programs and communication with ET's.

Phase Three—The Erosion of Executive Oversight of Clandestine Organizations

The election of Dwight Eisenhower in 1952, brought with it not only a Republican administration, but also a profound policy shift in how political management of the ET presence would be conducted—the formal involvement of Corporate America and the Council of Foreign Relations in managing the ET presence. Eisenhower had been supported in his Presidential campaign by the Rockefeller family and it was therefore no great surprise that he chose Nelson Rockefeller to be in charge of reorganizing the government. Rockefeller from 1953-59 was Chairman of the President's 'Advisory Committee on Government Reorganization.' In addition, he became the President's Special Assistant in Cold War Strategy (1954-55) and was critical in shaping the Eisenhower's views and responses to the ET presence.

The Rockefeller family derived much of its wealth and influence from the Standard Oil Company established by John Rockefeller which established a powerful monopoly in the US oil industry whose legacy continues today under the Exxon/Mobil/Chevron banners. [16] Nelson Rockefeller, the grandson of John Rockefeller, was a 'moderate Republican' who was a liberal in political issues and strongly supported the liberal internationalist idea of a global political institution, but conservative in the economic sphere. [17] In asking Nelson Rockefeller to advise him and reorganize government in general and the policy making infrastructure concerning the ET presence in particular, Eisenhower was giving Corporate America a prominent role in the way in which government attempted to address policy issues—a view consistent with the ideological underpinning of the US Republican party.

As far as the ET question was concerned, this meant that Corporate America would play a prominent role in the clandestine efforts to reverse engineer ET technology. [18] The immediate consequence was that the scientific laboratories that were previously directly funded by the Department of Defense were reorganized in terms of their location and funding base. These laboratories now received corporate funding though contracts awarded by military organizations, rather than being directly funded by the military as was the case during the Second World War and the Truman administration. Including Corporate America provided the important benefit of introducing a further layer of secrecy that could effectively keep prying Congressmen away from the truth about the ET presence. The Congressional oversight that was, in theory at least, possible for government/military funded scientific laboratories working on reverse engineering ET technology, would be impossible with corporations nominally in charge of the scientific laboratories working on the same clandestine military projects, using the same personnel, resources and funding. With Project Blue book underway and Congress attempting to discover what was really happening concerning the ET' presence, a thorough reorganization involving a prominent role for Corporate America, in Rockefeller's view, was needed if secrecy was to be maintained. The 'sleight of hand' involving Corporate America provided an important means of politically managing the ET presence—total secrecy could be maintained by simply invoking the mantra of private sector market forces, thereby ensuring immunity from congressional investigation.

Another important policy shift was the inclusion of the Council of Foreign Relations as the source for suitable recruits for a top secret policy committee whose exclusive task was to provide policy recommendations of the various political, economic, social, religious and legal issues concerning the ET presence. The Rockefeller family became important benefactors in the establishment of the Council of Foreign Relations in 1921 by making significant yearly donations of $1,500; making a large donation of $50,000 for the Council's new headquarters in 1929, donating the building that became the headquarters of the Council in 1945. [19] A measure of the Rockefeller influence could be seen in their support for individuals being appointed to powerful positions. In the early 1970's, for example, David Rockefeller, who eventually became Chairman of the Council, went against the wishes of a nominating committee to appoint William Bundy to the editorship of the influential journal, *Foreign Affairs*. [20] By bringing the Council of Foreign Relations to the center stage of how the Eisenhower administration would gain recruits for clandestine organizations designed to make policy recommendations concerning the ET presence, Nelson Rockefeller had maneuvered himself and his family to the center stage of how the ET presence would be politically managed.

The most significant institutional reorganization as far as political management of the ET presence was concerned was expanding and formalizing MJ-12 as an autonomous institution fully authorized by executive order to deliberate upon and make policy decisions on the ET presence. MJ-12 became formally embedded in the Covert Operations Committee of the National Security Council—Committee 5412, named after National Security Council Edict 5412. MJ-12's earlier existence as an ad hoc committee appointed by executive authority, was now transformed into a permanent subcommittee institutionally embedded within the most secret of all the National Security Council's committees. Evidence from whistleblower testimonies suggest that Truman's ad hoc committee, MJ-12, was reorganized so as to now comprise two layers. [21] The outermost layer was a group of up to 40 individuals who would form a Study Group (hence the names PI-40 and Special Studies Group also attributed to MJ-12) whose function was to provide specialized studies and policy recommendations concerning ET issues for a smaller decision making group (MJ-12) that would actually make official policy recommendations for implementation after gaining executive approval by Eisenhower.

The Special Studies Group/PI-40 formed under Eisenhower held their first meetings at Quantico Marine Base in Virginia and its 35 members were drawn exclusively from the Council of Foreign Relations. The Study Group had two directors Henry Kissinger and Zbigniew Brzezinski, and comprised prominent individuals as Dr Edward Teller, Paul Nitze, David Rockefeller, and McGeorge Bundy (later Kennedy's Special Assistant for Foreign Affairs). [22] According to Cooper, the Rockefeller's built a lavish retreat For the Study Group in an exclusive area in Maryland. [23]

MJ-12 comprised 19 individuals, reviewed the various studies and deliberated on policy issues concerning the ET presence on the basis of a qualified majority system of 12 votes being necessary for an issue to be passed. [24] According to William Cooper, who served on the Naval Intelligence briefing team for the Commander of the Pacific Fleet, this smaller group was headed by the President's Special Representative for Foreign Affairs, and its composition was determined according to the following formula. The President's Special Representative for Cold War Strategy (aka National Security Advisor—Nelson Rockefeller); the Director of Central Intelligence (Allen Welsh Dulles); Secretary of State (John Foster Dulles); Chairman of the Joint Chiefs of Staff (Admiral Arthur Radford); Director of the FBI (J. Edgar Hoover); six men from the executive of the Council on Foreign Relations ('Wise Men') and six men from a secret scientific organization called the JASON group (led by Dr. Edward Teller) that were all members of the Council of Foreign Relations. [25] It is likely that among the main organizational rules governing membership of MJ-12/PI-40 were that all appointments had to be approved by MJ-12; that individuals could not belong to both MJ-12 and PI-40, term limits applied to how long individuals could stay on the policy making body MJ-12, and no term limits applied in the case of the PI-40.

As with the policy coordinating role played by the National Security Council in providing advice to the President, MJ-12 had a similar function in coordinating policy concerning the different clandestine organizations involved in various aspects of the ET presence. As the organization developed to study specific policy issues concerning the ET presence, PI-40 had a significant role in framing policy issues and determining priorities that would influence the way MJ-12 made policy recommendations. As the architect of the institutional reorganization that led to the expansion of MJ-12, and as the President's Special Advisor, Rockefeller assumed the critical role of head of MJ-12. Furthermore, Rockefeller through his family's connections also could influence the selection of appointments from the Council of Foreign Relations and the JASON Group for PI-40. Accordingly, Rockefeller played a critical role in influencing the strategic principles and imperatives that would subsequently govern policy making on the ET presence. Rockefeller's influence gradually led to his estrangement with Eisenhower as a result of the latter realizing that executive oversight of the ET presence was being eroded due to Rockefeller's reorganization. Eisenhower's concern resulted from two main ways in which executive oversight was eroded; the role of Corporate America, and the way information was provided by MJ-12/PI-40 in dealing with the ET presence.

The shift from the Roosevelt /Truman administration models of government funded military-scientific laboratories that conducted clandestine military projects, to a model that made US corporations nominally in charge of these clandestine projects, led to a cooperation that Eisenhower believed became a threat to executive government. This was immortalized for the general public by Eisenhower's Famous warning in his January 1961, departure speech of the danger of the 'military-industrial complex'.

In the councils of Government, we must guard against the acquisition of unwarranted influence, whether sought or unsought by the Military Industrial Complex. The potential for the disastrous rise of misplaced power exists, and will persist. We must never let the weight of this combination endanger our liberties or democratic processes. We should take nothing for granted. Only an alert and knowledgeable citizenry can compel the proper meshing of the huge industrial and military machinery of defense with our peaceful methods and goals so that security and liberty may prosper together. [26]

At the end of his administration, Eisenhower evidently felt that the military-industrial complex had grown too powerful and had slipped out of the control of him and his principal advisors in how the ET presence was to be politically managed. Essentially, clandestine military projects with Corporate America receiving funding through contracts awarded by the military, meant that the President and his principal advisors, had lost control of what was occurring in the clandestine projects and organizations that formed an elaborate 'military-industrial complex' weaving through the various military and intelligence organizations that worked on different aspects of the ET presence. The loss of control of what was happening in military-corporate laboratories carried with it a loss of control of over the quality and accuracy of the intelligence information that found its way back to the President and his senior advisors. The 'military-industrial complex' evidently, was able to frame policy issues and contingencies concerning the ET presence in ways that dictated government policy to the extent that Eisenhower and his advisors felt frustrated and alarmed. This suggested that the MJ-12 and/or PI-40 had been compromised by the military-industrial complex, and was framing policy issues and imperatives in ways that eroded executive oversight of these clandestine organizations.

Nelson Rockefeller, the architect of the reorganization that included Corporate America in how the ET presence would be politically managed, resigned from his position as the President's Special Assistant for Government Reorganization in 1959 to successfully run for New York governor. Rockefeller's prominent role in both Corporate America and in MJ-12/PI-40, however, ensured that control of how the ET presence was to be politically managed increasingly lay with the MJ-12/PI-40 and the military-industrial complex responsible for reverse engineering ET technology, and the various intelligence agencies focused on the ET presence. What Eisenhower was alluding to in his departure speech was that, at least as far as political management of the ET presence was concerned, a political coup had occurred. [27] Eisenhower had been maneuvered into a role that merely gave constitutional validity to policy recommendations that were crafted on information that the President had no independent means of confirming. That Eisenhower felt this way is evidenced in reports by one of the military officers who directly served under Eisenhower, Brigadier-General Stephen Lovekin who wrote:

But what happened was that Eisenhower got sold out. Without him knowing it he lost control of what was going on with the entire UFO situation. In his Last address to the nation I think he was telling us that the Military Industrial Complex would stick you in the back if you were not totally vigilant . . . . And I think that he realized that all of a sudden this matter is going into the control of corporations that could very well act to the detriment of this county. This frustration, from what I can remember, went on for months. He realized that he was losing control of the UFO subject. He realized that the phenomenon or whatever it was we were faced with was not going to be in the best hands. As far as I can remember, that was the expression that was used, "It is not going to be in the best hands." [28]

The Kennedy administration marked an important milestone in the erosion of executive oversight of the ET presence. Kennedy was made aware of the ET presence when, as a young Senator serving on the Foreign Relations Committee, he was informed of the ET presence. Like his Republican predecessor, the new Democratic President, Kennedy, found that the political management of the ET presence was dominated by the clandestine military and intelligence organizations, in concert with MJ-12/PI-40, that released information on ETs in a way that was biased towards a particular outcome, the most obvious being the need for continued funding of their respective programs. If an information 'spin' was indeed occurring, and certainly that is what the Eisenhower experience suggests, then the ET presence would permanently remain in the category of a national security threat that required strict secrecy, with minimal government oversight and extravagant funding levels. The involvement of Corporate America in fulfilling military contracts meant that executive oversight would not succeed in discovering the true ramifications of the ET presence and what clandestine organizations were really up to. This problem of having no way of checking and confirming the information supplied by clandestine organizations that was suspected of being "spun" in a way that supported particular outcomes was certainly what concerned Eisenhower, and was a problem that Kennedy also confronted.

Kennedy and his most trusted senior advisors evidently labored hard to reestablish executive oversight and control but were similarly frustrated as was Eisenhower and his team of advisors. Kennedy's Special Assistant on International Affairs (aka National Security Advisor), McGeorge Bundy, and other cabinet members from the Departments of Defense and State, the Director of Central Intelligence and the Chairman of the Joint Chiefs of Staff were most likely members of MJ-12 but this did not apparently assist Kennedy in gaining the changes he required. According to a former steward aboard Air Force One, Bill Holden, he and Kennedy had the following conversation when flying to Europe in the summer of 1963. "What do you think about UFOs, Mr. President?" Kennedy became quite serious for a moment, and replied, "I'd like to tell the public about the UFO situation . . . but my hands are tied." [29] Rockefeller's institutional reforms made it impossible for one individual, even a sitting President, to take control of the policy making process concerning the ET presence.

Kennedy's efforts to reestablish executive control and overturn the disturbing reality that the military-industrial complex was acting with minimal executive oversight in dealing with the ET presence and, more importantly, influencing how the ET presence was to be politically managed, certainly led to an escalating series of confrontations. For example, Kennedy's initiative to improve relations with the Soviet Union under Nikita Khrushchev and cooperate more in responding to the ET presence certainly disturbed those clandestine organizations that held real influence in how to politically manage the ET presence. Documents have been found supporting this idea that Kennedy desired greater cooperation with the Soviet Union, and that this was opposed by the military-industrial complex. [30] Furthermore, it has been claimed that Kennedy issued an ultimatum to Majesty 12, that "he intended to reveal the presence of aliens to the American people within the following year [1964), and ordered a plan developed to implement his decision." [31]

The institutional restructuring under Rockefeller that made possible for Corporate America to participate in conducting highly classified programs with clandestine organizations embedded in military and intelligence departments, was not going to be overturned by an upstart Democratic President committed to a more transparent and cooperative national and international effort to politically manage the ET presence. It is likely that Kennedy's assassination was partly linked to his efforts to wrest control back over how to manage the ET presence. [32] Those responsible could have come from any of the clandestine organizations that felt their operations threatened by Kennedy's policies. An outcome of the crisis involving the Kennedy administration would have been that the formal policy making group, MJ-12, would have begun making policy choices without necessarily gaining Presidential approval. This marked a departure from the Eisenhower administration where, at least, Eisenhower had to give formal approval for major MJ-12 policy recommendations to be implemented.

Kennedy's assassination marked the culmination of a process that in all constituted a 'de facto political coup' where executive oversight of the ET presence came to an end in less than a decade. Eisenhower's Republican affiliation and choice of Rockefeller to reorganize government strategies and play a leading role in formatting how the ET presence was to be politically managed, was what effectively led to the erosion of executive oversight of the ET presence. Eisenhower became aware later in his administration that he had lost control, and that a "silent political coup" was occurring. Kennedy's unsuccessful effort to reestablish control and assassination marked a turning point in the erosion of executive oversight. The kind of executive oversight achieved under the Roosevelt and Truman administrations where the President and his senior advisors were fully informed and exercised firm control over the political management of the ET presence was now a distant memory. Real control over how to politically manage the ET presence had slipped into the hands of the clandestine military and intelligence organizations that operated secretly, with no executive oversight and lavish budgets. It is therefore understandable why, at least from a bureaucratic perspective if not a national security stand point, that clandestine organizations had a strong interest in maintaining the status quo and opposing efforts to yield to greater transparency and executive oversight.

As the key policy coordinating body, MJ-12/PI-40, would certainly have been aware of the advantages of such a 'de facto political coup' and most likely played a supporting, if not principal, role in the erosion of executive oversight of all aspects of the ET presence. MJ-I2/PI-40 was institutionally positioned to benefit greatly from this loss of executive oversight which meant that Presidential administrations knew less about what was really happening on ET issues, and would have to rely on MJ-12/PI-4O for accurate information on what was occurring within the military industrial complex. MJ-12/P1-40 could play its policy coordinating role with little real interference or scrutiny from Presidents and their policy advisors who simply did not have the means of confirming or challenging the information provided to them by the various clandestine organizations involved in dealing with the ET presence; and/or the policy advice provided by MJ-12/PI-40. The inability of Presidential administrations to gain independent and accurate information on the ET presence meant that MJ-12/PI-40 could put its own spin on the available information to produce policy outcomes inline with MJ-12/PI-40's priorities and needs. One of these needs was to ensure a degree of autonomy that minimized executive interference in affairs that MJ-12/PI40 probably decided were outside of the experience and abilities of Presidential administrations that were at best only temporary players in the need to politically manage the ET presence. The loss of executive oversight meant that MJ-12/PI-40 became the main player in determining how the ET presence was to be politically managed. [33] This led to the fourth phase in the political management of the ET presence—The Era of Autonomy and Impunity for Clandestine Organizations.

Phase Four—The Era of Autonomy and Impunity for Clandestine Organizations

President Lyndon Johnson, like Kennedy, was not trusted by MJ-12/PI-40 and was simply denied information concerning the ET presence. [34] MJ-12/PI-4O during the Johnson administration operated without executive oversight and politically managed the ET presence by coordinating between four main constituencies. The first was the various clandestine organizations embedded in the different military services that were part of the military-industrial complex involved in reverse engineering ET technology for weapons production The second constituency was the intelligence Organizations that attempted to gather information on ET activities; the ET agenda, establish channels of communication with the ETs; and which were embedded in the Central Intelligence Agency, National Security Agency, and the Defense Intelligence Agency. The third constituency was the President and his senior advisors who while not fully aware of the scope of the ET presence, were least aware of the existence of these clandestine organizations and of the policy coordinating role played by MJ-12/PI-40. [35] The fourth and last constituency was Congress and the general public who were most out of the information loop, and simply unaware of the extent of the clandestine programs set up to deal with the ET presence.

The Air force investigation begun at the end of the Truman administration, Project Blue Book, had been, according to Col Phillip Corso, who sewed in the Eisenhower administration and was also briefly the head of a secret Pentagon project to reverse engineer ET technology, "pure public relations from the start" that was designed to keep the general public focused on debating whether or not there was sufficient evidence for the existence of ET piloted UFOs. [36] The termination of Project Blue Book in 1969 represented the confidence of those politically managing the ET presence that numerous UFO sightings and public reports of contact with ETs no longer represented a threat to the official policy of nondisclosure of the ET presence.

In its role as the key policy coordinating body in the web of clandestine organizations that dealt with the ET presence, MJ-12/PI-40 now assumed firm control over how the ET presence was to be politically managed. Identifying the chief function of MJ-12/PI-40 and who its key players were is therefore central to understanding how the ET presence was politically managed, and how the erosion of executive oversight of the ET presence and the autonomy of MJ-12/PI-40 would impact on future Presidential administrations. The organizational function of MJ-12/PI-40 can be likened to that of a chess player who has to manage a whole range of pieces with different functions, values and strengths in order to achieve an ultimate goal—victory. This meant that MJ-12/PI-40's primary role was that of developing a grand strategy to deal with the ET presence in terms of the variety of ET races, their varying agendas and activities, and foreign national governments and clandestine governments on one side of the chess board (the opponent); and on the other side, the four different constituencies that made up the pieces of ones own side of the chess board. The architect of this strategic role for MJ-12/PI-40 was Nelson Rockefeller and the Council of Foreign Relations who essentially designed the institutional rules by which MJ-I2/P1-40 would interact with other constituencies involved in various aspects of the ET presence. There is strong evidence from whistleblower sources that the master strategist of the Special Studies Group that made up the outer layer of Mj-12/PI-40, was a key Rockefeller protégé, Dr. Henry Kissinger whose experience in managing the ET presence went further back than is commonly appreciated. [37]

The Prominence of Henry Kissinger as PI-40's Master Strategist

Their is significant evidence that Nazi Germany had partially succeeded in reverse engineering downed extraterrestrial craft that had been discovered by Nazi authorities in the mid 1930's. [38] The partly successful efforts by top Nazi scientists in understanding and revere engineering this ET technology was a major factor in Nazi Germany's advanced weapons technology program and prolongation of the war effort in order to fully deploy these new weapons systems. At the conclusion of the Second World War, a top secret effort to repatriate the same Nazi scientists in order to utilize their expertise was begun by US Amy's Counter Intelligence Corps. "Operation Paperclip" as this secret effort was called, involved the removal of hundreds of Nazi scientists to the well funded military-scientific laboratories created to produce weapons for the war effort. [39] A little known figure in "Operation Paperclip" was a young German speaking US Army intelligence officer with a German Jewish background—Henry Kissinger.

Kissinger was born in Fuerth Germany on May 27, 1923, and served in the Army Counterintelligence Corps from 1943-46. At the close of World War II, he stayed on active duty in occupied West Germany. He was assigned to the 970th Counter Intelligence Corps Detachment, among whose 'official' functions included the recruitment of ex-Nazi intelligence officers for anti-Soviet operations inside the Soviet bloc. [40] Kissinger's detachment, in reality, was playing a key role in "Operation Paperclip"—a role that would mark him out in military intelligence circles as someone who had the keen intellect and strategic thinking abilities that could handle the most important strategic policy issue facing the US—how best to respond to the ET presence. [41]

Kissinger returned to the US, and in 1947 began his university education as an undergraduate at Harvard University. Kissinger, however, retained his ties to the military, as a Captain in the Military Intelligence Reserves. This enabled him to continue to play a role in issues pertaining to the ET presence as the policy at the highest level of the Truman administration was being developed. By 1950, Kissinger was now a graduate student and was working part time for the Department of Defense. He regularly commuted to Washington—as a consultant to its Operations Research Office which was under the direct control of the Joint Chiefs of Staff. The Operations Research Office 'officially' conducted highly classified studies on such topics as the utilization of former German operatives and Nazi partisan supporters in CIA clandestine activities.

Kissinger's official duties were once again a cover for his role in coordinating the recruitment and utilization of former Nazi scientists in clandestine projects involving the reverse engineering of ET technology, and dealing with a range of intelligence and strategic issues surrounding the ET presence.

In 1952, after completing his PhD, Kissinger became a consultant to the director of the Psychological Strategy Board, an operating arm of the National Security Council for covert psychological and paramilitary operations. Thus Kissinger's role expanded to dealing with the extensive policy issues surrounding the ET presence. Kissinger's inside knowledge of Operation Paperclip and the ET presence, combined with his strategic thinking abilities, marked him as someone who would rapidly assume a prominent position in the decision making hierarchy surrounding the ET presence. As a member of the Council of Foreign Relations, Kissinger would undoubtedly have come to the attention of its most prominent members as someone who could provide leadership on how to respond to the ET presence.

In 1954, President Eisenhower appointed Nelson Rockefeller his Special Assistant for Cold War Planning, a position that officially involved the 'monitoring and approval of covert CIA operations'. This was a cover for Rockefeller's true role as head of MJ-12; and most importantly, directing US foreign policy in the wake of a secret 'treaty' signed between an ET race from the Orion Constellation and the US. [42]The 'treaty' has been a source of much speculation but its existence and content has been revealed by a number of former military and government intelligence 'whistleblowers'. [43]

In 1955, Kissinger became a consultant to the National Security Council's Operations Coordinating Board—the highest policy—making board for implementing clandestine operations against foreign governments. Kissinger's analytical and strategic skills were used not only for coordinating US policy in clandestine operations against foreign governments, but also for the clandestine operations against ET races. [44] Kissinger's role in the clandestine operations, his close relationship with Nelson Rockefeller, his intellectual abilities, all combined to lead to a steady increase in his influence. Rockefeller and others running clandestine organizations understood the danger in not coordinating clandestine policy towards ET races and reverse engineering, with the more conventional foreign policy issues that were the focus of public attention.

Coordinating the extensive range of issues and problems would require someone with the strategic thinking abilities to coordinate these two arenas. Kissinger's abilities marked him out in the mind of Rockefeller, the Executive Committee from the Council of Foreign Relations, and military intelligence, as the person best qualified for this critical role. Rockefeller was instrumental in appointing Kissinger as one of the two Directors of PI-40, the Study Group that would provide policy advice to MJ-12 in response to the Treaty signed with the ET race from Orion in particular, and the ET presence more generally. [45]

As a Director and key strategist of PI-40, Kissinger would certainly have been aware of the need to politically manage the ET presence through ensuring the autonomy of MJ-12 and PI-40 and to render efforts of executive oversight ineffective. More importantly, MJ-12/PI-40 had steadily grown in institutional authority and power to the extent that it could now exert political influence over the executive branch of government. Strongly influencing, if not outright control of, successive Presidential administrations was viewed to be a critical part of how the ET presence had to be politically managed, effectively dismantling the executive oversight that was such a prominent feature of the Roosevelt and Truman administrations.

What contributed to this need for MJ-12 to control/influence future administrations is the irony that while most policy national security officials, politicians, the news media and the public, believed that the Soviet Union was the primary threat to US Security, in fact the US was secretly cooperating extensively with the Soviet Union in responding to the ET presence This meant that beneath the official Cold War rhetoric and armed conflicts that consumed public attention and resources, clandestine cooperation was occurring against what was perceived to be a common threat. [46] In short, the US and USSR were strategic allies as far as addressing the ET presence was concerned, while simultaneously being strategic competitors in the geopolitics of the Cold War. This meant that much of the animosity that characterized the Cold War was a charade that helped divert the general public away from what was really happening. Such a charade could only work if the most senior officials within the Presidential administration were familiar with the ET presence, so as to moderate more bellicose policy makers who believed the Cold War was for real, and were fully ready to use nuclear weapons against the Soviet Union in response to a perceived attack.

Influencing successive Presidential administrations could be achieved by embedding key PI-40 members in senior policy positions of incoming Presidential administrations so as to ensure nondisclosure of the ET presence, and moderating Cold War hostilities. For the Kennedy/Johnson administrations, this individual was McGeorge Bundy, one of the original members of PI-40, who upon becoming National Security Advisor would have become the chair of MJ-12. [47] In the case of the future Nixon administration, this would be achieved by embedding within it an even more prominent PI-40 member who could control President Nixon when necessary.

For the Nixon administration, this person would be no other than Henry Kissinger who was plucked out of public obscurity in 1968 to be appointed National Security Advisor of President-elect Nixon The instrumental figure in Kissinger's appointment was Nelson Rockefeller who had lost to Nixon in the 1968 Republican convention, and subsequently arranged for his protégé to become part of Nixon's team.

[48] Kissinger was intent on centralizing foreign policy making in the White House and the National Security Council, thereby ensuring him a central role in shaping not only US foreign policy, but also clandestine policy towards ET races in his new role as Director of MJ-12. Given his long history as a Director of the Special Studies Group/PI-40 since its formation, Kissinger would have been the most experienced and powerful head of MJ-12 since Nelson Rockefeller. [49]

In Seymour Hersh's critical biography of Kissinger's political managerial style during the Vietnam era, what emerges is that Kissinger was intent on amassing as much power as possible in managing international affairs. [50] Kissinger systematically undermined the positions of others who could pose a threat to his control of international affairs, especially that of the new Secretary of State, William Rogers, and other key policy makers in the Nixon administration. [51] Kissinger emerges in Hershe's biography as a political figure paranoid about ceding power to others who in his view lacked the subtlety and acumen in dealing with critical foreign policy issues. Kissinger's managerial style was to ensure that all information passed through him as the principal filter for shaping Nixon's priorities and thinking on foreign policy. A passage from a former Kissinger aide, Morton Halperin, reveals Kissinger's political managerial style.

On January 25, 1969, five days into the administration, the NSC was convened for its first meeting. The issue was Vietnam, and Halperin, now clearly Kissinger's top aide, was assigned to summarize all the papers and prepare a covering memorandum for the President. He carefully listed the various options in the two- or three-page summary, leaving boxes for the President to initial his choices. The idea was to reduce the President's workload. If Nixon chose not to read the attached documents, he could merely review Halperin's summary (which, of course, came with Kissinger's imprimatur) and make his decision. Henry loved summary and thought it was terrific. But, 'Mort,' he said, 'you haven't told the President what options we should choose.'"

"I thought to myself," Halperin recalls, "we're not supposed to be giving positions; we're just supposed to send summaries of the options." Years later, Halperin would realize how naive he had been: "Henry had been publicly saying that we were just going to sort out the issues for the President. I didn't know that Henry wanted to give him the decisions be should take. I was surprised—because I still believed what Henry had said." The Kissinger summary papers, with their recommendations, would become the most secret documents in the Nixon White House. [52]

Kissinger's political managerial style while in government is very significant since it provides insight into how decision making in PI-40 was conducted under Kissinger as the Study group director, and later in MJ-12 when he become its head during the Nixon/Ford administrations. [53]

Kissinger's role in guiding US foreign policy was dictated by his philosophy of 'realpolitik'. Realpolitik was modeled after his favorite international statesman, 19th century German Chancellor, Otto Von Bismark, who skillfully managed international alliances and limited wars to transform Prussia/Germany into a great power without provoking an international alliance against Germany.[54] For Bismark, international politics was a grand chess board where morality and sentiment played at best a secondary role, and what really mattered was the skillful use of one's resources in achieving one's strategic objective of maximizing power.[55] 'Realpolitik' dominated Kissinger's approach to international politics as evidenced in places such is Laos, Cambodia, Chile and East Timor where morality and sentiment played no role in these countries treatment as pawns in the grand game of international chess where the US competed with the Soviet Union to maximize its geopolitical power, while simultaneously cooperating strategically in responding to the ET presence.

Little known to the general public1 however, Kissinger adopted the same role in steering US policy in how it would respond to the ET presence. Morality and sentiment would play at best a secondary role as the US gradually improved its resources in order to increase its strategic position vis-à-vis the ET races visiting Earth. The moral orientation of these ET races that interacted with humanity and the clandestine organizations that were aware of ET activities were not given great emphasis in Kissinger's realpolitik concerning the ET presence. What mattered was the extent to which ET races would provide resources for US clandestine organizations to improve their weapons technology and thus improve the US's strategic position vis-à-vis different ET races. Kissinger's realpolitik was the way in which the complex political, social, economic and environment issues would be managed vis-à-vis the ET presence. Kissinger's a role would be similar to his 19th century hero, Bismark, Kissinger would play a key role in transforming the US into the dominant global power that could deal with ET races as an equal, without sparking a damaging interplanetary war with one or more of the ET races that would spell the end for US sovereignty and freedom. Kissinger's close association with the Rockefeller family ensured that Corporate America would continue to play a prominent role in the political management of the ET presence.

With Kissinger, during the Nixon administration, simultaneously playing prominent roles in US foreign policy and its clandestine interplanetary policy' through MJ-12/PI-40, what emerges is that the political management of the ET presence was dominated by a few individuals intent on amassing as much institutional power as possible, and not delegating authority to those outside of MJ-12/PI-40 who were viewed to lack the necessary experience, political sophistication and intellect in dealing with the complexities of the ET presence. Eisenhower 'a warning that the political management of the ET presence was "not in the best hands" now appeared prophetic.

Political Impotence of the Carter and Clinton Administrations and the threat posed by Reagan

The election of Jimmy Carter in 1976, brought in a new Democratic President who had declared that he would reveal the truth about the ET presence once in office. Carter was the first US President who was on the public record as having witnessed a UFO. [56]

Carter, however, would find that as President, he would be unable to determine the full extent of US clandestine programs focused on the ET presence, far less have any power to influence how to politically manage the ET presence. Even though his National Security Advisor, Zbigniew Brzezinski was one of the first directors of PI-40 and would have now taken over the chair of MJ-12 from Kissinger, Carter and his principal advisors found that they were simply denied the necessary information on the ET presence making it painfully clear that executive oversight of the ET presence was nonexistent. [57]

A project funded by the Carter administration in May 1977 through the Stanford Research Institute (SRI) to explore Extraterrestrial Communication was terminated four months later through Pentagon pressure. The Pentagon simply threatened the directors of the SRI that it would terminate projects the Pentagon funded for SRI if the latter went ahead with the White House Center. [58] After the debacle over its Extraterrestrial Communication project, Carter and his senior advisors quickly recognized they were 'minor players' as far as the ET presence was concerned. Indeed, this lack of ability to politically manage ET affairs, may well have been a critical factor in the Iranian revolution that did so much to undermine Carter's reelection chances.[59]

The Republican electoral campaign of 1980 brought with it a new dimension to the political management of the ET presence. Ronald Reagan came into the campaign as a crusading anti-Communist with fixed views that only negotiating from a position of military strength was the means of countering the Soviet threat to global democracy. Privately, however, Reagan had a similar perspective on the ET presence and what he viewed as the need to negotiate from a position of strength vis-à-vis the 'ET threat to humanity'. [60] Like his predecessor, President Carter, Reagan had an encounter with UFO's. [61] Unlike Carter, however, he developed a strong belief that the ET presence was a threat to humanity that had to be militarily contained. In contrast, his opponent in the Republican Primaries, former CIA Director George Bush, brought with him a more moderate Republican ideology—an ideology that was more consistent with the views of MJ-12/PI-4O, which Bush had previously been a member of, and of Henry Kissinger who was by now the undisputed master strategist for MJ-12/PI-40 with nearly 40 years experience in dealing with the ET presence. The election of Reagan over Bush would certainly have come as a disappointment to Kissinger and MJ-12/PI-40 not only in terms of it bring in another 'outsider' who impacted the ability of MJ-12/PI-40 to politically manage the ET presence, but also because it allowed a dangerous element to emerge in the clandestine effort to manage this presence.

MJ-12/PI-40, under Kissinger, was fully aware of the complexities of the ET presence in terms of different races and orientations, and ensuring that interaction with the numerous clandestine organizations embedded in the different military organizations and intelligence agencies coordinated in a way that maintained a 'global balance of power'. What Kissinger and MJ-12/PI-4O were most concerned about was the danger of clandestine organizations in the US military and/or intelligence services engaging in a dangerous confrontation with ET races that could degenerate into a large scale hostilities leading to a 'war of the worlds'. As the master 'Bismarkian' strategist, Kissinger was concerned to maintain the 'balance of power' while simultaneously advancing the strategic position of the US vis-à-vis ET races PI-40, again under the leadership of Kissinger after having served his full term as head of MJ-12, was therefore intent on containing any 'military adventurism' on the part of clandestine organizations in the US military that were at best too confrontational, or at worst infiltrated by ET races intent on initiating global confrontation. [62] What most concerned MJ-12/PI-4O was the possibility that a Presidential administration could be unduly influenced by clandestine military organizations that either were prone to military adventurism and/or been infiltrated by ET races.

Soon after his election, Reagan demonstrated a rigid belief of the nature of the ET threat, and laced many of his public statements referring to the ET presence and its threat to humanity. [63] According to Dixon Davis, one of the two CIA agents appointed to brief Reagan when he was President-elect: "The problem with Ronald Reagan was that all his ideas were all fixed. He thought that he knew about everything—he was an old dog." [64]

Reagan's anti-Communist rhetoric and massive buildup of military forces was a cover for Reagan's true desire to militarily confront ET races. [65] His first major public comment on an ET threat occurred at a 1985 US-Soviet Summit meeting with Mikhail Gorbachev at Geneva when he said:

I couldn't help but—when you stop to think that we're all God's children, wherever we live in the world, I couldn't help but say to him (Gorbachev) just how easy his task and mine might be if suddenly there was a threat to this world from some other species from another planet outside in the universe. We'd forget all the little local differences that we have between our countries and we would find out once and for all that we really are all human beings here on this Earth together. Well I guess we can wait for some alien race to come down and threaten us, but I think that between us we can bring about that realization. [66]

It his unscheduled comment at a US-Soviet Summit were not itself provocative enough expression of Regan's views on the possible threat of an ET presence, then his speech to the Forty-second UN General Assembly of the United Nations on September 21, 1987, was even more provocative and disturbing in its implications:

In our obsession with antagonisms of the moment, we often forget how much unites all the members of humanity. Perhaps we need some outside, universal threat to make us recognize this common bond. I occasionally think how quickly our differences worldwide would vanish if we were facing an alien threat from outside of this world. And yet I ask—is not an alien force already among us? [67]

For Colonel Phillip Corso, and other conservative military officers, Reagan was a hero who knew how to best respond to the ET presence—a global defensive shield that could shoot down ET craft anywhere around the planet. [68] The Strategic Defense Initiative had little to do with shooting down ballistic nuclear missiles, and really was part of a planetary shield desired by clandestine organizations in the military wanting to militarily confront the ET presence.

Regan's conservative political philosophy and public statements on the need lot a massive military build up to the Soviet threat, were allusions to the perceived danger of an ET invasion. Reagan and his political advisors were considered by Kissinger, Brzezinski, and others in MJ-12/PI-40, a threat to the political management of the ET presence and to the tenuous peace that existed between clandestine organizations around the planet and ET races. Given the gravity of Reagan's fixed views and the implications for managing the ET presence, it is very likely groups responsive to the concerns of PI-40 played a role in attempting to have Reagan removed from public office and replaced by an MJ-12/ PI-40 member, George Bush the Vice President and former head of the CIA. The Hinkley assassination attempt in 1981 was possibly an attempt by organizations loosely linked with PI-40 to either remove or intimidate Reagan so as to prevent what could have been a disastrous unraveling of the covert global cooperation in managing the ET presence. [69] The eventual result of the assassination attempt was that the Reagan administration's militaristic impulses were sufficiently restrained so as to ensure that no military confrontation with ET races would spiral out of control.

The 1988 election of George Bush once again allowed MJ-I2/PI-40 to again dominate the strategic thinking of a Presidential administration. As a former member of MJ-12/PI-4O, Bush was all too aware of the need to politically manage the ET presence in the mold dictated by Kissinger during the Nixon administration. Indeed, Kissinger's support was critical in the appointment of Bush to become the Director of the CIA in 1975, and his 'promotion' to MJ-12 from PI-4O, not long after the Watergate scandal had begun to subside. [70] Public secrecy, monopolizing derision making power in MJ-12/PI-40, maintaining the balance of power, and continuing to reverse ET technology for weapons acquisition, and maintaining the prominent role of Corporate America in dealing with the ET presence, were the keys to politically managing the ET presence.

MJ-12/PI-40 was certainly content with its influence under the Bush administration and there is evidence that international events were managed in a way that would support the 1992 reelection of President Bush. The 'End of the Cold War' was certainly a 'gift' to the Bush administration that normally would have ensured a second election victory for an administration enjoying such a tremendous foreign policy success. [71] The successful outcome of the Gulf War in 1991 was similarly an event that would have normally secured a successful reelection campaign. The outcome of the 1992 Presidential election appeared so certain, that prominent Democrats decided not to run and viewed 1996 as the best time for a Presidential campaign.

The election of President Clinton was certainly a surprise development for MJ-12/PI-40 and once again had the effect of placing an 'outsider' in the White House. Clinton, like Carter before him, soon found out that he had minimal influence over the political management of the ET presence. Even more disturbing, his senior political officials including the Director of Central Intelligence, James Woolsev, and Secretary of Defense, William Cohen, had little knowledge of the ET presence. [72] Stephen Greer narrated the following exchange he had with a famous astronaut:

Recently I was in Washington meeting with a very famous astronaut. Everyone would know this person's name . . . This particular astronaut had during his career been in possession of a very specific piece of incontrovertible piece of evidence related to UFOs. It is something that if disclosed would be clear and definitive. This astronaut described how he had approached and worked directly with President Clinton's Secretary of Defense William Cohen to look into and retrieve from classified projects this specific piece of evidence—of that which he had all the specific details . . . the words used by this astronaut to me were "there was an inordinate large amount of money and personal lime by the Secretary of Defense William Cohen was spent to locate this evidence, and he was never given access to it." [73]

This suggested that many of those sitting in MJ-12/PI-40, were hangovers from the Bush administration, and Clinton's political appointments were not trusted to maintain secrecy. Clinton's efforts to extract information from clandestine organizations proved fruitless as evidenced in the following quote from William Laparl, who worked with the CIA in the early days of the Clinton Administration:

It was known among the high CIA people, and the people who had contact with these people, that the Clintons were on the prowl for UFOs. Bill Clinton had been asking anyone who would listen to him, to tell him the secret. You know, he would get some Admiral in there, and say "By the way, tell me the UFO secret." They would just look at him like "What planet are you from?" [74]

Clinton's interest and efforts to gain information on the ET presence and clandestine projects were a threat to MJ-12/P1-40 insofar as Clinton's initiatives threatened the veil of secrecy that had been existing since the 1940s. More importantly, Clinton's efforts may well have been viewed as the initial stages of an attempt to reestablish executive oversight. It as not to difficult to surmise that many of Clinton's political problems were a result of clandestine efforts to distract the Clinton administration, and ensure minimal support for his domestic policies. Clinton became resigned to serving his term with only minimal knowledge of the ET presence, and without having any serious impact on how to politically manage the ET presence. His remarks to a question from a Northern Ireland teenager in November 1995 testify to his political impotence on the ET presence:

I got a letter from 13-year-old Ryan from Belfast. Now, Ryan, if you're out in the crowd tonight, here's the answer to your question. No, as far as I know, an alien spacecraft did not crash in Roswell, New Mexico, in 1947. (Laughter) And, Ryan, if the United States Air Force did recover alien bodies, they didn't tell me about it either and I want to know. (Applause) [75]

The election of George W. Bush in 2000 once again led to an insider, or at least an insider's loyal son, to be in the White House. George Bush, Sr., would henceforth play a key role in steering his son, who lacked the kind of intellectual qualities to be a member of PI-40 or Council on Foreign Affairs, in his own right, but served as a useful figurehead that could gain the loyalty of the American public in ways that the more urbane and sophisticated George Bush Sr., and Nelson Rockefeller before him, never could. This set the stage for a new phase in the political management of the ET presence, the takeover of a foreign country for purposes exclusively to do with the strategic advantage this would provide in politically managing the ET presence.

Phase Five—The Political Management of Iraq and its Ancient ET Technology

In the 2000 Presidential election campaign, George Bush jokingly responded to a question concerning the ET presence by saying that that was Richard Cheney's area of expertise. [76] Cheney served as Secretary of Defense in the first Bush administration and according to the formula first used during the Eisenhower administration to determine the members of MJ-12, Cheney would have been a former active member of MJ-12, and was very likely a current member of PI-40. Bush would have been briefed at some point by Cheney and/or his father about the ET presence, and was simply deferring to the clear expertise Cheney had in the area. Other key administration figures in the new Bush administration such as Donald Rumsfeld, Secretary of Defense during the Ford administration, also would have served on MJ-12 according to the Eisenhower era formula, and would have subsequently sat on PI-40, before finding himself elevated once more to MJ-12. Other new Bush administration officials such as Paul Wolfowitz and Richard Perle, both of whom had earlier held prominent positions in the Department of Defense, were probably briefed on the ET presence, during there tenures.

What distinguished this senior group of officials in the Bush administration with backgrounds in the Department of Defense, was their 'hawkish' Republican world view that espoused such novel international theories as preemptive military intervention against rogue states and creating a global defensive shield against them. While such a theory ostensibly was focused on the threat such states posed if they possessed Weapons of Mass Destruction, the knowledge these officials had concerning the ET presence, suggested that they were merely continuing the threat based assessment of the ET presence that had earlier dominated the Reagan administration. The global defensive shield against rogue states using ballistic nuclear missiles against the US was again a convenient cover for the true intent to be able to militarily intervene against ET craft anywhere on the planet. Aside from the close ties this group of Bush administration 'hawks' had with the Department of Defense, some of them also had extensive ties with corporate military contractors that worked on various clandestine projects concerning the ET presence. The professional background and policies of the Bush administration 'hawks' suggests they formed the public head of a 'cabal' of senior military officials based in clandestine organizations that promoted a militaristic responses to the ET presence. [77] [Such officials would no doubt have been aware of clandestine organizations of other countries with whom they cooperated/competed with in terms of gaining maximum strategic advantage vis-à-vis the ET presence, and the strategic significance of different rogue states in the tenuous balance of power that existed on the planet between different ET factions and clandestine government organizations. A key rogue state in such a strategic struggle was Iraq under Saddam Hussein.

There is evidence that Iraq once hosted an advanced civilization that interacted extensively with ET races. [78] The likelihood that thus interaction led to the development of technology either left behind by these ETs or based on ET technology that lay buried in the Iraqi desert has been argued to be the primary motivation for the US led military intervention in the 2003 Gulf war. [79] Once the need to invade Iraq emerged for reasons that primarily concern the political management of the ET presence and the strategic importance of Iraq, it became important to manage this in a way that would minimize global chaos. This was especially important since clandestine organizations in the US were competing with similar organizations in France, Germany, Russia and even China that presumably already established some understanding of precisely what was available in Iraq, and its strategic significance for the global balance of power.

The influence of key 'hawks in the Bush administration, Donald Rumsfeld, Paul Wolfowitz, Richard Cheney, Richard Perle and others, was likely viewed by MJ-12/PI-40 as a useful political tool for gaining access to Iraq's strategic ET resources. This had to be managed in a way, however, that would maintain global consensus and nor precipitate a conflict that could easily spiral out of control and lead to the intervention of ET races. Like the Reagan administration before it, the more militaristic impulses of the Bush administration needed to be restrained by MJ-12/PI-4O, if Iraq's ET resources were to be gained over the objections of clandestine organizations based in France, Germany, Russia and China.

In August of 2002 when debate over a preemptive US attack was at its height, Kissinger released a significant policy statement that cautioned the Bush administration from alienating its historic allies in dealing with Iraq, and argued that "the notion of justified preemption runs counter to modern international law which sanctions the use of force in self defense only against actual rather than potential threats." [80] This combined with key statements from other former officials from Republican administrations, such as former Bush Sr., National Security Advisor and a likely PI-.40 member, Brent Scowcroft, led to President Bush moving towards a more internationalist agenda that embraced the role of the UN. [81] This culminated in Bush making a speech before the UN a day after the anniversary of September 11 where he emphasized the importance of confronting Saddam Hussein and the important role of the UN in playing a leading role in containing the threat posed by rogue states. [82] Kissinger and Scowcroft represented a very visible initiative by MJ-12/PI-4O to reign in the Bush Hawks insofar as the importance of global consensus was recognized as a key ingredient for the strategic goals of militarily intervening in Iraq, removing the Saddam regime, and gaining access to whatever ET technology was hidden in Iraq. The Republican hawks within the Bush administration had been temporarily outflanked by Kissinger, other Republican moderates and MJ-12/PI-40 members. The clandestine organizations in France, Germany and Russia therefore had more time to reach an accommodation with the Bush administration. Security Council Resolution 1441 was a triumph for Kissinger's approach to politically managing the ET presence.

By February of 2003, it had become clear that Germany, France and Russia were maneuvering to block the US intervention. Rather than this being purely an altruistic desire to preserve global peace in the face of an unjustified US attack to destroy 'nonexistent' weapons of mass destruction, these European states desired to keep the US out of Iraq due to the increased strategic power this would give to US based clandestine organizations. Kissinger's role is indicative of the strategic struggle occurring behind the scene over ET technology in Iraq, and the shifting alliances this caused. Kissinger subsequently came out with a key policy speech criticizing France and Germany as threatening the NATO alliance. [83] His speech indicated that the time was ripe for a US military invasion. Kissinger and MJ-12/PI-40 had given the blessing for an invasion which was now inevitable. Rather than regional devastation as was first feared, what occurred instead was a rapid collapse of the Saddam regime. US clandestine organizations had achieved their military objectives without precipitating regional and global chaos. The passage of a new UN Security Council resolution in May 2003 by a margin of 14-0 endorsing US administration and reconstruction of post-war Iraq by the UN marked a decisive victory for the Bush administration, and the influence of MJ-I2/PI-40 in politically managing the ET presence.

What was significant in the post-conflict administration of Iraq was the speedy departure of the Pentagon appointed civil administrator, Jay Garner, and his replacement by Paul Bremer, a former State Department Ambassador at large for Terrorism, and someone with strong connections to Kissinger, formally in charge of the civil administration of Iraq. This indicated that the more moderate policies backed by Kissinger in dealing with the ET presence, had prevailed over the more confrontational policies of the Bush hawks. A power struggle between moderate and hawkish factions of MJ-12/PI-40 was occurring behind the scenes, and the Kissinger backed moderates had been successful.

The policies implemented by Bremer in terms of setting back the schedule for the election of an interim Iraq administration, indicate that the US is set to remain in military control of Iraq for a number of years. Iraq's role in terms of being a host to ancient ET technology and analyses which suggest it plays a significant role in the possible return of ET races that sponsored Iraq's ancient Sumerian civilization makes Iraq a significant actor in the political management of the ET question.

Conclusion: Politically Managing the ET presence and the Challenge to Democracy and Liberty in America

Political management of the ET presence has evolved greatly since the Second World War era. Starting initially as a process firmly controlled by Presidential administrations that exercised executive oversight, thereby making it part of the democratic process despite its secrecy and lack of congressional participation, political management evolved to the point where Presidential administrations were not fully informed of and had no executive control over many aspects of the ET presence. This meant that the political management process had dubious constitutional validity and was controlled by a few actors who could be tied to the Rockefeller-Kissinger axis, and their respective ties to US corporations and elite foreign policy bodies such as the Council on Foreign Relations. The US-British-Australian intervention in Iraq suggests that political management of the ET process has evolved to yet another level. Now the US and its allies are prepared to militarily intervene in others in order to gain strategic goals vis-à-vis the ET presence. The most important of which are to maintain official secrecy of the ET presence, withhold from the general public the true nature of the historic role played by ETs in ancient civilizations, and to gain whatever military advantage possible from the reverse engineering of ET technology found in countries that, like Iraq, have been prominent sites hosting an ET presence. According to whistleblowers sources, there are numerous ancient ET bases that are being increasingly discovered around the planet. [84] Consequently, it is likely that the intervention in Iraq will set a precedent for similar interventions elsewhere across the planet for reasons that increasingly have to do with the political management of the ET presence. The strategic thinking of organizations such as MJ-12/PI-40 is based on the perception that the best analytical minds and strategic thinkers are employed in managing the ET process and that while this may not be acceptable from a democratic standpoint which emphases executive or congressional oversight of all government activities, it is acceptable from a national security prospective. What can be concluded here is that the view that indeed the "best minds" are in charge of the political management of the ET presence is misplaced. Information of the ET presence has been increasingly controlled and spun in a way that suggests that real decision making power has been in inexorably restricted to fewer and fewer individuals who reflect conservative political philosophies typically associated with the Republican party. While it is impossible to say exactly how many individuals exercise real influence in politically managing the ET presence, the history of the Rockefeller-Kissinger involvement and the prominent roles played by corporate America and the Council on Foreign Relations, suggests that this influence is restricted to very few. Eisenhower's warning about the "best minds" not being in control suggests that the elite club of "experts" that dictate how the political management of the ET presence is to be conducted, are overly influenced by corporate and elite interests sympathetic to world views associated with the military-intelligence communities. Introducing greater transparency into all aspects of managing the ET presence will make it possible to expand the restricted circle of power and influence that controls information concerning the ET presence in a way that does indeed make it possible for the best minds to be formerly in charge of politically managing the ET presence. The erosion of executive control over the political management of the ET presence has reduced Presidents to at best, rubber stamps of the MJ-12/PI-40 policies (this appeared to be the case in the Nixon, Ford, Reagan and both Bush administrations), or, at worst, to political impotence as appears to have occurred in the cases of the Carter and Clinton administrations. The policies of the present Bush administration indicate that the US Presidency is reduced little more than a vehicle for the realization of questionable policies concerning how to manage the ET presence. When combined with the blanket of security that has prevented the US Congress and the American public playing a meaningful role in the political management of the ET presence, the current situation is a profound problem for those truly committed to principals of democratic governance and liberty in the US and elsewhere on the planet. President Eisenhower demonstrated he became all to aware of the true problem confronting the US as a nation in dealing with the ET presence—a de facto political coup by interests closely allied with corporate America and the military-intelligence communities. It is time that the American public understood the true nature of his warning and begin comprehensive political reforms to address the threat to liberty Eisenhower was alluding to.

-END NOTES-

(The following paper is also written by Dr. Michael E. Salla and explains the E.T. history, the presence of a star-gate in Iraq, and the need for U.S. control over it.)—Sabre

Abstract

Most, if not all, criticisms of the Bush administration's motivation for launching a preemptive war on Iraq focus on a combination of the imperial world views of conservative politicians in power in Washington, D.C., and the corporate interests that drive the political agenda of the Bush administration. This study will provide a radically different political analysis of the Bush administration's motivation for going war, and of the explanations offered by his critics. This study provides an exopolitical analysis of the policy dimensions of an historic extraterrestrial presence that is pertinent to Iraq and a US led preemptive attack. It will be argued that competing clandestine government organizations are struggling through proxy means to take control of ancient extraterrestrial (ET) technology that exists in Iraq, in order to prepare for an impending series of events corresponding to the 'prophesied return' of an advanced race of ETs. The Columbia Space Shuttle may well have been a high profile victim of such a proxy war intended to send a message to US based clandestine organizations over the preemptive war against Iraq.

In conducting this analysis, this study examines the available evidence of an historical ET presence in Iraq, and then applies this evidence to better understand the contemporary political situation in Iraq. The study will then analyze the motivations of the main political actors in the prospective US led preemptive war against Iraq. The study concludes by making some policy recommendations concerning how to respond to the legacy of an ET presence in Iraq and its contemporary political relevance.

About the Author

Dr. Michael E. Salla has held academic appointments in the School of International Service, American University, Washington DC (1996-2001), and the Department of Political Science, Australian National University, Canberra, Australia (1994-96). He taught as an adjunct faculty member at George Washington University, Washington DC., in 2002. He is currently researching methods of Transformational Peace as a Researcher in Residence in the Center for Global Peace (2001-2003) and directing the Center's Peace Ambassador Program which uses transformational peace techniques for individual self-empowerment. He has a PhD in Government from the University of Queensland, Australia, and an MA in Philosophy from the University of Melbourne, Australia. He is the author of *The Hero's Journey Toward a Second American Century* (Greenwood Press, 2002) and author/coeditor of three other books: *Essays on Peace* (Central Queensland University Press, 1995); *Why the Cold War Ended* (Greenwood Press, 1995); and *Islamic Radicalism, Muslim Nations and the West* (1993). He has authored more than seventy articles, chapters, and book reviews on peace, ethnic conflict and conflict resolution. He has conducted research and fieldwork in the ethnic conflicts in East Timor, Kosovo, Macedonia, and Sri Lanka. He has organized a number of international workshops involving mid to high level participants from these conflicts. He has a personal website at www.american.edu/salla/.

Introduction*

In his 2003 State of the Union address President George W. Bush declared "the gravest danger facing America and the world, is outlaw regimes that seek and possess nuclear, chemical and biological weapons."

(1)   In his speech, President Bush eloquently expressed his main motivation for launching a preemptive war against Iraq in order to prevent "a day of horror like none we have ever known." Critics of President Bush's preemptive policy, including the political commentator, Robert Fisk, argue the upcoming US led war against Iraq "isn't about chemical warheads or human rights: it's about oil."

(2)   According to another prominent political commentator, Michael Lind, the motivation lies in the preemptive military doctrine

championed by 'neo-conservatives' such as Deputy Defense Secretary Paul Wolfowitz, whose policy views were given more prominence after the September 11 attack.

(3)   Most, if not all, criticisms of the Bush administration's motivation for going to war focus on a combination of the imperial world views of conservative politicians in power in Washington, D.C., and the corporate interests that drive the political agenda of the Bush administration. This paper will provide a radically different political analysis of the Bush administration's motivation for going war, and of the explanations offered by his critics. It will be argued that the focus on either the factors supporting a preemptive war against an Iraq possessing weapons of mass destruction; or on criticisms against US imperialism and corporate interests, are not so much wrong, but simply reflect a limited political paradigm for understanding the motivations behind US foreign policy. The political paradigm to be used in this paper is based on 'exopolitics'.

(4)   This paradigm starts with the premise that there exists an extraterrestrial (ET) presence on Earth which clandestine government organizations have been withholding knowledge of from the general public and elected public officials. Rather than being an unsubstantiated 'conspiracy theory' with little relevance to contemporary policy issues such as a preemptive US war against Iraq, it will be argued that an exopolitical analysis can provide a more comprehensive understanding of what motivates the Bush administration in launching a preemptive attack against Iraq. Exopolitics as an emerging field of public policy is primarily based on the evidence provided by a range of sources supporting the idea of an ET presence that is known by clandestine government organizations that suppress this from the general public and elected political leaders.

(5)   The most important evidence comes from former military and government officials who have come forward to give 'whistle blower' testimony in a number of non-governmental initiatives to promote disclosure of the ET presence.

(6)   While many disagree over the plausibility of the available evidence and take various positions either for or against the existence of an ET presence and government non-disclosure of this presence, exopolitics is based on the premise that such debate ought not preclude discussion of the implications of such a presence among policy makers and the general public. Therefore when one examines contemporary

international issue such as a US led preemptive war in Iraq, one can explore the viewpoints offered by those using an exopolitical analysis, and consider the plausibility of these for a more comprehensive understanding of foreign policy, irrespective of the ongoing debate over the persuasiveness of the available evidence. This is necessary since those actively involved in a non-disclosure program will be use disinformation, intimidation and other strategies to deter witnesses, distort evidence, and deter public attention from the ET presence and how it pertains to a range of contemporary policy issues. Finally, a clandestine campaign of non-disclosure needs to be considered in conducting exopolitical analysis in terms of the likelihood of this influencing and/or compromising available empirical evidence that otherwise would confirm an ET presence. Therefore, exopolitical analysis has some key differences from more conventional approaches to political analysis which are based on a traditional social scientific method of a value free, objective analysis of available processes, institutions and actors in the public policy arena.

What follows is an exopolitical analysis of the policy dimensions of an historic ET presence that is pertinent to Iraq and a US led preemptive attack on the regime of Saddam Hussein. In conducting this analysis, I will first examine the available evidence of an historical ET presence in Iraq; then apply this evidence to better understand the contemporary political situation in Iraq; I will then analyze the motivations of the main political actors in the prospective US led preemptive war against Iraq; and finally conclude by making some policy recommendations.

What's the Evidence for an historic ET presence in Iraq?

The strongest available evidence for an historical ET presence in Iraq comes from cuneiform tablets directly recording the beliefs and activities of the ancient Sumerians whose civilization began almost overnight in 3800 BC. Most of these cuneiform tablets relate stories of the Sumerians interacting with their 'gods'. Most archeologists initially accepted that these were merely myths and attached little importance to them other than giving insight into the mytho-religious beliefs of the ancient Sumerians. That viewpoint received a major challenge in 1976 when the Sumerian scholar, Zecharia Sitchin, published the first of a series of books on his translations of thousands of Sumerian tablets.

(7) Rather than treating the stories of the gods as myths that had little empirical relevance, Sitchin interpreted the tablets as literal descriptions of events as they occurred in the time. translations of Sumerian cuneiform tablets revealed precise information on a range of topics that he argues could not have been possible for a civilization at the initial stages of its development with no obvious predecessor civilization to borrow from. According to Sitchin, the Sumerians had detailed knowledge of all the planets in the solar system, understood the precession of the equinoxes, and also had an understanding of complex medical procedures.

(8) As to where they could have gained this detailed knowledge, Sitchin's translations suggest that the Sumerians provided a clear answer for its ultimate source. They revealed in their tablets that all their knowledge came from a race of extraterrestrial visitors the 'Anunnaki' ('those who from heaven to Earth came') who were not only teachers for the Sumerians, but also played a role in the creation of the human race. The origin of this ET race was a planet called Nibiru that had a long elliptical journey around the sun, and returned to this region of the solar system every 3,600 years.

(9) When Sitchin's innovative work was first published, it raised great controversy and intense debate between those either for or against his main thesis of an historic ET presence in Sumer that was responsible for starting the remarkable Sumerian civilization. Among those responding favorably to Sitchin's thesis included established popular authors such as Erik Von Danniken who had himself written in 1969, the best selling book, *Chariots of the Gods*, that proposed an historic ET presence in different parts of the planet.

(10) Another popular author, David Hatcher Childress, outlines with great detail the ET technology possessed by ancient civilizations in his various books.

(11) Less well known authors such as William Henry have similarly published books supporting Sitchin's thesis.

(12) To support Sitchin's thesis, many authors typically cite biblical texts that make reference to the 'gods' that resided on Earth and interacted with humanity. The biblical text most referred to is the apocryphal book of Enoch.

(13) While the Book of Enoch was excluded in most versions of the Old Testament, it was nevertheless part of ancient Hebrew scholarship and indeed is included in the Ethiopian and Slavic versions of the

Old Testament. The Book of Enoch describes a rebel group of angels, the 'Nephilim', who numbered 200, who settled the Earth and interbred with the human population before being recalled and punished by their superiors, the 'Elohim'.

(14) The Book of Enoch provides contextual detail for mysterious verses in the Book of Genesis which describes a time when the 'sons of gods', the Nephilim/Anunnaki, interbred with humanity, and created a race of giants/heroes that ruled over the rest of humanity: "The Nephilim were on the earth in those days-and also afterwards-when the sons of God went to the daughters of men and had children by them. They were the heroes of old, men of renown."

(15) Supporters of the Sitchin thesis say that this is part of the biblical evidence that an advanced ET race did in fact exist on Earth, had a long interaction with humanity, and even played a role in the creation of the human race. In addition to the growing number of authors, independent archaeologists and biblical commentators supporting Sitchin's thesis, there are a burgeoning number of individuals who claim to be in telepathic communication with ET races who reveal information on the historic ET presence in Sumer. One of these ET 'channels', is Jelaila Starr who claims to be in touch with beings from Nibiru itself, and she regularly releases online information from the Anunnaki themselves on her website.

(16) Another is Sheldan Nidle who 'channels' an ET race from the star system Sirius gives extensive information on the historic presence and influence of the Anunnaki in his books and website.

(17) There is great controversy over the extent to which such a disparate collection of evidentiary sources can substantiate the Sitchin thesis of an historic ET presence in Sumer. Despite the controversy generated by such sources, they provide a wealth of information that merits closer examination in terms of their public policy implications. Given that exopolitics is based on the premise of an ET presence that is subject to non-disclosure by clandestine government organizations, it is possible to provide an exopolitical perspective on Sitchin's thesis despite the ongoing debate over the consistency and accuracy of the available evidence.

How Does the Historic ET presence relate to US policy in Iraq?

An independent archaeologist that discusses a direct link between the ancient ET presence in Sumer (southern Iraq) and current US focus on the regime of Saddam Hussein, is William Henry. Henry's main thesis is that there existed in Sumerian times a technological device which he describes as a 'Stargate', that the Anunnaki/Nephilim used to travel back and forth from their homeworld and the Earth, and also how they travel around the galaxy.

(18) Henry focuses on the following scene described by Sitchin's interpretation of a cuneiform tablet of an Uruk ritual text: Depictions have been found that show divine beings flanking a temple entrance and holding up poles to which ring-like objects are attached. The celestial nature of the scene is indicated by the inclusion of the symbols of the Sun and the Moon . . . depicting Enlil and Enki flanking a gateway through which Anu is making a grand entrance.

(19) Rather than a simple temple scene involving the chief Anunnaki of the Sumerians, Anu and his two sons, Enlil and Enki, Henry proposes that the above scene represents a transportation device used by Anu and others from the elite Anunnaki. If so, then such a device is most likely located in Sitchin's the Sumerian city of Uruk which was the founding city of the Sumerian civilization and the home of Gilgamesh, the famed king of the Epic of Gilgamesh.

(20) Sitchin, and authors such as David Childress who discuss the various technologies used by ETs and ancient civilizations, missed the significance of the Stargate in their own translations of the above texts and investigations of ET transportation. Both operated in a conventional paradigm where transportation occurs through rocket propelled vehicles.

(21) Sitchin focuses on rocket propelled spacecraft in his description of the Anunnaki and their various trips to and from the Earth. For example, describing the transportation used by the Anunnaki in moving between their earth and space based locations, Sitchin wrote: "The texts reveal that three hundred of them—The "Anunnaki of Heaven," or Igig—were true astronauts who stayed aboard the spacecraft without actually landing on Earth. Orbiting Earth, these spacecraft launched and received the shuttle craft to and from Earth."

(22) It therefore can be concluded that there were two forms of transportation used by the Anunnaki. One was a form of rocket technology familiar to us which was used by the resident Anunnaki on Earth who Sitchin, described as the "rank and file Anunnaki" who administered the Earth and humanity according to the dictates of their space based compatriots.

(23) Another transportation technology was the Stargate which presumably was used only by the highest class of Anunnaki, who dispensed the tasks of harvesting the Earth's resources to the resident and space based Anunnaki (lesser gods/rebel angels). Interpreting the Babylonian Epic of Creation, one gains an idea of the way tasks were allotted and the hierarchy of the Anunnaki in the way the 'supreme god' Marduk, dispensed tasks to his subordinate Anu, chief of the Anunnaki: Assigned to Anu, to heed his instructions, Three hundred in the heavens he stationed as a guard; the ways of Earth to define from the Heaven; And on Earth, Six hundred he made reside. After he all their instructions had ordered, to the Anunnaki of Heaven and of Earth he allotted their assignments.

(24) Thus the Anunnaki operated outposts both on Earth and in Space to maintain their control over the planet. Given the strict hierarchy of authority described by Sitchin in his detailed analysis of the Anunnaki and their interactions with one another and humanity, it is likely that the Stargate would have been revered and a subject of awe by the resident Anunnaki and humanity who could only observe its operation but were not allowed to use it themselves. As such, there would have been only a limited number of Stargates around the planet, with the Sumerian Stargate being located in the most important of the ancient Sumerian cities—the most likely being the ancient capital of Uruk, home of the ancient kings, which is located in Southern Iraq. Significantly, after a 12 year lull in excavations, a team of German researchers in 2002 resumed excavations in the buried city of Uruk. Using a magnetometer which is able to detect the presence of man-made objects beneath the soil, and a powerful computer system in Germany, German geophysicists were able to map out the buried structures of the sprawling ancient capital of 5.5 km2 that was where Sumerian civilization began.

(25) An important event in the Sumerian descriptions of the Anunnaki, was the latter's final departure from the planet during a series of cataclysmic events that culminated in the period 1800-1700 BC.

(26) Indeed, conventional archeologists support the view that there was a regional cataclysmic event that occurred at that time.

(27) If in fact there were two modes of transportation used by the Anunnaki, when most of the resident/lesser Anunnaki left by conventional rocket ships, elite Anunnaki most likely left by the Stargate and closed it down. Predictably, given the reverence and awe surrounding the Stargate, it would not have been left unprotected in the interim period between their departure and its reactivation with the prophesied return of the Anunnaki. A wide number of sources describe the present era in terms of a 'prophesied return of the gods/Anunnaki'. The notion of a 'prophesied return' in the context of the former Anunnaki presence varies in meaning according to three different perspectives. The first perspective is based simply on the idea of the gods or 'Anunnaki' physically returning to resume a prominent role in influencing human affairs and overseeing the use of resources of the planet.

(28) In such a scenario, the first wave of Anunnaki would arrive to create the favorable conditions for the anticipated return of the Anunnaki elite. This would involve the lessor Anunnaki first returning and activating the Sumerian Stargate that would be required for the return of their leaders. Presumably, this would be heralded as a sacred event that should be celebrated by all humanity. The authors, Clive Prince and Lyn Picknett, argue that there has been an identifiable chain of global events involving key religious and political actors preparing humanity for just such a return.

(29) The second perspective on the "return of the gods/Anunnaki' is the return of their home world, the planet Nibiru. Indeed, there has been much interest generated by a range of books and online web sites devoted to the topic of the return of Nibiru, the home world of the Anunnaki. Many authors cite a variety of astronomical evidence supporting the idea of a tenth planet that has long been speculated to influence the orbits of Uranus and Neptune ever since it was found that Pluto (discovered in 1930) could not account for these perturbations. In the late 1970's two astronomers, who were both from the US Naval observatory, Tom van Flandern and Richard Harrington, begun publishing a series of papers, supporting the existence of a tenth planet.

(30) widely cited reference to support the thesis of a tenth planet that is known but not released to the general public, is a series of press

releases by the astronomical team that were searching a part of the sky which calculations by van Flandern and Harrington suggested would be where the tenth planet was. In December 1983, the chief astronomer in charge of the Infrared Astronomical Satellite (IRAS) run by the Jet Propulsion Laboratory, Dr Gerry Neugebauer, announced possible confirmation of such a planet. The Washington Post Reporter summarizing the announcement wrote: "[a] heavenly body possibly as large as . . . Jupiter and possibly . . . part of this solar system has been found in the direction of the constellation of Orion by an orbiting telescope . . ."

(31) After a total of six major newspapers covered the announcement, there was a retraction of the announcement and public silence by astronomers on the possible existence of a tenth planet. Rumors began to emerge of an active campaign of suppression of information and intimidation by clandestine government organizations.

(32) For example, one of the astronomers at the US Naval Observatory, Dr Richard Harrington, spoke publically and wrote articles on the hypothetical planet X and there has been speculation that this directly contributed to his untimely death by 'natural causes' in 1992.

(33) A third perspective on what the 'prophesied return of the gods/ Anunnaki' means can be found in authors who focus on the significance of the upcoming end of the current Mayan 5,200 year cycle. According to John Major Jenkins, the Mayans were aware of the way the solar ecliptic plane comes into alignment with the galactic plane on periodic basis. This makes it possible for more intense cosmic energies to reach the earth from the galactic core. There is then a corresponding increase or decrease in the level of human consciousness, i.e., parts of the brain either go online or off-line, as these Mayan cycles go through their different phases. According to Jenkins, the year 2012 corresponds to the end of the current Mayan cycle and will lead to a rapid transformation of global consciousness.

(34) Numerous authors refer to this as a New Age of more enlightened global thinking and increased human potential.

(35) The 'prophesied return of the gods' may therefore signify a rapid in human consciousness as dormant parts of the brain come on line when the solar ecliptic comes into alignment with the galactic plane. In this explanation, humanity itself would develop 'god-like' powers which spontaneously become accessible to large numbers of humanity. A number of authors, for example, have been describing

the amazing abilities and psychic powers of an increasing number of children world wide.

(36) In sum, the available information on the 'prophesied return of the gods' can be understood to signify an important milestone in the growth of human civilization. The 'prophesied return' can be interpreted either literally or metaphorically to mean either a physical return of the 'gods'/Anunnaki; the return of a mysterious 10th planet to the solar system; or a rapid growth in the consciousness of humanity as the solar plane comes into alignment with the galactic plane. Despite the controversy over what precisely such a 'prophesied return' signifies, the factors in such a return that most pertain to the current political situation in Iraq and preemptive military intervention by the US can be identified and analyzed. The first factor is that an ET transportation device, a Stargate, or some other important ET artifact, may lie buried in the desert of Southern Iraq which presumably will play a role in the 'prophesied return of the gods'. Second, it is possible that there will be a return of a tenth planet that plays a critical role in return of the Anunnaki and/or which significantly impacts on the global environment. Finally, there is the potential for a rapid acceleration of human consciousness as the end of the Mayan Calendar, 2012, approaches.

Exopolitical Analysis of US policy towards Iraq

a Stargate existing in Southern Iraq that will play a role in such a 'prophesied return of the gods', then it is most likely that clandestine government organizations that greatly influence or control the Bush administration, are aware of the existence and the role of this Stargate. Iraq's President Hussein is most likely also aware of such a Stargate's existence as might be inferred by his architectural projects intent on reviving the grandeur of early Mesopotamian civilizations, and cementing his place as the restorer of Iraq's past glory.

(37) More significantly, his permission for a German team of archaeologists to resume excavations in the Sumerian city of Uruk after detailed underground mapping, suggests that this may be the location of the Sumerian Stargate. This knowledge of a buried Stargate, may also be part of the reason why the German government has been publically opposed to a US preemptive war against Iraq. If in fact both the

Hussein regime and the Bush administration believe that a Stargate lies buried in the sands of Southern Iraq, then there most likely exists a race to gain access to it and to control it. William Henry's thesis is that this is indeed the political underpinning of the continuing military conflict in Iraq.

(38) From the perspective of the Bush administration, control of the Sumerian Stargate would enable clandestine government organizations to continue their global campaign of non-disclosure of the ET presence.

(39) This is strongly implied by the Bush administration's penchant for secrecy and overturning many of the Freedom of Information initiatives from the earlier Clinton administration. Control of the Stargate, in addition to any other Stargate's that may have been established in the capitals of other civilizations, e.g., the Egyptian, Incan and Aztec; would presumably give clandestine government organizations greater leverage with ET races that are presently interacting with the planet, or are predicted to arrive on the scene in some event associated with the 'prophesied return of the gods'. At the very least, control of the Sumerian Stargate would allow clandestine government organizations to dictate the pace of global transformations that ET races promise to introduce to the Earth with their advanced technology, superior knowledge and heightened psychic abilities.

On the part of Hussein Regime, control of the Stargate would allow him to activate it and to fulfill prophesy by facilitating the return of an advanced race of ETs, the elite Anunnaki. President Hussein probably imagines that in return for his loyalty to the elite Anunnaki, he would be rewarded with a position of great global authority. Perhaps he would even see himself as some kind of human savior facilitating the return of the gods who would solve all of humanity's problem, and end the rule of clandestine government organizations perpetuating non-disclosure of the ET presence. Significantly, European governments such as Germany, and perhaps even France and Russia, may be giving themselves greater leverage in the future control of the Stargate by offering diplomatic cover for the Hussein regime as a *quid pro quo* for allowing the resumption of archaeological digging in Uruk. These governments and the clandestine organizations associated with them that have access to knowledge over the ET presence, most likely have deep suspicion over the willingness of the US to share information and control over the future of the Sumerian Stargate and any other ET technology discovered in Iraq.

Sitchin's thesis of an ancient ET presence in Sumer combined with the notion of a variety of ET transportation devices described by other authors in their research of ancient civilizations, and resumption of archaeological excavations of the first Sumerian capital Uruk in 2002, give support to William Henry's thesis of a Stargate that lies buried in the sands of southern Iraq. This provides important contextual information that is helpful in understanding the true motivations of the Bush administration in launching a preemptive attack on Iraq.

It may be argued that the Bush administration and the Hussein Regime are both in a race against time to gain access and control of the Stargate in the ruins of Uruk or some other location in Iraq, before the prophesied return of the Anunnaki. At the moment, a stalemate exists. Hussein controls the ground in Southern Iraq, and is permitting the German led excavations in Uruk, while the US led coalition controls the sky and is monitoring the situation. The Bush administration wants control of Iraq territory to take control of excavations of Uruk to uncover its buried Stargate, and closely monitor and control it. In contrast, Hussein wants to find and activate the Stargate for his greater glory and presumably the benefit of humanity.

The primary evidentiary support for the above discussion is admittedly thin for conventional public policy experts and may sound better suited to a fictional thriller than serious public debate. From a conventional perspective, a scattered assortment of independent archeological authors, radical exegetical interpretations of biblical texts, the writings of 'channels' of ET knowledge, speculative papers from astronomers hardly constitute a persuasive source of information for understanding the motivations of US foreign policy. The prevailing explanations of a Bush administration as either devoted to eradicating Weapons of Mass Destruction, and/or being driven by oil interests and imperial ambitions would predictably prevail for those unconvinced by the above sources.

However some important circumstantial evidence which lends plausibility to the Henry thesis of a Stargate as the true focus of the Bush administration or at least key interest groups behind it. The first piece of circumstantial evidence, are the overwhelming whistleblower testimonies confirming the existence of clandestine government organizations responsible for suppressing public knowledge of an ET presence, and which controls all official interaction with ET races.

(40) From an exopolitical perspective, then, the clandestine suppression
of a *contemporary* ET presence supports the conclusion that there
is also an active clandestine suppression of an *ancient* ET presence
which also has significant public policy implications.

The second piece of circumstantial evidence is the powerful diplomatic
support given by Germany and France to the Hussein Regime in warding
off a preemptive military strike. So powerful has been this support that the
US Secretary of Defense, Donald Rumsfeld, disparagingly referred to them
as the "Old Europe" in response to a reporter on January 22, 2003:

"You're thinking of Europe as Germany and France. I don't. I think
that's 'old Europe.' If you look at the entire NATO Europe today, the center
of gravity is shifting to the East. And there are a lot of new members. And if
you just take the list of all the members of NATO and all of those who have
been invited in recently—what is it, 26, something like that? [But] you're
right. Germany has been a problem, and France has been a problem."

(41) Rather than backtracking on what was a diplomatic bombshell, the
Bush administration has instead continued to go to extraordinary
lengths to isolate the German and French positions on Iraq. For
example, the administration encouraged the leaders of Spain,
Portugal, Italy, the United Kingdom, Hungary, Poland, Denmark
and the Czech Republic, to write a letter to the Wall Street Journal
on January 30 that said ``the Iraqi regime and its weapons of mass
destruction represent a clear threat to world security."

(42) Rather than merely an intense diplomatic debate over different
policy positions on Iraq, the striking language and positions taken in
this debate suggests a more fundamental conflict over issues hidden
from the public view. It is likely that there exists a factional struggle
between clandestine government organizations set up to deal with the
ET presence in the US, with rival organizations created in Germany,
France and also Russia.

The third piece of circumstantial evidence is the resumption of
excavations of the first capital of Sumer, Uruk, by a German archeological
team in 2002. Given the prominence of Uruk and its likelihood as the site
for a Sumerian Stargate, then resumption of excavations raises questions
over why they were resumed at this time and what is being sought. Given
that political tensions in Iraq had not significantly diminished in 2002

with it being a likely source of future military conflict with the US, it can be suggested that there are powerful hidden motivations for what on the surface appears to be a purely scientific dig of an ancient Sumerian capital.

The fourth piece of circumstantial evidence was the destruction of the Space Shuttle Columbia during its final descent on February 1, 2003, at an approximate height of 38 miles and travelling at Mach 18. One of the astronauts was the first Israeli in Space, Colonel IIan Ramon from the Israeli Airforce. Col Ramon reportedly played a role in the Israeli attack on Iraq's nuclear facility in Osirak in June 1981, and there has been speculation that his mission involved intelligence gathering over Iraq during the Shuttle's orbits. The destruction of Columbia occurred 16 minutes before touch down when its fuel tanks would have been virtually empty. A likely source of the Shuttle's destruction, given the speed and height of the Columbia, would have been some form of attack from an organization or state possessing military capabilities well beyond any terrorist groups and indeed most nations. The likely cause would have been a clandestine government organization that desired to send an important message to its US rivals over the threatened preemptive attack on Iraq.

When all the primary and circumstantial evidence is put together, what emerges is a very plausible case that supports Henry's thesis of a power struggle that goes to the heart of the ET presence and the continued clandestine suppression of ET related information and its full implications. The interpretations of the motivations of the Bush administration in launching preemptive war on Iraq in terms of the concerns raised in Bush's 2003 State of the Union address, or the corporate and imperial interests suggested by his critics such as Robert Fisk and Michael Lind, can all be described as part of the surface layer of motivations driving the Bush administration. At a deeper level, it is likely that there is great anxiety by clandestine government organizations in terms of what would happen if Hussein, with the support of the German and other European governments, gained access to the Sumerian Stargate or other ET technology buried in Uruk, or if the Stargate were to somehow reactivate without clandestine government personnel present to monitor and control the Stargate. President Bush's State of the Union address outlining the need for a preemptive attack on Iraq, in most likelihood masks a hidden agenda to gain access to the Stargate or other ET technology in Uruk and elsewhere in Iraq. Such access would presumably perpetuate clandestine government control over global resources and information at a time of increased ET activity and influence.

Conclusion: Policy Implications and Recommendations

If the exopolitical perspective is a more accurate description of the motivations driving the Bush administration in pushing for a preemptive war on Iraq, then the following policy recommendations can be made.

(43) First, the quality of evidence substantiating an historic ET presence and clandestine government cover up has a significant degree of credibility and persuasiveness. This supports the creation a new field of public policy, exopolitics, which would study the historic ET presence in terms of its implications for contemporary public policy.

Second, there is a need to promote official government disclosure of the historic ET presence and/or the impending return of these ETs; and to make more representative the policy making process that has evolved in government responses to such information.

Third, evidence suggests that the present military preparations for a war against Iraq have little to do with weapons of mass destruction, but are designed to perpetuate US clandestine government control of information concerning the historic and present ET presence. Such a preemptive war should therefore be stopped and a resolution between the US, Iraq and interested European governments should be encouraged.

Fourth, evidence suggests that the Iraq conflict and the destruction of the Columbia Space Shuttle mask a deep factional struggle between clandestine government organizations associated with different national governments that were initially created to deal with the ET presence. It is recommended that there is public disclosure of these organizations and their efforts in monitoring and responding to the ET presence, and that these organizations become accountable to elected public officials.

The final policy recommendation is that there needs to be more effort in determining the extent to which congressional and legislative oversight is required for organizations created in different countries to deal with all aspects of the ET presence, both past and present, and on the implications of a projected return of a race of ETs associated with the birth of human civilization.

This paper suggests that the best mechanism for responding to the existence of ancient ET technology in the ancient Sumerian capital of Uruk and/or elsewhere, is a willingness by major world governments and associated clandestine organizations to share information and control over these ET assets. A preemptive war conducted largely for the control of a 'Stargate' in Uruk which pits the US and its allies, against an Iraq which is tacitly supported by key European nations, could be calamitous if indeed the 'prophesied return' signified an actual physical event involving the ancient ET race that played a role in the start of human civilization. Competing clandestine government organizations struggling through a proxy war over the control of ancient ET technology in order to prepare for those events corresponding to the 'prophesied return of the gods', would hardly send the best example of a mature humanity responsible enough to continue to exercise sovereignty over the Earth's resources. The Columbia Space Shuttle may well have been a high profile victim of such a proxy war intended to send a message to US based clandestine organizations over the preemptive war against Iraq. Human sovereignty may therefore be at stake at the very time where there exists an opportunity for a rapid movement forward in the evolutionary growth of human consciousness. It is up to all humanity to decide how we respond to the challenge posed by clandestine organizations struggling over Iraq's historic resources to further their respective secret agendas.

-END NOTES-

 -FAMILY BRIEFING DOCUMENT-

-PUBLIC INFORMATION-
"The Disclosure Project"
A compilation By Rogue Sabre

The Disclosure Project, (*www.disclosureproject.com*),
began in 2001 and was spear headed by Dr. Stephen Greer.

Dr. Greer assembled 400 Government and Military personnel, to include Generals and Ministers of defense, as "Whistle Blowers" and witnesses to come forward to Congress and swear under oath that they were directly involved in, among other things, covering up and keeping from the public, the fact that "WE ARE NOT ALONE." Swearing, indeed, that we have made direct contact with Extraterrestrials' and have been working with them, in some cases, to back engineer their technology. These individuals have given their statement to congress and all of the interviews can be seen for free on youtube.com thanks to Dr. Greer and Project Camelot.—Sabre

THEIR STATEMENT:

The Disclosure Project is a nonprofit research project working to fully disclose the facts about UFOs, extraterrestrial intelligence, and classified advanced energy and propulsion systems. We have over 400 government, military, and intelligence community witnesses testifying to their direct, personal, first hand experience with UFOs, ETs, ET technology, and the cover-up that keeps this information secret.<>

On Wednesday, May 9th, 2001, over twenty military, intelligence, government, corporate and scientific witnesses came forward at the National Press Club in Washington, DC to establish the reality of UFOs or extraterrestrial vehicles, extraterrestrial life forms, and resulting advanced energy and propulsion technologies. The weight of this first-hand testimony, along with supporting government documentation and other evidence, will establish without any doubt the reality of these phenomena.

"Never doubt that a small group of thoughtful committed people can change the world; indeed, it is the only thing that ever has."
—Margaret Mead, anthropologist

"Few men are willing to brave the disapproval of their fellows, the censure of their colleagues, the wrath of their society. Moral courage is a rarer commodity than bravery in battle or great intelligence. Yet it is the one essential, vital quality for those who seek to change a world which yields most painfully to change."
—Robert F. Kennedy 1966 Speech, US Democratic Politician

BACKGROUND BRIEFING POINTS

FOR CONGRESSIONAL HEARINGS & LEGISLATION

- The Disclosure Project is a non-profit research effort that has, since 1993 when it began as Project Starlight, been identifying top-secret military, government and other witnesses to UFO and Extraterrestrial events.
- To date, several hundred such witnesses have been identified throughout the world and spanning every branch of the armed services, the NRO, DIA, CIA, NASA, the former USSR, and other agencies and countries. Over 100 have been videotaped, thus far; 70 have been transcribed into edited testimony. A four hour videotape summary of testimony and an over 500 page briefing document is available that contains excerpts of this historic testimony.
- The weight of this testimony, along with supporting government documents and other evidence, establishes beyond any doubt the reality of extraterrestrial life forms, UFOs, or extraterrestrial vehicles, and advanced energy and propulsion technologies resulting from the study of these vehicles.
- The testimony and evidence proves that these vehicles have been tracked on radar on many occasion, have landed and/or crashed on terra firma, and have been retrieved and studied by specialized and compartmentalized projects. Advanced technologies which have been identified from the study of these vehicles, once disclosed, will replace currently used forms of energy generation and propulsion. These technologies will enable the Earth to attain a sustainable civilization without pollution, energy shortages, or global warming. These technologies are already fully operational. They have been developed within super-secret, unacknowledged special access projects. In short, the definitive solution to the world's energy, pollution, and poverty problems exists within compartmentalized projects that need planned disclosure and relevant legislation.
- The programs controlling this issue are operating outside of legally required Congressional oversight. Even Presidents have been left out of the loop, deliberately deceived, and denied access. Therefore, urgent action is needed on the part of Congress, the White House, and other institutions to obtain the necessary oversight and control

of these operations to ensure that these now-classified technologies are prepared for disclosure and the eventual near-term application for world cooperative energy generation and propulsion.

- A clear and on-going threat to the national security and world peace has arisen because of unauthorized covert actions that have led to the targeting and downing of these extraterrestrial vehicles and to related covert plans to weaponize space. Since it can be proven that we are sharing space with other civilizations, it is critical that a full disclosure of this long suppressed information take place, and that the National Missile Defense System (NMD/BMD/SDI.) be re-evaluated by policy makers in light of these revelations.

- There is no evidence that these extraterrestrial civilizations are hostile to humanity or the Earth. Rather, the testimony shows that they are very concerned about nuclear and space-based weapons systems, and human warfare. Therefore, a cooperative world policy and law must be immediately established to prohibit the targeting and striking of these vehicles.

- Urgent Congressional, White House and UN action is needed to allow any and all witnesses to testify under oath so that a full, honest and open disclosure may occur this year, 2001, including witnesses with high level security clearances.

- A US Presidential Executive Order is needed to protect these military, government, and other witnesses, and to declassify secret projects and their related technologies.

- The world community needs to research and develop diplomatic programs and protocols, laws and treaties to address this issue and to interface with these civilizations in a manner that is peaceful, non-violent and mutually beneficial.

WE, THE PEOPLE, CALL ON THE U.S. CONGRESS:

- To hold open, secrecy-free hearings on the UFO / Extraterrestrial presence on and around Earth.
- To hold open hearings on advanced energy and propulsion systems related to the subject that, when publicly released, will provide solutions to global environmental and other challenges.
- To enact legislation which will ban all space-based weapons.
- To enact comprehensive legislation to research, develop and explore space peacefully and cooperatively with all cultures on Earth and in space.

President Eisenhower's farewell address to the nation, January 1961

"In the counsels of Government, we must guard against the acquisition of unwarranted influence, whether sought or unsought, by the Military Industrial Complex. The potential for the disastrous rise of misplaced power exists, and will persist. We must never let the weight of this combination endanger our liberties or democratic processes. We should take nothing for granted. Only an alert and knowledgeable citizenry can compel the proper meshing of the huge industrial and military machinery of defense with our peaceful methods and goals so that security and liberty may prosper together."

(This speech meant, now that the projects had been privatized by companies that had been opened up by the Rockefellers' and were being funded by Government grants, that the President himself could no longer over-see these projects because there were just going to be too many to follow, that it was up to those personally involved to make sure the right thing was done on behalf of humanity, and it was up to the rest of us out there to keep a suspicious and watchful eye over what was going on in this complex as a whole.)
—Sabre

(Another agency founded and operated by Dr. Steven M. Greer is CSETI, *www.cseti.com* It is this agencies aim to actually educate people about the ET presence, to provide them with a short training course and along with Dr. Greer and his team, go to known UFO sighting locations and attempt to "make contact".)
—Sabre

THEIR STATEMENT:

The **Center for the Study of Extraterrestrial Intelligence** (CSETI)
is an international nonprofit scientific research and education
organization dedicated to the furtherance of our understanding of
extraterrestrial intelligence. **CSETI** was founded in 1990 by Dr. Steven
M. Greer, who is the International Director. CSETI's projects include
the *CE-5 Initiative* and the *Disclosure Project*.

CSETI is especially interested in the ETI-Human relationship and in the
peaceful furtherance of this relationship. A thorough review of existing
data and documents concerning "UFO's" indicates that the Earth has
been visited by ETI and Extraterrestrial Spacecraft (ETS) for decades, if
not centuries, and that this contact has intensified since 1947.

There is a great deal of compelling evidence for the ongoing contact
between ETI and humans. This evidence includes the existence and
authentication of top secret documents confirming the retrieval of crashed
ETs and their occupants, the numerous photographs and films showing
ETS, the reports of thousands of credible civilian and military observers,
and the presence of associated landings.

CSETI is committed to the thoughtful long-term development of bilateral
ETI-Human communication and exchange, and open public education on
the subject. We regard this endeavor as the most important scientific and
educational project of the next thousand years and invite you to join us.

CSETI is supported through the voluntary efforts and contributions of
its friends around the world, and **we welcome your involvement now!**

CSETI trainings and workshops are designed to facilitate in-depth
analysis on all aspects of ETI studies. Topics include: human-initiated ETI
contact; origins and motives of current ETI visitors; the ETS' make-up;
peaceful applications of ETI technology and 'new' physics; non-harmful
disclosure for all peoples of the world, security and diplomacy issues;

world peace and unity as a prerequisite for extensive ETI contact and universal peace.

The CE-5 Initiative

Definition: CE-5 is a term describing a fifth category of close encounters with Extraterrestrial Intelligence (ETI), characterized by mutual, bilateral communication rather than unilateral contact. The CE-5 Initiative has as its central focus bilateral ETI-human communication based on mutual respect and universal principles of exchange and contact. CE types 1-4 are essentially passive, reactive and ETI initiated. A CE-5 is distinguished from these by conscious, voluntary and proactive human-initiated or cooperative contacts with ETI. Evidence exists indicating that CE-5s have successfully occurred in the past, and the inevitable maturing of the human/ETI relationship requires greater research and outreach efforts into this possibility. While ultimate control of such contact and exchange will (and probably should) remain with the technologically more advanced intelligent life forms (i.e., ETI), this does not lessen the importance of conscientious, voluntary human initiatives, contact and follow-up to conventional CE-s types 1-4.

CSETI is the only worldwide effort to concentrate on putting trained teams of investigators into the field where 1) active waves of UFO activity are occurring, or 2) in an attempt to vector UFOs into a specific area for the purposes of initiating communication. Contact protocols include the use of light, sound, and thought. Thought—specifically consciousness—is the primary mode of initiating contact.

(Below is an open letter from Dr. Greer, as the CSETI representative, to MJ-12/PI-40, aka the Majestic 12)—Sabre

Open Letter to PI-40*

By Steven M. Greer, M.D.
12 September 1996
Copyright CSETI 1996

From Chapter 42 of the book "Extraterrestrial Contact—The Evidence and Implications", available for order at *http://www.disclosureproject.org/shop.htm*

* Note: PI-40 is the acronym last mentioned to me as the code for the group controlling covert operations dealing with the UFO/ET issue. It is not known if it is accurate or currently in use.

Dear PI-40,

The time has come for us to discuss directly and openly plans related to a disclosure on the existence of ET life forms.

As you know, we are working with important world leaders regarding a near-term disclosure on this matter which is intended to be factual, evidence-based and non-alarming. We have become aware of a number of your operations, both historically and currently, related to ET reconnaissance, research and reverse-engineering. Since capability on your part for disclosure has been reached, we feel it important to proceed with a dialogue with those controlling current covert operations so that an optimal and safe outcome occurs.

Certainly, you are aware of the profound and far-reaching implications of this subject, and it is our desire for this information to come out in as positive and in the least-destabilizing manner possible. Ideally, this disclosure would occur:

- With the cooperation, whether explicitly or tacitly, of your operations
- Without recriminations or vilification of your operations or people, provided non-interference is assured
- Without the application of 'false or deceptive indications and warnings' related to this matter
- While avoiding either the demonization or deification of the ET life forms

I have asked the President to provide for amnesty for your group and its personnel, provided there is cooperation and non-interference, and restraint on your part is observed. Neither the United States, nor the world, can afford a conflict over this issue.

But we are aware of certain projects of your group which we feel run counter to a safe and effective resolution of this problem. We are requesting that a 'stand-down' order be issued regarding these types of destabilizing and dangerous projects, including:

- Tracking with intent to target, and targeting with intent to destroy, ET assets
- Covert human-controlled 'abduction' programs and related genetic, biological and psychological disinformation projects
- Development and implementaion of 'false and deceptive indications and warnings', which intend to stage false or hoaxed ET events of a violent nature, for the purpose of consolidation of control and other manipulative purposes
- The continued general disinformation and denial control points in the scientific, political, academic and media communities

There are a number of areas where I feel healing and resolution are needed, especially the conflict and estrangement between:

- Various ET groups and your entity
- Your operations and those of the traditional military and governmental operations and leadership, as well as the public
- The ET groups and humanity in general.

How may we help?

Personally, I am exquisitely aware of the time in which we live and the unprecedented risks and changes of the coming one to two decades. And while in the short term I know we are bound to encounter the general turbulence and chaos of a global phase transition, in my heart I am certain that the intermediate and long term future is exceedingly hopeful. I feel we must look to that far-horizon, while yet being mindful of the current exigencies.

There is a Chinese proverb: Unless we change directions, we are likely to end up where we are headed.

I feel where you are headed with this issue is currently very dangerous, and it will take vision, courage and some faith to change it. Not every power which can be acquired should be, and not all acquired power should be used. It is time for restraint and wisdom; perhaps together we can find the safest path to the next century.

Presently, we are aware of certain covert projects which, if pursured further, could result in a worst-case scenario and outcome for the entire planet. There is no point in pursuing a scorched-earth policy, or should

I say an interplanetary version of the old Mutually Assured Destruction (MAD) policy of the Cold War.

The misunderstandings, fears and suspicions of the past need to be suspended, for the sake of all of us. The recent targeting and destruction of ET spacecraft, while no doubt demonstrating a certain sophistication and prowess, may lead to catastrophic consequences in the future. Can we work together to find a safer and more sustainable way to deal with our differences with other life forms?

We should begin working together now, throughout the world, to establish the foundations of global and universal peace. The means for doing so are available to us; it is our responsibility to act and implement this plan.

With the establishment of universal peace, then and only then will we be able to disseminate the infrastructural and technological means of creating a high technology, sustainable civilization on this planet. A civization which will be using advanced, zero-point and related technologies for peaceful purposes, and which will no longer need to destroy this extraordinary planet on which we all depend for our survival.

Surely, we cannot establish world peace, and continue to have conflicts with our planetary neighbors. And yet without world peace, and the peaceful use of these new technologies, we lack a civilization capable of surviving yet another century.

Ultimately, then, we are in a crisis of spirit and vision, and only secondarily of outward circumstance. I would like to offer to you to help work on this issue, to come together to find a resolution to this most challenging problem.

We would like to help you in creating a dialogue with the ET peoples which would lead to resolution and peace. And to bring about a truthful and hopeful disclosure on this subject which would serve as the catalyst to the establishment of global and universal peace. And then to plan the eventual dissemination of the technological means to create a truly sustainable civilization on this planet.

I know this can be done. Yes, there will be some turbulence along the way, but it will be worth it. It will not be easy, but the rewards are so great, no effort can be too much.

I would invite you to contact me, at any time and at any place of your choosing, to pursue this further. We have many people on our team who can help us create a future civilization on Earth which can grow in peace for hundreds of years. This future is waiting for us to create it.

My roots are humble here in North Carolina; I am no saint, and lay claim to no nobility. By my father was part Cherokee, and there is a Cherokee teaching which says that all that we do should be done with consideration for seven generations to follow.

For the sake of our children, and our children's children's children, I hope we can work to establish a time of peace on Earth, and beyond.

I look forward to hearing from you soon.

Sincerely,

Steven M. Greer, M.D.
Director of CSETI

-PUBLIC INFORMATION-
Disclosure Project
"Whistle Blowers"

-PUBLIC INFORMATION-

This information has been sourced from *www.projectcamelot.org* Project Camelot is diligently driven, managed and operated by Bill Ryan and Kerry Cassidy. This is a web based site that has charged itself with the highly esteemed duty of personally video interviewing all of the "Deep Black" Operators willing to come forward, who were directly involved in the Military Industrial Complexes Black Budget Operations, dealing directly or indirectly with the back engineering of Extra-Terrestrial space craft, their propulsion systems, technical capabilities, and weaponry. All of the Black Op's Operators who have come forward as "Whistle Blowers" have given their statements to Congress and their amazing testimonies have been video recorded by Project Camelot. All of these videos' are shown in their entirety and can be viewed for free at *www.youtube.com* by typing Project Camelot in the search box followed by the name of any individual of interest.—Sabre

THEIR STATEMENT:

Welcome to Project Camelot

Overview and mission statement:

- To provide researchers, activists and 'whistleblowers' with access to all forms of media in order to get the truth out.
- Our focus includes but is not limited to the following:

  - extraterrestrial visitation and contact
  - time travel
  - mind control
  - classified advanced technology
  - free energy
  - possible coming earth changes
  - revealing plans that exist to control the human race.

- To establish 'safety in numbers' and unite these disparate factions under an umbrella of protection for activists and 'whistleblowers' who may have concerns for the safety of themselves and their loved ones.
- To provide a tribute to all activists in paradigm-challenging fields who have worked for the benefit of humanity . . . and who have suffered or been silenced for speaking the truth.

In Tribute:

- Many courageous, free-thinking individuals have suffered for their commitment to help humanity.
- We honor them here, and offer protection and support for those who follow in their path.
- We are dedicated to getting the truth out.
- *Click here* for our Tribute Page in which we honor a number of exceptional men and women who have paid the highest price for speaking the truth.

Project Camelot . . .

- Enables activists in paradigm-challenging fields to make a firm statement about their work, their intentions, and their positive state of mind.
- Provides access to all forms of media. The objective is to get the truth out . . . and facilitate appropriate, secure publicity for all those with information to share, but who fear reprisals for their stand.

We shall prevail.

Being of sound mind, heart and spirit, we each declare the following to be true:

- We have no intention of ending our own lives.
- We will not tolerate suppression of our truths, our ideas, our freedoms, or our work.
- We stand together to support others in the expression of truths and freedom to speak out . . . no matter how radical those ideas may seem.
- Standing for freedom takes courage; together we shall be strong in the face of all odds.
- If it is ever claimed that we have committed suicide, disappeared, been institutionalized, or sold out financially or in any other way to self-interested factions, we declare those claims false and fabricated.
- We testify, assert and affirm without reservation, on behalf of all those who have dedicated their lives to the ending of secrecy and the promotion of freedom of thought, ideas and expression . . . that we shall prevail.

First they ignore you, then they laugh at you, then they fight you, then you win.
Mahatma Gandhi (1869-1948)

MEN WANTED FOR HAZARDOUS JOURNEY. LOW WAGES,
BITTER COLD, LONG HOURS OF COMPLETE DARKNESS.
SAFE RETURN DOUBTFUL. HONOUR AND RECOGNITION IN
EVENT OF SUCCESS.
Ernest Shackleton, Antarctic explorer (1874-1922)
(The advertisement above, placed in a London newspaper in 1912,
inspired nearly 5,000 replies.)

Project Camelot
SUPPORT FOR RESEARCHERS AND WHISTLEBLOWERS

Project Camelot is a website formed in support of researchers and 'whistleblowers' who challenge current paradigms. Our aim is to provide safety in numbers and thereby establish a climate of trust and support across many disciplines on behalf of disclosure of truth with an emphasis on the freedom to explore.

Bill Ryan and Kerry Cassidy
PROJECT CAMELOT
8 January 2009

All of the interviews contained below are available to watch for free on www.youtube.com by typing project Camelot followed by the appropriate name of the individual of the desired view in the search box.

I highly recommend the Dan Burisch video as well as the John Lear interview. (Sabre)

- **VIDEO INTERVIEWS**
- **VIDEO INTERVIEWS**
- **SPECIAL REPORTS**
- **SPECIAL REPORTS**
- **AUDIO INTERVIEWS**
- **AUDIO INTERVIEWS**

—click below for each interviewee—
—click below for each interviewee—

Benjamin Fulford
Benjamin Fulford
Dr Bill Deagle
Dr Bill Deagle
Bill Hamilton
Bill Hamilton
Bill Holden
Bill Holden
Bob Dean
Bob Dean

Boriska
Boriska
Sgt. Clifford Stone
Sgt. Clifford Stone
Dan Burisch
Dan Burisch
Dan Sherman
Dan Sherman
David Corso
David Corso
David Wilcock
David Wilcock
Duncan O'Finioan
Duncan O'Finioan
Gary McKinnon
Gary McKinnon
George Green
George Green
Gordon Novel
Gordon Novel
'Henry Deacon'
'Henry Deacon'
James of WingMakers
James of WingMakers
Jessica, Crystal Child
Jessica, Crystal Child
*** NEW *** Jim Humble
*** NEW *** Jim Humble
Jim Sparks
Jim Sparks
John Lear
John Lear
Leo Zagami
Leo Zagami
Luca Scantamburlo
Luca Scantamburlo
Marcia Schafer
Marcia Schafer
Michael St. Clair

-PUBLIC INFORMATION-
Rogue Sabre
2012

-FAMILY BRIEFING DOCUMENT-

The following information is a personal briefing I wrote to my cousin.
I am adding this in as it makes this file easier to comprehend when told
from the beginning of time, and gives the reader a point of reference as
to where it is this whole journey for mankind begins.
—(Sabre)

Well Bob,

The briefing that I have is huge, and the story that this planet and its'
world leaders have to tell, and have been covering up, is the biggest secret
and the greatest story ever to be told in the history of mankind.

The fact that it hasn't been told, is because it's just too big. I mean
too big in more than just long, I mean it is almost unbelievable. Quite
frankly, this has been kept a secret for so long, solely on the fact of its'
sheer unbelievability. Deep Black Op Agents want the story to be told,

insisting that it must, and most of them want out. Some have even tried to leave only to wake up "patched".

"Patching" is what they do to guys who are deep black operators who try to run or leave their respective program sectors. What a "patch" was, was a surgical implant of a bio-regulation apparatus that is placed right into the center of your chest that you or anyone else could actually see, and was full of a serum that keeps you alive. (Now days I'm sure it is as simple as a trans-dermal patch) This medical procedure integrates this "patch" into your body as part of its biological life sustaining systems and if you don't "check in" every couple of days to get your patch refilled with serum you'll die. It's a life support insurance policy and a way for the agency to control those who need this type of control. Some of the projects that were developed are still very much under way and proceeding regardless of anything, and they are so compartmentalized only those directly involved even know of their existence.

This is why Obama is going through every last single program that is federally funded with a fine tooth comb. When he took over office he sent out his emissaries' to all of the companies and agencies receiving federal funding for complete disclosure and a Presidential update on the projects funding usage and their current status in an attempt to bring them all back under direct control of congress and the presidential office. In some instances those messengers were not received well.

NASA told him he didn't have a need to know and Cape Canaveral had shut the door on his investigative team on more than one occasion. Now it is thought he might be eliminating all projects that refuse to comply with full disclosure of its operations.

(Down played news coverage aired like this on FOX-40), "Apparently there's *also* something about what's going on with *NASA and NASA Administrator Mike Griffin* and an investigation by Lori Garver who is heading up a 'space transition team' of Obama's to check into how money is being spent."—And it was never mentioned again.

As a result there is a scramble going on by these agencies/companies and their clandestine department leaders to secure funding by whatever means necessary.

George Bush Sr. was the head of the CIA before being President, due to that, he is the only President since Truman to know the entire truth, and the detailed history, of this planets interaction, and projects with, or related to, the Extraterrestrial and trans-dimensional beings' that we have made contact with.

Have you noticed over our short written history that everyone has been obsessed with ruling the Middle East?

Everything I am about to tell you has been deciphered off of the walls of every temple on the planet, including the pyramids, by archeologists' from every country from the most prestigious learning institutions on this globe over the last 200 years and tracked all the way back to Sumer. The science is all in, and for the most part, is incontestable. The man who is responsible for bringing us a great deal of this information is Dr. Zacharia Sitchen. READ HIS FIVE BOOKS, starting with "The 12th Planet". His discoveries and writings are not contested and are still to this day read at the greatest learning institutions and Universities' on the planet. Many of the tablets containing even more information on this subject, written by this civilization are housed in the largest underground library on the planet at the VATICAN.

This story is as old as our solar system and believe it or not begins with the creation of it, as well as our planet, and eventually us.

Everything is the manifestation of sound. All things are made up of atoms with sub-atomic properties that make them unique from each other, each with a different number of protons and electrons. These differences make atoms vibrate at different frequencies. Some atoms are sub-atomically structured to create or be seen as wood, steel, granite, etc . . . The manifestation of light has a sound. Each color in the spectrum of light vibrates at a certain speed or frequency and therefore creates its own unique sound. Believe it or not, it's the musical octave scale. 8 to be exact, and each one of our planets in this solar system vibrates to a note in this musical scale, C, D, E, F, G, A, B, C, and there are 3 harmonic frequencies known to most as the "Station Identification" musical chime used by none other than the National Broadcasting System (NBC). This all must be told for you to fully understand Bob, so don't think I'm just rambling on here. THOUGHT FREQUENCY AND THE MANIFESTATION THEREOF, IS ONE OF THE SECRETS TO THE PHYSICS BEHIND THE CREATION OF EVERYTHING AS WE KNOW IT.

As this solar system was creating itself, it started from the sun outward. Once it had gotten to the big planets, Jupiter and Uranus, the gravitational pull of these monsters pulled in a planet that was not (as we have been told by Zacharia Sitchen as read in ancient Sumerian texts) a creation of this solar system. It was a planet from somewhere else, but none the less, was pulled into the orbit of this solar systems sun due to the pull from our big planets. Our planets circle the sun counter clock-wise, but not this one, it rotates clock-wise. This one was 4 times the size of our current Earth and had a moon of its own, then Earth was known and documented by the Sumerians as being named Tiamet. Long story short, this moon belonging to this intruding planet ran into Earth/Tiamet and smashed it in half, knocking the unbroken half of Earth/Tiamet into its current orbit and then turning the other part of Earth/Tiamet into the great meteor belt, aka—"The hammered bracelet", this is written and understood in Sumeria by those who wrote it as truth as told to them by the Annunaki.

Now as these two planets passed each other there was a transfer of water and life. The both of which came from the intruding planet, Nebiru—Planet of the crossing, or planet X, or the 10th planet.

This planet was inhabited by a race of beings called the Annunaki. Which, when translated means,—(Those who from heaven to earth came)

Now the story on these temples walls, and written in these ancient texts, is that these Annunaki came here for gold, to repair the atmosphere of their planet, which was damaged somehow. And after doing the work themselves to mine this planet for its gold for a long time, they decided to fashion a worker or a slave to do the work for them, and eventually created mankind by mixing their DNA with the DNA of a creature that was already here on Earth.

Now these ETs are who taught mankind everything we needed to know to get us rolling along.

So, these creatures/astronaughts created man and mined this planet for gold and its mineral resources. They also created the most beautiful temples on the planet in commemoration of themselves and as a way to utilize and harness the power of the stars, this planet, and/or other planets. This is a physics and science that we have been trying for a long time to understand and at this point we've just about got it figured out. It is written that the first Man, the Adam, was taught and understood all of this knowledge and his understanding was even compared to that of his creator.

Ok, now before I get ahead of myself, I must reiterate, READ ZACHARIA SITCHENS WORK. It will inform you about everything that happened with these astronauts prior to the history of man as well as mans history up to the great flood recorded in the King James Version of the Bible and beyond as well as bring clarification of it to you. Now let me get to what has been going on around this planet with its governments for about the last 60 years.—(Sabre)

The following information is sourced from *www.projectcamelot.org*

The Big Picture:
A hypothesis

The following is a detailed summary of our tentative, current understanding of the core issues related to the presence of the visitors, and the possible earth changes that may be ahead of us.

By its very nature it is subject to change and revision at any time, as the Big Picture comes into focus and becomes clearer. It is also intended to provoke thought and to invite readers to do their own research and analysis. It would be extraordinary if every detail below were correct.

Much of this information is also elsewhere on this site (in *interview* form or on the *Henry Deacon pages*) and elsewhere, but the summary below may be a useful synopsis for the purposes of discussion.

The Roswell Catastrophe

The visitors who crashed at Roswell were future humans[1]. They were not from another planet[2], but from a future Earth—stepping (which may be a better word than *traveling*) back in time to 1947 to attempt to deal with serious problems which had occurred in their history[3]. Their mission was to try to change their past by creating an alternative branch of their own timeline, so that particular events—about to happen to us in our very near future—would not actually occur.

The Roswell visitors were on a purely altruistic mission. They did not have to do this, but chose to . . . out of compassion. But the mission went disastrously wrong—not just because they crashed (an accident caused by high-powered radar—later the military realized this and made use of radar as a weapon), but because they had a device with them which was their only means, as an orientation device in time and space, to get them home and back to their own time.

The device was a little box, highly advanced and multifunctional in nature, and was far smaller than the "Looking Glass" that Dan Burisch[4] and Bill Hamilton[5] describe as being subsequently utilized by military scientists in various experiments. When the box was acquired and investigated by the military, this became a catastrophe in itself. It made the timeline problem many times worse, because this both introduced time manipulation technology to us at the wrong time . . . and also told the military what lay ahead.

It can not be stressed too strongly how totally calamitous for us all the Roswell incident was. It was a major, major setback, right at the start of the future humans' project to help fix the problem. Acquiring a device such as that which the Roswell visitors carried would immediately alter the timeline which the future humans were trying to change in the first place . . . so there would then be two timelines that need fixing, and not just one.

NOAA, the Dark Star, and Global Warming

A small organization within NOAA (the *National Oceanic and Atmospheric Administration*) is aware of what scientists there sometimes call the "second sun". This is a massive astronomical object, possibly a brown dwarf[6], which is on a long elliptical orbit around our own sun on an inclined plane to the rest of the planets. To align with other researchers[7], we'll refer to this as the *Dark Star*. (This planet has now been identified as NEBIRU, the 12 member of our solar system)

The Dark Star is now approaching, and is causing resonance effects on our sun in various ways[8]. This is the cause of the warming of *all* the planets, not just the Earth[9]. This information is classified, but has been known for a number of years[10].

This issue is connected with the Roswell catastrophe described above. The problems the future humans were attempting to address were multiple, but principally featured a possible event triggered by a massive 'spike' of solar activity at some point in our currently near future.

We emphasize most strongly that this event is only possible (having been observed in Looking Glass devices in a possible future) . . . and, importantly, is now evaluated to be unlikely[11].

The increase in solar activity is caused only in part by the Dark Star, multiple factors being at play. These are complex. Some of them are on a galactic scale[12], and are associated with natural, periodic events which the Earth has suffered through a number of times previously. What makes this particular time completely unique for our planet is that there is a convergence of serious factors—such as global warming[13], overpopulation, and our propensity for choreographing war[14]—all of which combine with these major, cyclic and solar events to simultaneously threaten the well-being of ourselves and the biosphere.

The large-scale events are unstoppable. It's also unclear when the 'spike' of solar activity is due to be—though our understanding is that this is imminent, and could occur at any time in the next ten years or so. Although the issue has been considerably hyped, it's impossible not to observe that the year 2012[15] is right in the middle of this bell-curve of probability.

What is possible, however, is to minimize the effects of the solar event. Evaluated Looking Glass data concludes that there is a 19% probability of the worse case scenario occurring, with 85% confidence that that 19% figure is correct[16]. It seems we're off the hook . . . although no matter what timeline one is on, significant problems lie ahead with the man-made crises that surround us (exacerbated by solar activity).

Insurance: Underground Bases, Project Preserve Destiny, and the Martian Colony

Readers may by now have made the connection with the trillions of dollars[17] spent in recent decades by various military agencies on underground bases in a number of different countries[18]. Given the possibility of an imminent near-ELE (extinction level event)[19], a cynic would argue that it's a smart use of our tax dollars to ensure that at least some humans survive[20].

Other readers will connect this potential scenario with the story told by Dan Sherman[21], in straightforward, unassuming and sober fashion, in which he was trained as an IC (Intuitive Communicator) as a preparation for an unknown future event that would include all electronic communications being rendered unusable[22]. The project he was part of was called *Project Preserve Destiny* . . . the title of which gives us a very heavy clue to what this was all about[23].

Researchers who have looked into the claims of the infamous TV program *Alternative 3*[24] conclude in the main that the show was a hoax. There is clear evidence that the show was a dramatic portrayal of arguably fictional events. However, it would appear that it may have been accidental disinformation rather than a hoax *per se*. Certain events portrayed in the program[25] are markedly similar to the scenario presented here.

Among those is the existence of a substantial Mars base, established in the early 1960s and supplied by a combination of stargates[26] and an advanced, classified space program codenamed SOLAR WARDEN[27]. The Mars base apparently has a number of functions[28] but among those would certainly be the ensured survival of the human species should anything untoward happen on or to our home planet.

What Can Be Done?

As has been written above, the major, cyclic large-scale events that are set to occur are unstoppable by us or any other race, no matter how advanced their technology. But what can be mitigated are certain effects. Much remains under our control, although as every day passes the clock is ticking: carbon emissions are not helping[13]; choreographed and orchestrated war[14] adds to the chaos; and new diseases and possible pandemics[29] may be controllable if we act in harness.

This is no different a warning than has been stated by many other messengers in recent years, but we place this in a context where the stakes are rather higher than usually recognized. It IS possible to work together to make a difference in a situation in which the backs of the human race may be up against the wall.

On an individual level, more is possible. Michael St Clair[30] says simply: find a safe place, and do it now. Many others agree. It may be smart to avoid living on the coast, on low-lying land, on a fault zone, in a major city, or on the sides of a volcano. Fresh water is likely to be a major issue. Some analysts consider the southern hemisphere (Australia?) to be safer than the north, as the craziness of orchestrated war threatens.

Remember that global warming will really start to bite in the near future. The Atlantic Conveyor[31] may conceivably shut down, plunging Europe into a deep freeze. The warmer the Earth gets, the more evaporation there will be from the sea; so expect Katrina-scale hurricanes (in the west) and cyclones (in the east—they're pretty much the same thing), and disrupted and extreme weather patterns everywhere.

There is a metaphysical approach, too. Lynne McTaggart[32] in her new book *The Intention Experiment: Using Your Thoughts to Change Your Life* argues as she did in her previous work, *The Field*, that consciousness is a factor that can and does affect macro-scale events. We would wholeheartedly agree. Morphic fields[33]—templates or patterns of manifestations yet to unfold—are as powerful and influential as they are invisible; and morphic resonance is also amenable to the power of thought[34].

It's even possible to bootstrap oneself into an optimum (least worst) timeline . . . though there's no instruction book on a mechanical means to do this. Spiritual methodologies such as the many forms of Meditation and other Yogic practice, Buddhism, Sufiism and Shamanism can assist one here. (This is far from an exclusive list, and may include some of the major organized religions. Prayer, if well understood and applied by a mature and aware spirit, is certainly capable of working miracles.)

Times are changing, and global awareness (actually, a large morphic field) is gradually being heightened. The real question is, how fast? The way to put one's shoulder to the wheel is to add one's own strong and optimistic intention to the concept of the positive outcome. And like all such mental or spiritual gymnastics, this has to be real: it cannot be faked.

Footnotes:

1. This information has been reported by Dan Burisch. See our *interviews* page and also *Dan Burisch's website*.

    Not all visitors are time travelers. There are a great many races currently visiting this planet; many of these are genuine extraterrestrials. Reported numbers vary, but the respected witness *Sgt. Clifford Stone*, for example, says he knew of 57. Other witnesses have reported numbers of a similar order of magnitude.

    The motivations of the visitors are likely to vary as much as human agendas do. We can safely assume that some are benevolent and altruistic; some are evil and self-serving; and others may have agendas we can only guess at and may be beyond human understanding.

2. According to Burisch, some of the subsequent visitors were future humans who were (at that future time) based on other planets, and no longer on the future Earth.

3.   This matter is complex. According to Burisch, some future humans are benevolent, while others are actually seeking to ensure that history is *not* changed; see note 23 below.

4.   Dan Burisch has made detailed technical drawings, from memory, of the 'Looking Glass' device he was familiar with. We'll publish these as soon as we have clearance. Henry Deacon has said that he never worked with or encountered the form of Looking Glass that Burisch described, and therefore is not in a position to confirm it.

5.   Bill Hamilton described the Looking Glass in some detail on *this page* of his website. (Note: the image shown is not the real thing; this is taken from the movie *The Time Machine*.) The text is also reproduced *here*.

6.   See Andy Lloyd's excellent website *here*, and the Binary Research Institute's website *here*. For a series of excellent graphics depicting the Dark Star's possible orbit, click *here*.

7.   Researchers differ in some details, but the consensus is that this is probably a *brown dwarf*. Deacon only referred to it as a massive astronomical object, causing serious gravitational and other effects. Burisch has stated in personal correspondence that his best recall is that he had been told that this object was a small black hole.

8.   The mechanism is unclear, but appears to include electromagnetic, gravitational, and other resonance effects.

9.   This information is in the public domain. It has been observed, for instance that *Mars*, *Jupiter* and *Pluto* are all warming up. There are many other references. *Click here* to read David Wilcock's and Richard Hoagland's comprehensive and detailed research into 'climate change' affecting every planet in our solar system.

10.   A possible contender for the companion star was announced in 1983 by the Infrared Astronomical Satellite (*IRAS*) team, reporting a Jupiter-sized object at a distance of 550 astronomical units (550 times the Earth's distance from the sun). The report was withdrawn soon afterwards, though IRAS have always claimed this was not a cover-up. Deacon reports that the small organization within *NOAA* has known about the "second sun" for a decade or possibly much longer, but has never made any public reference to it.

11.   Burisch has stated on record that according to the most recent computer-analyzed Looking Glass data, there is a 19% chance of the worse case scenario occurring, with 85% confidence in that 19% figure.

12. Burisch and Deacon both report that the principal factors at play are large scale and galactic in nature, linked with long-term, recurring cycles of influence on the Earth. Read *this article* about the correlation between cosmic ray activity and global warming, for instance.

13. Al Gore, in his acclaimed documentary *An Inconvenient Truth*, makes the case for the harmful effect of increased carbon emissions with considerable impact. However, the real cause of global warming is very likely to lie with other larger scale causes (see note 9 above).

14. *The Report from Iron Mountain*, among other documents in the public domain, references the organized and orchestrated requirement for war in order to maintain macro-scale economic and social stability.

15. 21 December, 2012 is when the famous *Mayan calendar* comes to an end. Much has been written in speculation about the reasons why the Mayans saw no need to extend their calendar beyond that date. Deacon and Burisch state that this apparently exact date is not a precise prediction and that the solar events in question could occur at almost any time in the window 2007-2016.

16. See note 11 above. Because (according to Burisch) steps were taken to decommission certain devices which were in danger of triggering massively amplified earth changes at the time of the solar 'spike', Burisch is confident in this calculation. Deacon states the the situation is continually subject to change, and is less sure of the figures.

17. See *this site* for some details, and also many others.

18. There are persistent reports of underground bases in the US, the UK, Puerto Rico, France, Germany, Norway, Canada, Australia, South America and Antarctica.

    Click *here*, *here* and *here* for three interesting photos of giant tunnel boring machines.

    There are also *undersea bases*, confirmed by Deacon and researched by *Dr. Richard Sauder* and others.

19. Burisch reports that according to Looking Glass data, in the worse case scenario up to 94% of the Earth's population might be killed over the period of a small number of years following the catastrophe.

20. Any high-level military scenario planning would be based on the premise that not everyone could be saved if the world's population were under serious threat.

21. Sherman's story is told in his excellent e-book, *Above Black*.

22. It's accepted by scientists that one of the possible effects from a sufficiently major solar 'spike' is that on Earth all electronic communications might be rendered unusable for some time.

23. The title *Project Preserve Destiny* is chilling, implying that the project may be based on the premise that a cataclysmic future event has been observed through very high technology time-portal devices . . . and that it is to be ensured that what has been observed must occur. Burisch reports that a certain group of future humans, whom he calls *The Rogues* and who come from approx. 45,000 years hence, are committed to making sure that events in their history do indeed come to pass in our timeline and are not averted.

24. *Alternative 3* was shown by UK's Anglia TV on 20 June, 1977— although it was initially scheduled to be premiered on 1 April. Anglia TV made a statement the day after the program that it had been intended as a joke.

    Click *here* for to see a streaming video of the program itself (54 mins).

    Click *here* for a newspaper cutting in which Leslie Watkins, who authored the *book* of the same name, reveals that although he intended it as fiction in support of the TV program, he appears to have accidentally hit on something very real with the premise he portrayed.

25. The premise described in the TV program (and the *book*) was that to deal with the twin problems of overpopulation and global warming, there were three alternative solutions:

    Alternative 1: to drastically reduce the human population on Earth;
    Alternative 2: to construct a network of underground bases to safeguard a small, elite population;
    Alternative 3: to establish a 'Noah's Ark' colony of humanity's best and brightest off of the planet—on Mars.

26. According to Deacon, stargates are routinely used for transport to distant locations, 'travel time' being instantaneous.

27. Deacon reports that SOLAR WARDEN encompasses a small fleet of large, highly classified spaceplanes.

28. According to Deacon, the Mars base is multifunctional and has a large population of around 670,000 (not all of whom are present-day humans). Situated underground on the bottom of an ancient dried-

up sea, it has, apparently, been in existence for thousands of years but was recently re-established in the early 1960s by an international team. Its functions include stargate access to more distant locations still.

29. HIV, SARS, Avian ('bird') flu, and other modern scares—whether orchestrated or not—come to mind.

30. See our *interviews* page for more on *Michael St Clair*, a well-known visionary and astrologer who predicts major problems in the coming few years and who advocates prudently identifying safe places to live. (St Clair has already moved to live in a small community in a remote location.)

31. Click *here* for a summary of the importance of the Atlantic Conveyor, a critical component of the oceanic circulatory system. If the Atlantic Conveyor were to shut down (as might be triggered by a massive amount of cool water flooding into the North Atlantic from, for instance, the melting of Arctic ice or Greenland glaciers), then the warm waters of the Gulf Stream would no longer reach European shores and Europe would experience a much cooler climate . . . despite global warming.

32. Lynne McTaggart's book, *The Intention Experiment: Using Your Thoughts to Change Your Life*, is available *here*. She is also the author of the ground-breaking book, *The Field*.

33. The British Biologist *Dr. Rupert Sheldrake* introduced the concepts of *Morphic Fields and Morphic Resonance* in his revolutionary 1981 book *A New Science of Life: The Hypothesis of Formative Causation*. Simply put, Sheldrake argues that species create Morphic Fields— templates based on accumulated experience that have an an effect on the behaviors, attitudes and abilities of subsequent members of the same species—which influence others encountering the same situations at the same or a later time.

34. The relevance here is that humans can consciously create Morphic Fields, which influence others once a critical mass of intention is gathered. It's one way in which a small group, powerfully and positively focused, can gradually change the world.

-FAMILY BRIEFING CONTINUED-

In brief, what does all this mean? It means that on a certain day here on our planet there is supposed to be a huge cataclysm that wipes out 2/3rds of the worlds population, from 6 billion down to 2 billion caused by men on this planet who are messing around with "jump gate" technology. Future humans came back and explained to us that on Dec 21st 2012 at 11:11 am, Earth will be in the exact galactic center of the Universe. At this precious moment all the planets in our solar system line up like a big clock to the 12 o' clock position signifying the beginning of a new time, or era. This is a really good thing, as this time is to bring the spirit of God back into the Earth. Not the return of Jesus, but the return of the spirit of Jesus in the form of vibrations from the center of the galaxy retuning everything to the new frequency of space that we are traveling through. The planets have a huge influence over us. They bring us luck or lack of it. Have you ever heard the expression "Follow your lucky star and you will receive good fortune your whole life?" Well it's true and the Government is aware that we are on the cusp of our evolution. Part of this evoloution will give us the ability to traverse the dimensions and the closer we get to galactic center, gates to these dimensions will be opening up on this planet allowing you to physically travel from one planet to the next and I'm assuming one dimension to the next as well. The physics behind this is complex but it is explainable to a certain degree. I do not fully understand its function or process, but it is explained to a point by the ETs in a document called "The law of one—Harvest".

Anyway, as our planet gets closer to this galactic plane it begins to move through this super energetic part of space called a Photon Belt. And what the time traveler told us, was unbeknownst to us, is that this natural creative energy of the universe is flowing in right now and bringing with it a cleansing and in the process, our evolution. But men had artificial star-gates open during the arrival of this energy and it caused the planet to increase its rotation speed just a little bit. Now due to this increase in speed and the weight of the ice cap at the North Pole, it forced the world to act exactly like a spinning bowl of water, and it tipped over, instantly.

All of the Continents float on either water or magma, and the force of the flip caused all the Continents to slide, drift and/or sink, partially or completely. This is why they came back, to tell us to shut down the star-gates to avoid this catastrophe.

This information was given to Truman at Wright Patterson AFB during a meeting with the ETs and it is thought to have given him his first heart attack. Since then the military has taken no chances, and since 1950 they have been spending trillions of dollars to build huge underground bases that are completely self sufficient for a select contingency of humanity to escape to when the time is near. The Operation is code named "Project Preserve Destiny" and it is the most highly classified secret on the planet. It involves all the countries of the world. Russia built a huge underground seed bank that houses every single seed for food on this planet so they can be replanted after the event. These are arks for humanity, built by our governments' to save as many as possible. There are places above ground that look like concentration camps with rail road tracks that run right next to them. The trains are AIR FORCE Trains and are marked as NOAH Trains, NOAH 1, 3, 7, etc . . . They are there to take those chosen to the Underground Arks. You can see these concentration type camp rally points on *www.youtube.com* by typing "Concentration camps in the US" in the search box.

The "Looking Glass" that the travelers brought with them is a device that allows the holder to see into the future. But the future is not set, so the device shows you a possible of 3 probable outcomes. Depending on which path you took you could possibly achieve this outcome. This devise was being passed around to all the leaders of the world so that they could see the cataclysm for themselves and choose the best way to prepare for the coming catastrophes'. This devise, albeit small and hand held is a star-gate.

The original technology to open star-gates was discovered on ancient cylinder seals in Sumer.

Here are the documents on star-gates provided by MAJ Dan Burisch, PhD the retired MJ-12 Operations Team Captain who worked at Area-51 in a clean sphere with a captured alien life form.

You MUST WATCH HIS DISCLOSURE INTERVIEW provided by Project Camelot on youtube.com, where he discloses his complete personal story up to his retirement and the release of the alien. Conformation of the underground bases follow as well.
—Sabre

Source material below is courtesy of Project Camelot

Project Looking Glass

The following text is copied from Bill Hamilton's *website*:

The Commentary that follows is from my [Bill Hamilton's] source that linked with inside sources and took notes on Project Looking Glass and Time Travel experiments:

"With regard to LG (Looking Glass): As I understand it, this device (at least 3 to 4 years ago) could not focus on a detailed sequence of activities in the future. In other words, you could not see exactly what would happen, like a series of events. I was told to consider the multiverse idea combined with work by Richard Gott on *cosmic strings*. The multiverse apparently is accessed when the forward mode is set. I was also told to consider the views provided by LG as one of many potential realities (at least in the future view mode).

I have also been told that recently there has been an effort made to outfit videotape recorders to be sent forward through the apparatus, thereby allowing the dark project people to gain some insight into what may take place.

When I heard about this several questions came to my mind. The most pressing of which was: if a camera were sent forward in time/space, would it be able to record anything other than what was immediately in front of its lens? I mean, what if LG were located in the middle of the Groom Lake facility, and the operators wanted to gain insight into the outcome of a conflict, say in the Middle East. How could a videotape recorder, set to record what was right in front of its lens at that location gather any data on the Middle East if it were still stuck in the middle of the Mojave desert when it got to the future??? Hell, something important could be happening right behind the camera and it would miss it—a couple of degrees change in camera direction allows one set of events to be seen while another set is completely overlooked, much less events half a world away.

To answer this question, my contact was not specific, saying only that cameras did not move, as mass does not change in its perspective to space time. However, such an item placed into the injected atmosphere, might experience a different time, if only briefly. And cameras could film within the gas or see images in the injected atmosphere as though it were a lens reflecting events in and around the column. I was given to understand that the tilt or positioning of the electromagnets would allow different views or positions in the environment to be reflected in the gas column.

(I feel confident that at least two rings of electromagnets are employed and that the rest of the device is composed of a barrel and the gas* injected into the barrel. (Two different sources have indicated that these are the basic components.) These magnets spin in different directions, creating a charge of some kind. Then the gas is injected into the barrel. Depending on the direction of the spin (I am sure speed and tilt and a bunch of other factors must also have an effect) time space can be warped forward or backwards by long or short distances relative to the present. I have reason to believe that the scientists have completed a map of the exact positions and speeds of the magnets necessary to reach targeted times both forward and back.)

Apparently, images of the events at different places, relative to the location of the device can be picked up and in essence reflected off the gas, causing it to behave like a teleprompter or crystal ball, for lack of a better example. But I am not entirely sure that mass does not move, or that mass is not affected, Since I was also told many years ago about an experiment that went very wrong in the early years of the LG project, involving a test subject of some kind. As I understand it there was significant movement of mass during that experiment, and it ended up with a rather gruesome death for the poor test subject. (I originally thought it was a monkey, but I found out that there were many test subjects that got sent through, so I am not certain what kind was involved in the experiment that went bad. However, in my typical reverse-logic search for corollaries, this tells me that there must have been many test subjects that made it through just fine. So

I am certain that any errors that were made or any miscalculations have long since been corrected).

I wish I could offer you more information. For what its worth, my sources have confirmed the presence of electromagnets and a barrel-like device which is injected with some kind of gas *[an independent source has stated that the gas concerned is argon. Project Camelot]* . . . these components seem necessary for LG to function as a viewing device. And as for any changes in mass, or movement within time-space . . . . I really don't know since my information sources would only tell me 'so much' about what they saw or experienced at the time they were involved. But it can be reasoned, based upon what they did say that there were significant experiments in the movement of mass back and forward through time, many of which were successful. I am sure much has been discovered and/or refined in the process since then."

The following images were created and supplied by Dan Burisch.

Note: the unattributed image on Bill Hamilton's website is a movie prop from the the movie "The Time Machine".

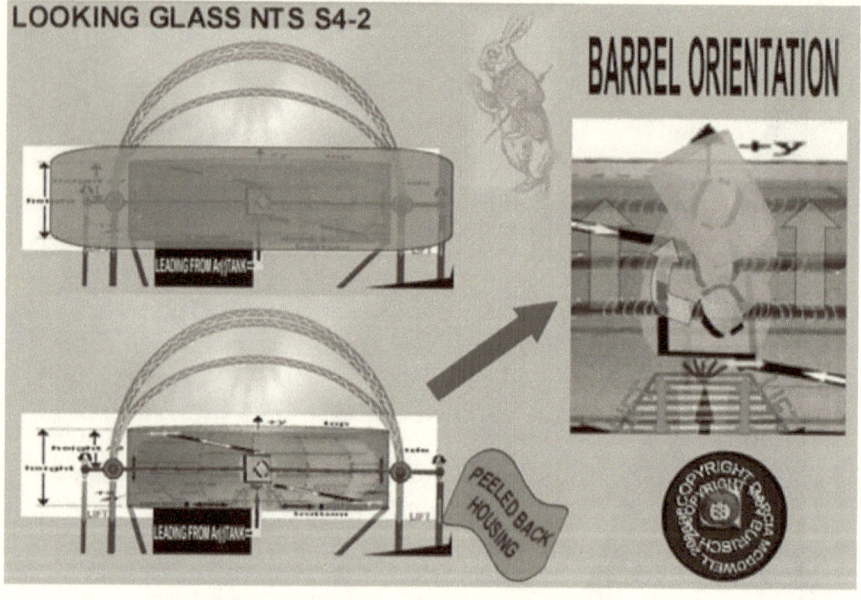

GENERAL BARREL ORIENTATION

-END NOTES-

Here is more "Star-Gate" testimony, courtesy of *www. projectcamelot.org* from another Black Ops Operator by the name of Henry Deacon confirming the previous information.
—(Sabre)

Stargates

Regarding Montauk, Henry said most of Al Bielek's information is correct. There are apparently several kinds of star-gate, notably: (a), The kind where you step through a portal and leave the device behind, and (b) the kind where you take the device with you.

The latter he said was a bit like "Think about where you want to go, and you're there." The mental interface is significant, apparently. He confirmed that as far as he knew Dan Burisch's information about star-gates was 95% correct, but the missing 5% was that he had no knowledge of the large-scale *Looking Glasses* Dan described. (We showed him *Dan's diagrams* and also *Bill Hamilton's source's text*.) Henry emphasized that this didn't mean they didn't exist, because there was so much compartmentalization—but simply that he had no knowledge or experience of those devices.

We showed him Dan's diagrams in person. We watched him while he examined them carefully. Then he suddenly said: "Did he tell you about the one in Iraq?" We asked him whether had he not heard *that part of Dan's interview*. Henry told us that he'd not finished watching the videos. We asked him what he knew. He said the Iraq star-gate was what the Iraq war was really all about, that its location was one of the biggest secrets, and that the war was at least partially about control of it. We asked him how he knew all this—did he read it in a briefing document? No, he said, not a briefing document. The only thing he would say was that it was "first-hand knowledge".

-END NOTES-

Now that you have a basic understanding of what has been happening I am going to provide you with some underground base information. What you are going to read next is a letter from an actual Politician —(Sabre)

Source material below is courtesy of Project Camelot

Since we published this report, our mailbox has been full of messages from all over the world.

We have been completely unable to reply to them all, but we have carefully read everything we have received.

After a week's silence, we heard from our source again. He is well, but is keeping his head down for understandable reasons. He has invited us to forward any relevant correspondence to him from others, and we will do that on a selective basis. He has also promised that he will draw the 'alien' symbols he saw, and fax them to us.

To those who have asked understandable questions about his veracity: we have done everything we can to confirm his identity and authenticity and are certain he is exactly who he says he is.

Meanwhile, *click here* for additional confirming information from another source, in a detailed report including maps and photographs.

The following message was received by us shortly after we posted our *tribute page* to Benazir Bhutto. Although we cannot attest personally to the information given here, we feel in good conscience that we cannot and must not keep this from publication.

The person who wrote this message to us did so under their own name (which we have checked), and enclosed a number of photographs of them with the Prime Minister of Norway and (separately) with Benazir Bhutto. We are certain of their bona fides, but you will understand that we cannot reveal their real name or publish any photos without consent.

The content of this message, if true, could hardly be more important.

[Edited for grammar and spelling errors where needed to ensure full clarity. As a preface to the message, the writer apologized for their English, which is not their language.]

I am a Norwegian politician. I would like to say that difficult things will happen from the year 2008 till the year 2012.

The Norwegian government is building more and more underground bases and bunkers. When asked, they simply say that it is for the protection of the people of Norway. When I enquire when they are due to be finished, they reply "before 2011".

Israel is also doing the same and many other countries too.

My proof that what I am saying is true is in the photographs I have sent of myself and all the Prime Ministers and ministers I tend to meet and am acquainted with. They know all of this, but they don't want to alarm the people or create mass panic.

Planet X is coming, and Norway has begun with *storage of food and seeds* in the Svalbard area and in the arctic north with the help of the US and EU and all around in Norway. They will only save those that are in the elite of power and those that can build up again: doctors, scientists, and so on.

As for me, I already know that I am going to leave before 2012 to go the area of Mosjøen where we have a deep underground military facility. There we are divided into sectors, red, blue and green. The signs of the Norwegian military are already given to them and the camps have already been built a long time ago.

The people that are going to be left on the surface and die with along the others will get no help whatsoever. The plan is that 2,000,000 Norwegians are going to be safe, and the rest will die. That means 2,600,000 will perish into the night not knowing what to do.

All the sectors and arks are connected with tunnels and have railcars that can take you from one ark to the other. This is so that they can be in contact with each other. Only the large doors separate them so that the sectors are not compromised in any matter.

I am very sad. Often I cry with others that know that so many will learn too late, and then it will all be over for them. The government has been lying to the people from 1983 till now. All the major politicians know this in Norway, but few will say it to the people and the public—because they are afraid in case they too will miss the NOAH 12 railcars that will take them to the ark sites where they will be safe.

If they tell anyone, they are dead for sure. But I don't care any more about myself. Mankind must survive and the species must survive. People must know this.

All the governments in the world are aware of this and they just say it is going to happen. For those of the people that can save themselves I can only say reach for higher ground and find caves up in the high places where you can have a food storage for at least five years with canned food and water to last for a while. Radiation pills and biosuits are also advisable if your budget allows it.

For the last time I say may God help us all . . . but God will not help us I know. Only each person individually can make a difference. Wake up, please . . . !

I could have written to you using another name but I am not afraid of anything any more. When you know certain things, you become invincible and no harm can come to you when you know that the end is soon.

I assure you 100% that things will happen. There are four years to prepare for the endgame. Get weapons, and make survival groups, and a place where you can be safe with food for a time.

Ask me anything and I will answer as much as I know about the Norwegian connection to all this. And just look around: they are building underground bases and bunkers everywhere. Open your eyes, people. Ask the governments what they are building, and they will say "Oh, it's just storage for food", and so on. They blind you with all the lies.

The marks of the alien presence are also there, and I often see the Norwegian elite politicians are not what they say they are. It's like they are controlled in every thought, and what they have to say is just as they are told to do things in such manners. It is clear for me who they are, and who they are not. You can see it in their eyes and in their minds.

Remember that those who are going to be in and around the city areas in 2012 are those that are going to be hit first and die first. Later the army will purge the rest of the survivors and they have a shoot to kill order if there is any resistance to bring them into the camps where every one will get marked with a number and a tag.

I also see that Benazir Bhutto is spoken of on your site. Her death was tragic. I have met Benazir, as you can see. You will also see from the photographs that I have met with a number of other notable politicians and world leaders.

The public will not know what happens till the very end, because the government does not want to create mass panic. Everything will happen quietly and the government will just disappear.

But I say this: don't go quietly into the night. Take precautions to be safe with your family. Come together with others. Work together to find ways to solve all the many problems you will face.

Kind regards

[Name and proof of identity supplied]

From immediate further correspondence, and in response to our specific questions, we can add the following clarifications in our source's own words:

I have been to several underground bases *[number given]*. We used the railcars to get around. Only a few special people were selected to be shown around. Those that run with the elite know of this.

I have evidence of my claims. I trust my sources 100%, but they are afraid to tell what they know. People are afraid for their lives and that is how it is. I just also want the public to know what the hell is going on. I am not afraid of death or any other thing.

All the elite politicians in Norway know of this. They also know that if they reveal anything they will be removed from office and will be denied access to the different underground bases when the time is up.

The NOAH 12 railcars are transport railcars between the different bases underground. They have a support system of these all around from one base to another. They are mainly used by the military and they control all of them. There are orange triangle symbols in each base and the check-ins are a kind of energy field that everyone has to go through.

The future for my children is all I think of—and that for all the other children growing up in the new world. We have to make a difference for them so that they grow up knowing what their parents did for them, such as giving this information to people like yourselves.

In 2009 the government of FRP will come into power and Siv Jensen will be elected Prime Minister. This is already known. It's important to understand that. The elections are all fake and the same persons and power elite get elected each time in turn. Look up the political history of Norway, and the people that run the country now.

Please share this information with the rest of the internet. When the time comes people will survive because of the information they have learned from the different sites on the net.

I will not get anything but trouble from posting this, and I have no need to mislead anyone. I do this only to expose what is to happen in my country and that maybe some people will survive what is to come.

Kind regards

[Name supplied]

After posting the above, we presented more questions to our source, who replied with the intriguing information which follows.

The paragraphs indented and marked with an asterisk (*) constitute further supplementary information received after we had asked a number of additional clarifying questions.

When I was in the military I was in the [name of service given]. At one point we were given a task to get something out of a base and deliver it to another base.

We were told: "DO NOT ASK ANY QUESTIONS. JUST DO YOUR JOB." Later, when we landed outside the base, we were taken by trucks to outside the base where there were large doors heavily guarded by other military personnel.

Or it seemed like they were military, but they had different suits on them: orange and black suits with the orange suits having a golden triangle on them and the black suits having a green triangle.

*As far as I can remember the triangle went downwards, like a pyramid but facing down and had some weird kind of sign it it. To me it looked like the letter 'E', but the lines were not connected in the 'E' like we write the letter 'E'. It was shaped like the 'E' and in the middle the letter 'E' was pointing inwards . . . not as in any language I can read or understand, and certainly not in Norwegian.

* The signs were as far as I can remember not on the arms because I saw them clearly . . . they were at the left side of the black suit just over the chest area and on the caps they had on. The signs were not that big—just as a regular sized patch is, but clear enough to see them.

We went through the large doors. I was thinking *what the hell is this*, and I felt a bit scared at first. It was like something out of a science fiction movie. This was the first time I was in such a base.

We then came to a 500 meter long tunnel and there were more of these military personnel waiting with guns and transport for us. We were divided up in groups. Some went another way, and I and my group were asked to come with the black suited guards so they could take us to another location. When we came to the end we were asked to put on some masks "for our own protection".

I was thinking *for our own protection . . . aren't we already protected by being inside this huge underground complex and by guards with weapons?*

We were then asked to step inside a railcar . . . and this is what I know of the railcars. They are run by some kind of blue crystal energy, I think, or at least that is how it appeared. Then we sat in the cars and I asked one of the guards "What is this?" He replied, "You don't need to know this, sir."

* At the front where the operator sits there was a box with a window just besides him, and just when before powering up you could see the large purple-blue crystals emitting a purple-bluish light . . . not blinding you but quite beautiful to watch indeed. I have never ever seen such energy light or crystals anywhere. I was thinking that must be the power source.

* Later on in the base I saw that some people were working on these purple-blue crystals. They were larger than the ones I saw in the railcar—ca. one meter in length and they were lined up one after the other. They were taking some light through them. They were in fact purple-blue and when the light went inside they turned more blue and had a stronger color—the

people had white masks on and goggles standing away from them when the light was going inside the crystals. I was about 20 meters from them and we were quickly rushed along when they said "Move on now."

* I also think that the energy fields that we went through before entering the check-in were powered by these crystals, because it was the same kind of light—or so it seemed to me. If I remember more I will let you know.

I could see that there was a tube-like system and the other railcars were just going so fast you just saw a light going by. I think this was a vacuum tube system where there is no drag.

* The railcars were just like the tube itself but inside. The tube was a bit larger. The main car or transport shuttle was I think ca. 12 meters in length and had a pointed shape in front and back and had seating for 10 people with the operator. You could drive them both ways and it was not necessary to turn it around. It was sealed from the side after you go inside and was quite okay . . . but the speed was too much, and I got sick after the ride.

* There was also some room for some cargo that you could take with you, but not much. There were a lot of these railcars around, as it seemed, and they just went passing by just like a light passes by in a flash. But you don't get to see the other rails either, because it's so fast that you forget to look and rather concentrate on your stomach when inside one of these.

Later on, after I went into politics, I found out what was inside of the rest of the base and what the bases are for, which I have already told you. I know that when this railcar moved—and it moved *fast*—I had never seen anything like this before. Later when we arrived at the end station I was feeling sick, and the other people that were there also felt the same way. One of the guards said it happens the first time to everyone.

When we came outside we were given goggles and asked to go through a security check. This is where it all gets weird.

There were guards with weapons all over the place . . . and remember I told you about the energy fields that one has to go through. I was thinking *I should not be here at this place*, and I was a little scared.

Then we got through this energy field and came to another room. I saw that there was a screen on the side of the wall that said HUMAN—NOT HUMAN—PURE—NOT PURE.

After I saw this I was thinking: *Are there 'NOT HUMANS' too ??*

* At the energy field check-in there was a screen, which I mentioned before. There was a weird language on the screen. I have never seen this before. Underneath it stood letters that were like the 'E' I told you about—but the only thing I could read was HUMAN—NOT HUMAN—PURE—NOT PURE.

[Project Camelot note: we are seeking further clarification on this]

The guards stopped us and told us to change clothes inside another room and come with them. When we had done so they said that it was time to go further down. Again, I was thinking *how large is this place?* We just came out of a rail system that runs for miles and miles . . . and then there is more?

We were then taken to a lift system, with seats, that was going to take us down . . . or, this is what I thought it was going to do. But it went sideways for about three minutes. In this place, time was not known to me because we had no way of telling what the time was. They had taken everything away from us at the check-ins.

* At the lift I remember I saw a letter that look like a headphone—like the hearing phones you have on your head when listening to music. It was just bent like that if you see it from the front side and pointing down.

All I can say is that when our job was done I was thinking that the world is not as it seems to be, and that many things are hidden from the public. It makes me sad and scared.

Later, when I got into politics, I began digging into this because I needed more answers. What I found out was that these bases were Arks for the government and some of the people and military to survive inside. There was a threat from outside that was going to be in the year 2012 and that the human species had to survive.

The 'Planet X' I learned about is from all what I have seen till now. The government knows this and is keeping it from the public. They have been tracking this object for a long time now and were given the first warnings from the USA.

I know that 18 bases exist in Norway. I don't know what many of the dangers are because I am not a scientist. But what I know is that before

2012 the different governments are going to leave for the bases that they have built for the last 40 or 50 years.

If this object goes by, there will be a lot of problems on the surface of the Earth. That is all I know. This is why they go underground.

If such an event comes, they have made sure that five years or more underground is going to be what they need to avoid this. When they know it's safe to surface, they will rebuild again. We were just told that we have to leave before 2012 and that there is something in space that is going to cause much destruction.

* I don't know if there is a threat from the sun itself. I am not so much into the science of things. I am just telling what I have seen and nothing more.

I can say that I have already said too much, but the people are now warned about this . . .

I have no need to make this up or create mass panic. I just want to tell the public what is to come, and I have done what I can from my side.

There are things in this world that are not known to the public. And there is one thing I can say about all this:

Be ready and have faith in yourself. There is no help in trusting the governments. Trust only yourself.

Kind regards
[Name supplied]

———————————

Comment from Project Camelot:

There is no 'hard' scientific data in our source's warning. The information does not explain exactly how or why 'Planet X' is such a threat (although we may guess). There is also no specific information about the exact need to go underground.

Our source is saying exactly what they know, and no more. It is very plausible that that as a politician the message's focus is on vital sociopolitical issues rather than science.

For more information about 'Planet X' and what it may or may not be, please visit Andy Lloyd's excellent website here, or Marshall Masters' website here. We were told by Henry Deacon (and this has also been widely reported by others) that the South Pole Telescope has been built to track what Henry called the 'second sun'. We urge readers to do their own research.

Between now and 2012 and beyond, many differing views, channeled messages, and insider revelations will come to the fore. Project Camelot is working hard on a documentary that will cover both positive and not so positive projections of our future.

It goes without saying that it is our sincere hope, prayer and full intention that the events portrayed above will not come to pass.

We are all co-creators of our world, from moment to moment. We urge each and every one of you to participate in consciously co-creating a bright and positive future for humanity, and to work with us in waking the sleeping and bringing to light the truth—wherever it may lead.

Since our *recent post* of the letter we received from a Norwegian politician about underground bases and that country's preparation for a future disaster, our mailbox has been full of messages from all over the world.

We have been completely unable to reply to them all, but we have carefully read everything we have received.

After a week's silence, we heard from our source again a few days ago. He is well, but is keeping his head down for understandable reasons. He has invited us to forward any relevant correspondence to him from others, and we will do that on a selective basis. He has also promised that he will draw the 'alien' symbols he saw, and fax them to us.

To those who have asked understandable questions about his veracity: we have done everything we can to confirm his identity and authenticity and are certain he is exactly who he says he is.

Independent Confirmation

Many friends and correspondents in Norway and Sweden have been in touch, some offering intriguing details. To many of them, this information was not new and confirmed what they had known from other sources. We received specific confirmation from two scientists and one ex-intelligence source that these facilities exist (and that both Sweden and Switzerland have the facilities to take their entire population underground if need be). A number have asked to be put in touch with the politician who wrote to us.

One correspondent confirmed a great deal of detail. His message is so important that we have published it here in full (having edited out personal information which may have identified him to the authorities). His family has first hand experience of the facilities which our source reported.

Dear Project Camelot team,

I just read your article about the Norwegian politician and his letter and information to you regarding underground bases, 2012, etc.

I must say I found this quite fascinating. I already know of the Svalbard storage place, which when it was announced made the signals go off for me thinking it had to be some specific reason for this for something they know will happen soon.

Regarding the rest, I am also Norwegian, I am from a military family, and have been told by my mother who served a long time in the military and other governmental institutions about these underground bases.

I know the location of one of these bases and have seen it myself. It's within a big mountain in my old city where I used to live and grew up before I moved from Norway a couple years ago. The place is called Baneheia.

My mother refused to tell any details about it, except for that it is created for an emergency. Only specific personnel will be allowed, such as herself and other military people, and other important people, politicians, scientists and other civilian people they consider important.

She has also ben to another base in an area called Evje, and this underground facility is very huge, she said, much bigger than the Baneheia facility. She also said there is a base in the Stavanger area that she knows of.

She also supplied me with a photo of a base in Sørreisa, and said this is another underground base but she is unsure of the size of this base. I attach the photo here. This is the top of the mountain base, and she told me these 'spheres' are probably powerful radar systems that monitor a huge area.

She told me other interesting things, such as that that the big electronic corporations of the world in reality are run by the same people at the very top level, but they just create different names and brands and 'packaging' of the same products, or rather products that are developed and owned by people high up in society that run these electronic manufacturers with total control.

Regarding Baneheia, my mother said that the rest of our family would NOT be allowed to enter there in case of an emergency.

She also said she could not say any more about what was inside or how it looked, as she had sworn and promised to her officials to not tell anything more, and she is a woman of her word and keeps to her vows.

On another occasion, I briefly got to talk to another person from the military who had also been there. He gave me a little more details. He said the facility is huge and has many separate rooms with different purposes, but he was not allowed to enter these rooms and was confined to a specific room where he worked with computer related systems for the military. But he said that, from what he could see, this facility could house several thousand people, and it was extremely huge and highly technological.

It was also reported that of seven men who worked together at some project in this facility, four of them died of cancer. This was published in Fedrelandsvennen, a Norwegian newspaper from the city of Kristiansand in 2001. Unfortunately, the article is no longer online.

One of my friends, who worked for NSA for several years, has also provided me with some great and astonishing information regarding underground bases and other activities of the governments.

He confirmed that there are plenty of these bases around and that in some of them they are doing some very strange experiments and research, related to things such as time travel, journeys to other dimensions, alien technology etc, and he himself was in the process of recreating a spaceship and working on computer simulations for this purpose, and he showed me complex computer generated material related to this as well as illustrations of this spaceship.

He said that at the moment the governments are in contact with Mars, and even have such 'Stargates' connected with Mars so that they can travel there in a few seconds instead of using spaceships. He also claimed they are currently working on terraforming Mars—this means making Mars a suitable and possible place for humans to live. At the moment, Mars is more or less dead on the Surface, though there seems to be some life underground.

He suddenly disappeared and for the last couple years I've been unable to locate him or get in touch with him.

I can mention however, that he was adopted, and according to himself raised up and being prepared to work for the governments since his early childhood due to his high intelligence and other 'abilities'. He was indeed an extremely bright individual, personally I would say a genius, and he was involved with extremely complex mathematical and scientific material that he showed and discussed with me on several occasions.

I will first of all present you with some maps and photos of the area where this main facility I've mentioned is located.

The photo on the top right of the composite image below is what is called Gimlemoen, which used to belong to the military and was restricted for unauthorized personnel. I was there several times due to my mother working for them.

Recently they officially closed down the military operations in Gimlemoen, and turned it into a university campus instead.

However, some of the old bunkers and underground systems still remain in that area. I have personally visited some of them and been deeply underground. At one point I ended up in a large room that had a huge iron door with a big metal wheel on it and several locks. I could not open this door so nor do I know what was behind it.

There is also an aerial photo of Baneheia, where the current underground base in the mountain is located. This is at the top left of the image below. I've made a purple circle at the location of the entrance.

This rectangular structure is the entrance which leads into the mountain. The white structure to the right of it is some other buildings I assume is related as they got tons of antennas on top and satellite dishes etc and other strange devices.

As you see, it's a large mountain, with lakes on top, so how deep inside it goes into the mountain can only be speculated.

[Click on each of the four images to enlarge them:]

[Google Earth or Acme Mapper N 58.16, E 8.00]

Also notice that Gimlemoen and the Baneheia mountain are very close, only separated by the river. If any underground connection ever was established between the facilities in Gimlemoen and the facility in the Baneheia mountain below the river I have no idea, but as I mentioned

I did find a 'bunker' going deep down underground leading to that room with a huge iron door that it was impossible to open. This was in Gimlemoen, and the room was covered with half a meter of water on the floor.

I've discovered something that seems like air vents in the middle of the woods on top of this mountain, which I assume provides air into the base below. The vents are placed in rather inaccessible areas and far away from any paths. They look like this, except that they are more modern.

I have on several occasions seen heavily armed military personnel outside the entrance of the mentioned facility. And this is strange, as Norway is not a country where the military or police normally walk around with guns. In fact, they are not allowed to carry guns unless it is an emergency situation. It's actually the only time I've seen the military carry guns in public while I lived in Norway.

Things should be pretty safe over here, I mean Norway is not exactly like Iraq having any specific needs for such bases and military work unless there is really something they know that is coming that the rest of us do not know.

-END NOTES-

The following is some activity involving the NAVY, confirming the existence of underwater Arks. This just goes to show how many billions of dollars have been spent, and how far they have gone to "Preserve Destiny".

I have not been able to do any research with anyone from the NAVY as of yet. Other than my father, who was formerly NAVY Personnel, now deceased, and my two uncles, I rarely get an opportunity to chat with anyone Navy.—(Sabre)

Additional information from Dr. Richard Sauder

Further Official U.S. Navy Documentation

Is there any more documentation from the U.S. Navy with regard to undersea bases? As it happens, there is. In 1972 the U.S. Navy published another report that discussed "undersea ports". The Report is entitled, "Subsurface Deployment of Naval Facilities." The document cites live sorts of facilities which the future Navy might situate underground. The reasons for going underground range from the tactical to the practical— many Navy bases face a real estate squeeze from the surrounding civilian communities and sea. They are hemmed in on all sides. The logical solution is to build underground, since the surface possibilities for expansion are constrained. So the Navy planners envisioned placing the following sorts of facilities below the surface:

(1)   Administration buildings
(2)   Medical facilities
(3)   Aircraft maintenance facilities
(4)   Ammunition storage facilities
(5)   Miscellaneous storage facilities '81

Most interesting for the premise of this section of the book, however, is the following statement:

Underground facilities may someday play a greater role in Naval operations because of future developments such as.

Improving effectiveness of satellite surveillance systems which could clearly require the subsurface deployment of' any system this nation desired to keep a secret.

The emergence, because of' its continuing invulnerability of the sea-based strategic missile system as our first line of deterrence against nuclear attack; and the importance of protecting its supportive basing and communication system, which may dictate the need for underground or undersea emplacement of key supporting elements of this force.

The increasing vulnerability of surface fleets which could lead to the need for an all-submarine Navy, including cargo and troop transport vessels, supported by undersea ports.

There you have it, the authors envision a submarine Navy, replete with "undersea ports" and "undersea emplacement of key supporting elements." This report appeared in 1972, the very year the secretive Glomar Explorer set sail. If the Pentagon has built "undersea ports" or manned facilities beneath the sea floor along the North American continental shelf or well out to sea, perhaps even in other regions of the world, it has done so clandestinely. I would caution that the absence of public evidence for such projects says absolutely nothing at all about whether they exist or not. in a dual system such as exists in today's United States with a purportedly open, allegedly constitutional government side-by-side with a parallel clandestine, "invisible" or "shadow" government that is hugely funded by the black budget, quite a lot is possible. Remember that tens of billions of dollars are available in the Pentagon's and intelligence agencies' so-called black budget each year. Considering the entire body of evidence, I think it is possible that highly secret, clandestine facilities beneath the ocean floor have been constructed by the American military-industrial complex.

Submarine UFOs and Closing Thoughts

I want to say something about the many submarine UFOs that have been seen over the years. Going back for decades, UFO witnesses have seen unexplained objects leaving and entering the world's oceans, seas, bays, rivers and lakes. In recent years, submarine UFO activity has been observed in the coastal waters off Puerto Rico and Iceland. And in late 1999 1 spoke with the host of a radio show in the Midwestern region of the United States who told me of recent sightings of UFOs seen entering and leaving the waters of Lake Erie. In light of observations such as these, I hypothesize that at least some of the observed submarine UFO activity could be related to clandestine sub-sea (or sub-lacustrine) bases made by terrestrial humans. This is a possibility that I believe researchers should consider when carrying out their investigations. Just because a phenomenon, or suite of related phenomena, are highly strange or bizarre, does not mean of itself that they must necessarily be alien, or extraterrestrial. Indeed, from published accounts in the popular aerospace press of recent years, it is clear that the American military-industrial complex has covertly developed a new generation of technology that has been carefully kept out of the public eye. In my view, it is possible that some of the objects witnesses have seen coming and going from the ocean depths may be a new generation of technology secretly developed and funded, and deployed from secret facilities and installations, some of which may be highly clandestine facilities buried beneath the ocean floor or beneath other bodies of water. I cannot prove this hypothesis, but the evidence I have uncovered logically leads me to raise the issue for further discussion and investigation.

—Dr. Richard Sauder from his book, ***Underwater and Underground Bases***

 -FAMILY BRIEFING DOCUMENT-

At this juncture Project Camelot has given us a rough look at the big picture. A UFO crashed, we retrieved it, capturing one live alien and a crashed space ship. The alien was taken to the Groom Lake Facility (Area 51), and worked with there by MAJ Dan Burisch of MJ-12/PI-40. MJ-12/PI-40 was a group of the most brilliant and influential men on the planet.

The almost 5 hour personal video interview of MAJ Dan Burisch telling this amazing story in great detail, is available, and can be seen for free at *www.youtube.com* under Project Camelot Dan Burisch, parts 1-4. This video MUST BE SEEN. MAJ Burisch issues a very commanding, standing order, to all Operators involved in the projects affiliated with, or under MJ-12/PI-40's command and control that must be heard by all who fight for humanity and its continuing existence.

This really is the biggest story ever covered up in the history of man-kind, and soon, there will be a lot of books about it. A lot of people are coming forward and with that, soon after, the President is going to have to disclose publicly.

I say this because right now the Chinese have a satellite orbiting the moon doing nothing but taking pictures. India has just announced its intent to launch a space probe to the moon to do the very same thing. And when these pictures are put into books and released, the things that we will engineer to echo the architecture up there is going to spark a whole new era of creativity on this planet.

> We have also disclosed that humans were engineered by a highly advanced race of beings visiting Earth as astronauts from a planet trapped within our solar system. These beings are our relatives and were the ones from whom we learned everything.

After studying ancient temples and sacred sites, the Germans located an abandon, crashed, or a left behind UFO from these visitors. Hitler's team began research until the end of WWI at which point, the project and its original science team, was brought over under "Project Paper-Clip" to continue research in America.

From this project, the "Philadelphia Experiment" was launched to create "Stealth Technology" in a project on a battle ship known as the "Eldridge". The operation was a success, not only did the ship disappear from radar it disappeared completely, not just from sight but from this time space continuum and slipped into hyperspace. There were problems with the physics and obviously no understanding of the electromagnetic hyperdimensional geo-planetary physics and the "gates" that could be opened with the utilization of this technology. So, due to these problems, the NAVY scrapped the project and sunk their money into Albert Einstein and he invented the H-Bomb and we won WWII.

In 1947, the US downed a UFO of our own, in Roswell New Mexico.

From this point multiple projects were under way, and in order to keep those projects classified, the president and his advisors created "Private" enterprises, and provided the Federal Grants for Funding to keep them secret. This is where the President lost his power of keeping them under control. These companies made huge advancements in multiple fields, PAC-Bell and General Electric led the way.

Somewhere during all of this, the sole survivor of the Roswell Crash was being worked with at Area-51. The survivor of the crash, who was referred to as "Captive", was forthcoming with information vital to mankind. "Captive", relayed the impending tragedy that befalls the planet in the future and backs up his testimony by showing man-kind this future with a time portal viewing device, (now known and referred to as a televisor), that he was in possession of.

Since this point in time the Military has been going mad, spending money on the construction of Deep Underground Military Bases (DUMBS). This has been a global effort with every nation in on the secret. These are huge cities underground that are literally miles square. The entrances to these cities are usually gained through a "Bolt-Hole" accessible on Military installations and at some agency HQ's. That's why the US started closing posts all over America and abroad, everyone went underground to new bases and the above bases were no longer used. All of these underground bases are connected by a vacuumed tunnel system that a maglev propelled bullet train moves through at mach speeds. This train only goes in one direction, counter clock-wise, to move you to the appropriate place on the globe in minimal time, and its network IS COMPLETELY GLOBAL. This train and the system it moves through is known as "The Terra-Drive", NAVY personnel who are in positions to "Know", refer to the system as just "The tubes".

Since 1950 teams of "Special Projects" groups have been at work to try and understand the circumstances surrounding the catastrophes' that are apparently just over the horizon and expected to occur around the year 2012. This brought the Mayan Calendar into the research, as its "End-Date" for the current "Time" is, and has been amazingly plotted by them, at a December 21, 2012 date at 11:11AM.

The ACTUAL END DATE IS UNKNOWN. It seems that there are a few things that culminated all at once and caused the planet to experience the problems that it is supposed to endure. These anomalies' have been the focal point of most all of our special interests science and physics teams put together by the US government.

Accompanied by think tank study groups for the last 60 years, they have been dissecting the circumstances to first, understand what physics was at work behind star-gate technology, not to mention the universe itself, before they could even attempt to provide a solution.

Now with a 60 year head start, and the assistance of ET technology, they have made a great deal of headway, and are even optimistic about the problems and the final outcome.

There are somewhat uncontrollable contributing factors responsible for what brings on the cataclysmic global events that "Captive" warned about and were confirmed by Project Looking Glass Operators.

1— The Earth passing through the "Photon Belt" in the "Galactic Center" of the Universe, in combination with open man-made Star-Gates in operation, posed the most serious problem. With an electromagnetic-hyperdimensional planetary geo-physics we had yet to understand, it caused the Earth to increase its rotation speed inducing a physical planetary pole flip.

2— The return of Nebiru back into the midst of the planetary solar grid, accompanied by its size, the 4 times larger than our Earth planet brings its electromagnet gravitational pull with it. This planet passes between Mars and Jupiter every 3,600 years and can be seen from Earth in the day time sky because it looks like a second sun. It causes great discord while it passes through our planetary solar grid on its orbit around the sun and SOMETIMES wreaks great havoc on the Earth itself. These events have all been documented by one civilization or another on the planet when they occurred. And Nebiru may or may not have had imposed an effect of its own on top of the star-gates being opened on Earth in 2012 during our trip through the galactic center, but "Captive" did make reference to that planet having an effect in some sort of way.

3— Now on top of all of this the sun itself lends a hand in the disruption of things by having a huge coronal mass ejection of electromagnetic radiation that thumps the entire solar system with this Electromagnetic Radiation Pulse. The pulse is the largest ever to be recorded, and although it is NOT harmful to humans, it IS HARMFUL to all types of electronics, COMMUNCATIONS, and electrical power transformers. Everything electronic will go down, including cars.

This was the "icing on the cake", as they say, because it completely wipes out the ENTIRE GLOBAL ELECTRICAL GRID ON THE PLANET INCLUDING THE SATELITES. Currently every electrical company on the planet has been made aware of the impending solar spike and has tasked their electrical engineers' to help develop some kind of insulation or shielding to protect against the threat.

No one on the planet at this point has any idea how to stop this event from taking place and the electrical engineers are being stressed by governments who are scrambling to come up with a solution. Following is the official report from NASA.—(Sabre)

A Giant Breach in Earth's Magnetic Field

Written by science.nasa.gov

Friday, 10 April 2009

Dec. 16, 2008: NASA's five THEMIS spacecraft have discovered a breach in Earth's magnetic field ten times larger than anything previously thought to exist. Solar wind can flow in through the opening to "load up" the magnetosphere for powerful geomagnetic storms. But the breach itself is not the biggest surprise. Researchers are even more amazed at the strange and unexpected way it forms, overturning long-held ideas of space physics.

"At first I didn't believe it," says THEMIS project scientist David Sibeck of the Goddard Space Flight Center. "This finding fundamentally alters our understanding of the solar wind-magnetosphere interaction."

The magnetosphere is a bubble of magnetism that surrounds Earth and protects us from solar wind. Exploring the bubble is a key goal of the THEMIS mission, launched in February 2007. The big discovery came on June 3, 2007, when the five probes serendipitously flew through the breach just as it was opening. Onboard sensors recorded a torrent of solar wind particles streaming into the magnetosphere, signaling an event of unexpected size and importance.

"The opening was huge—four times wider than Earth itself," says Wenhui Li, a space physicist at the University of New Hampshire who has been analyzing the data. Li's colleague Jimmy Raeder, also of New Hampshire, says "1027 particles per second were flowing into the magnetosphere—that's a 1 followed by 27 zeros. This kind of influx is an order of magnitude greater than what we thought was possible."

The event began with little warning when a gentle gust of solar wind delivered a bundle of magnetic fields from the Sun to Earth. Like an octopus wrapping its tentacles around a big clam, solar magnetic fields draped themselves around the magnetosphere and cracked it open. The cracking was accomplished by means of a process called "magnetic reconnection." High above Earth's poles, solar and terrestrial magnetic fields linked up (reconnected) to form conduits for solar wind. Conduits over the Arctic and Antarctic quickly expanded; within minutes they overlapped over Earth's equator to create the biggest magnetic breach ever recorded by Earth-orbiting spacecraft.

A computer model of solar wind flowing around Earth's magnetic field on June 3, 2007. Background colors represent solar wind density; red is high density, blue is low. Solid black lines trace the outer boundaries of Earth's magnetic field. Note the layer of relatively dense material beneath the tips of the white arrows; that is solar wind entering Earth's magnetic field through the breach. Credit: Jimmy Raeder/UNH.

The size of the breach took researchers by surprise. "We've seen things like this before," says Raeder, "but never on such a large scale. The entire day-side of the magnetosphere was open to the solar wind."

The circumstances were even more surprising. Space physicists have long believed that holes in Earth's magnetosphere open only in response to solar magnetic fields that point south. The great breach of June 2007, however, opened in response to a solar magnetic field that pointed north.

"To the lay person, this may sound like a quibble, but to a space physicist, it is almost seismic," says Sibeck. "When I tell my colleagues, most react with skepticism, as if I'm trying to convince them that the sun rises in the west."

Here is why they can't believe their ears: The solar wind presses against Earth's magnetosphere almost directly above the equator where our planet's magnetic field points north. Suppose a bundle of solar magnetism comes along, and it points north, too. The two fields should reinforce one another, strengthening Earth's magnetic defenses and slamming the door shut on the solar wind. In the language of space physics, a north-pointing solar magnetic field is called a "northern IMF" and it is synonymous with shields up!

"So, you can imagine our surprise when a northern IMF came along and shields went down instead," says Sibeck. "This completely overturns our understanding of things."

Northern IMF events don't actually trigger geomagnetic storms, notes Raeder, but they do set the stage for storms by loading the magnetosphere with plasma. A loaded magnetosphere is primed for auroras, power outages, and other disturbances that can result when, say, a CME (coronal mass ejection) hits.

The years ahead could be especially lively. Raeder explains: "We're entering Solar Cycle 24. For reasons not fully understood, CMEs in even-numbered solar cycles (like 24) tend to hit Earth with a leading edge that is magnetized north. Such a CME should open a breach and load the magnetosphere with plasma just before the storm gets underway. It's the perfect sequence for a really big event."

Sibeck agrees. "This could result in stronger geomagnetic storms than we have seen in many years."

 -FAMILY BRIEFING DOCUMENT-

As if what you just read wasn't enough, I totally overlooked an issue concerning one of the events that is to join in on the calamity of these Earth changes that we have been discussing and the Government has been preparing for. And that is the passing by, or "drive by", of a rogue meteor next to Earth. I don't know the particulars, how big, how close, etc . . . But what I do know is that it is part of the American Native Hoppi (Hope-ee) Indian tribe's prophecy. They say that this rogue rock is supposed to be blue in color and quite a spectacle as it makes its appearance. Some astronomers from NASA as well as a small contingency from the Astronomical Society claim that this comet or meteor has already made its pass and everything is clear. I don't know what the truth is concerning this matter, but I am currently looking into it and cross referencing Hoppi Indian Legend to see if they have an actual hard date that this is supposed to occur, and/or if they are confirming weather or not they have already witnessed this event as having taken place themselves. Also I am not sure exactly what type/s of havoc this passing shall bring to the planet but from what I have gathered it is somewhat responsible for the oceans and lakes turning blood red. It's due to some type of fall-out from this meteor landing on the planet and how it lands on this planet, I'll bet, is directly related to the article above. If there is no electromagnetic shield around our planet to deflect these types of charged energy particles they could, depending on whether we are positively or negatively charged, be magnetically attracted right to us. And if they are expecting this to happen I'm betting the charge is the exact opposite of ours.

Another revelation I wanted to let you in on, that is under study, is that if you are born in space your soul will remember it's passed lives. That's speculatively due to the absence of the Van Allen belt. If your soul has to come to Earth to inhabit your body it will not remember who it is because of the transfer of information that the Van Allen belt has an electromagnetic effect over, stopping or stymieing this outflow once the soul has crossed over the belt and onto Earth. Therefore that is why you don't remember who you are, or should I say were, in your passed life. Anyway, I think, that's why there are projects studying the effects of babies being born in space and the effect the Van Allen belt has over the human species. I wonder if they are under the impression that after 2012 the Van Allen belt will no longer be of hindrance to the soul, (as the magnetosphere is failing), and those who will be born after 2012 will "Know Themselves". Will you remember your passed lives and remember who you were? And will that cause us to evolve past this point of childhood that we have been stuck in? Well I would like to at least see one of these theories play out. We would really become far more intelligent than what we currently are.

These are some pretty interesting people, the Hoppi. Our government has sent teams to them to actually hear their prophecy and write it down. Not only that, but the Hoppi already have a huge underground facility that their tribe will retreat into at the time in question and are quite prepared to ride this thing out. Our government was astonished to see just exactly how prepared these people were when they got there. From my understanding it is the Hoppi Indian Reservation that the Archuletta Mesa is on Bob. This is the site of the joint Alien/Human base I was telling you about earlier that I want to go and see for myself. The Indians know where the entrances to this facility are and are just really scared to even try and go near them. They are scared of the super-natural or the spirit world and such. Actually the creatures that we are working with or against are multi- or what they call trans-dimensional beings. So the Hoppi are not far off of the mark, but none the less, they are just really scared to go poking around in the Archuletta Mesa.

This is a crop circle that was drawn up by one faction of ETs (and there are many) warning us that the sun is expanding, during which at some point a rogue meteor will pass by our planet, accompanied by the return of planet X, creating a great deal of problems.

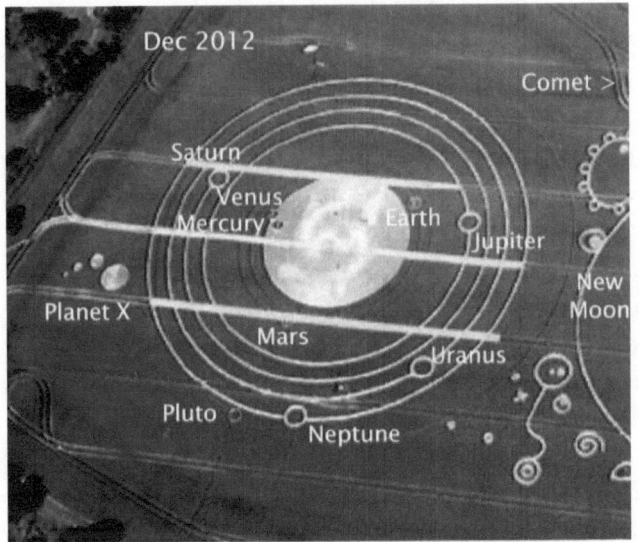

Listed here are a couple of reasons for the demise or fall of man-kind. Basically this type of death could possibly be avoided if one was prepared for the confrontation.

An earth quake could potentially kill anyone, what is not known is that a large enough earth quake, like the ones expected, will be so loud that the sound of them will kill 10's of thousands more than just the jolts themselves. Thunder from lightning can not hold a candle to the sounds produced by the mantle of this planet being ripped apart from centrifugal forces created by an accelerated spin of this planet, the sound itself would flatten trees.

But the ring of fire could be survived if one had a good gas mask, a fire retardant suit, and a place underground to retreat to that had enough food & water to last 6 months or more. Just long enough for the hot ash to settle or cool in the stratosphere and begin its long journey back to the planet. Unless we were lucky and due to the fact that the electromagnetic shield could be down the volcanic dust might be swept out into space by the vacuum of the cosmos. That would be a break, however unlikely.—(Sabre)

The following information is sourced from projectcamelot.org

- There is some renewed earthquake and volcanic activity on the West Coast of the US to be aware of:

Salton Sea earthquake may mean trouble for San Andreas Fault
Scientists are concerned that the recent quakes in the Salton Sea area may trigger the 'big one' in the Southern California area.

Quakes shake California's southeastern desert
Thursday, March 26, 2009
"Scientists are closely watching the increased earthquake activity because it is near a section of the San Andreas Fault that has not broken loose in more than 300 years."

March 26, 2009: Alaska Redoubt Volcano erupts twice—with more expected
This indicates growing activity along the Ring of Fire and has links to the San Andreas fault which runs the length of California—and could be triggered further as a result of these eruptions.

   -FAMILY BRIEFING CONTINUED-

---

What's provided next is testimony from Black Ops Operators who have come forward as "Whistle-Blowers". Some have sworn under oath, and some are still deep inside and coming forward to legitimize information that has been released, and provide more detail's to reveal important information concerning the status of some of the projects that have been disclosed and some of the projects that are still very much under way.

As I stated before Bob, you must see the Dan Burisch video on youtube.com and then watch "John Lear tells all". He is an X-CIA pilot whose father invented the Lear Jet. John was tossed out of the CIA after coming forward to tell the public that the space shuttle was not being used to repair satellites', it was being used to shuttle parts and equipment to the base on the moon. He will show you photos of the bases up there and much more. Watch HIS VIDEO TOO. After the CIA tossed him he came forward to spill the beans.

Another brilliant guy to watch is Richard Hoagland; he worked for NASA for 20 years. He will show you more pictures of the huge bases on the moon and reveal more about bases on Mars.

The speculative reason we are in Iraq is to protect a real live, soon to be active, STAR-GATE. It's the original one, the one that the first astronauts opened when they were here, and there is more than one. But this one, I believe, is a direct link to Nebiru, I cannot confirm this, but one thing is for sure, and that is that when Nebiru comes back around we want this gate open, and we want to be in control of it if the need be to access it.

It is written in the King James version of the Christian Bible that when GOD returns EVERY MAN will fall to his face to bow down before the presence of the Lord to worship him forever and ever. I think our species now sees this as a form of slavery and has been preparing for this return. I think that is why we are moving to gain control of these sacred sites or "Star-gates", and I think that is another reason for the underground bases.

I myself have a couple of questions about this gate and our abilities that I'm going to go out on a limb here to voice. First I will state that most would try to say that the Governments would want control of this gate so that they can control who is coming through and control what types of information these beings who do come through are allowed to divulge or share with the general populace of man-kind. To wit I would say, To think that obtaining power over one or the other with the intension of controlling either, let alone both is ludicris, if they want to come through and communicate I don't think we could stop it. Second, if the Galactic Alignment does open mans mental ability to access or exercise any kind of "God-Like" powers then why would we try to stop or control something with-in ourselves that could ultimately save us, protect us, or even set us free from something or someone who could possibly possess the ability to present a threat to us?—(Sabre)

The information below has been sourced from Project Camelot

6 April 2009

- ***Future Threat to Global Electric Grid:*** We have been contacted by the wife of an electrical engineer who has been officially briefed about a serious major threat to infrastructure in a few years' time.

We believe this information is credible and important but the contact may now have been lost. Therefore we are publishing what we have—with many questions unanswered. This is the first message we received, on 27 March:

Hello B & K
*   My husband is an electrical engineer for a national power company. He has just been told that they are expecting an event in the next 3-4 years that will render every transformer in the world useless. They are desperately trying to find a solution to the problem. If they don't, the entire global electrical system will go down. I know this fits in with some of your research. Have you heard about this?*

In answer to my questions, she then replied:

Hi Bill: My husband is working away from home at the moment on this problem, and I have no way of contacting him. He is not allowed to carry a cellphone on the job. I'll clarify with him when I speak to him next, but the impression I got was that it was definitely going to happen. He mentioned something about electromagnetic clouds in space that the Earth is drifting into. They are under incredible pressure to find a solution. My husband is an engineer, a man of rational logic. He is not prone to flights of fancy, but this news has disturbed him no end.

An Eye Opening Interview with 'Henry Deacon'

This interview was transcribed from video as the interviewee expressed a wish to remain anonymous ('Henry Deacon' is a pseudonym, prompted by his similarity to the likeable and creative polymath on the *Eureka* TV series). Certain details have been deleted and/or amended in order to ensure that his identity remains concealed, and the transcript has been "cleaned" of most expressed natural hesitations and the like. Meanwhile, it is most important to note that none of the factual information disclosed has been altered or amended in any way.

Henry's name and employment details are known and verified and we were able to meet with him personally more than once. He was understandably a little nervous but definitely wanted to talk with us. In conversation, he sometimes responded with silence and meaningful glances or enigmatic smiles rather than words. He was entirely disarming, in a very quiet way, and was not always certain about what he should or shouldn't say. At times, however, he took great pleasure in revealing the truth about some key matter in a way that could not be traced back to him. One or two supplementary details were provided by e-mail after the interview.

The most important piece of additional information comes at the very end of the transcript, where Henry confirms Dan Burisch's testimony.

Readers are welcome to distribute this freely with the proviso that no part of it must be altered or deleted. Unaltered extracts can be quoted if the context is made clear. We consider this interview to be extremely important and in our opinion the information revealed should be made widely known.

For further information from Henry, updated in February 2007, please click here. However, it's probably best to familiarize yourself first with the material presented below as an introduction to Henry and his testimony.

Click here for a further update (May 2007)

We have heard nothing from Henry Deacon for the last five weeks. Prior to our last communication, in the last week of March 2007, he had told us he was being 'coerced'. Up to that time he had been in very regular communication with us. We now feel certain he has been forced into silence.

While this newest update contains fascinating further information on a number of topics (including more information on Mars and some details of the inside story of 9/11), we continue to believe that the most important information he has shared is to be found in the second update. However, as stated above, new readers are advised to read Henry's testimony in sequence.

Start of interview

Please tell us a little bit about yourself—as much as you feel you can.

I'm a current employee of one of the three letter agencies [he plays a little word game with us until we guess the right agency, which he then confirms]. I'm probably taking quite a risk by speaking to you like this, though I don't intend to reveal any information that in my judgment is both classified and specific to national security. I've been involved in many projects with many different agencies over many years.

To jump right in at the deep end, I believe I was a walk-in around eighth grade. I have memories of coming from another planet, and these are all mixed up, all mingled with human boyhood memories. It's very weird, and hard to explain what it feels like. I've never had any problems intuitively accessing complex scientific information and I've often found myself understanding complex systems with no detailed briefing or training. I work essentially with systems. I don't mean to be arrogant, but I do know a great deal of advanced information, scientific and otherwise. I just seem to know it. I can't say more than that right now.

Can you give even any clues about which agency you work for?

No, not on public record. I just can't afford that.

What information do you feel you have that's important for the world to know?

There's so much that it's hard to know where to start.

I knew about 9/11 two years before the event, for instance. Not in specific terms, but certainly in general ones. It was talked about, an event like that, something that would change the game, let's say.

I know that there's a planned war between the US and China scheduled for late 2008. This is also geopolitical and not concerned with Black Ops as such. These were both just events that I got to hear about in passing. I have no detailed information about them.

You mean that China and the US are working together to stage a war?

The Pentagon started the planning in 1998. You have to understand that China and the US are hand in glove with everything. This war is a joint op between the US and China. Most wars are set up that way and have been for a while.

You want something else that's just unpleasant to hear? I also heard from someone who was serving in a unit that worked with missiles deployed for testing in the Pacific and the Far East. The missiles were shipped to the test location in very tightly sealed containers, very secure, hermetically sealed. After the tests, the container would be shipped back, sealed the same way, but empty, supposedly empty. On one occasion, this guy was present when a container was opened. It wasn't empty. It was filled with bags of white powder.

Cocaine?

I leave you to draw your own conclusions. I doubt it was sugar. Let me just say that, hypothetically speaking, and let just me say that to protect myself, if such a plan was in operation, it makes perfect logistical sense as it's a totally secure way to get around all security, customs, international boundaries and ports, and all those checks. It's perfect, like the way guns and ammunition used to cross borders in diplomatic bags going between consulates. This happens all the time.

Would you call yourself a physicist?

Yes. I cover other specialties as well, but yes, I'm a physicist. And I specialize in systems. Livermore is a good place to be, everyone's very professional there. They don't, you know, they don't play games there.

What can you say about the current state of physics in the military-industrial complex?

It's dozens of years ahead of mainstream physics which is published in journals in the public domain. There are projects dealing with subjects beyond the belief or experience, beyond the imagination, of many public domain physicists.

Can you give us any examples?

[long pause]
There's a project called *Shiva Nova* at Livermore which uses arrays of giant lasers. These are *huge* lasers, huge capacitors, many terawatts of energy, in a building built on giant springs [extends his arms to show the size], all focused on a tiny tiny point. This creates a fusion reaction which replicates certain conditions for nuclear weapons testing. It's like a nuke test in lab conditions, and there's very powerful data collection focused on that point where all the energy is focused.

The problem is that all extremely high-energy events like this create rips in the fabric of spacetime. This was observed back in the early Hiroshima and Nagasaki events, and you can even see it in the old movies. Look for what looks like an expanding energy sphere, and I can send you a link to show you. The problem with creating rips in spacetime, whether they're big or little, is that things get in that you don't want to be there.

Things get in?

Things get in. Things that we all know about that are discussed on the net a lot. Beings, and influences, and all kinds of weird stuff, and I can tell you they've created *big* problems.

What kind of problems?

[pause]
The problem of their presence and then what happens next. The other problem is that if you're creating rips in spacetime you're messing with time itself, whether you mean to or not. There have been attempts to fix that, and it all results in a complicated overlay of time loops. Some ETs are trying to help, and others, others are not. When predicting futures, we can only talk about probable and possible futures. This is all extremely complex and very highly classified. Basically, it's just a huge mess. We've opened Pandora's Box, starting with the Manhattan Project, and we haven't yet found a way to deal with the consequences.

The problem of multiple timelines sounds like the information reported by Dan Burisch . . . can you comment on that?

[shakes head] I don't know about any of that.

OK. We'll send you the links so that you can see the interviews. But what you're saying also corroborates the information reported by "Mr X" on the Camelot website. Have you seen or read those interviews?

No, what does he say?

"Mr X" is an archivist who for a six-month period had the opportunity to work with classified documents, films, photos and artifacts back in the mid-80s when he was working on a special project with a defense contractor. He says that he read that the principal reason for the ETs' interest in us was because of nuclear testing and the general threat of nuclear weapons.

That sounds about right. Except only one or two ET groups are concerned about nuclear weapons, not all of them.

OK. What else can you tell us about the timeline problem?

Just that it's unresolved. The risk is, you see, that each time we try to fix it, it adds to the problem. It just gets worse all the time.

Are the aliens—or some of the aliens—time travelers? Dan Burisch states this.

Yes.

Do you know about the Montauk Project?

That caused a huge problem, and generated a . . . created a 40-year loop.
I don't know about Al Bielek. I believe some of his information is suspect. But something like that definitely did happen, the Philadelphia Experiment, too. John Neumann was very involved in all of that.

And Tesla, and Einstein?

Don't know. But Neumann . . . [nods head]

Montauk was real?

Yes. That was a real mess. They created a time split we're still unable to mend. Now, understand this also relates to Project Rainbow, the Stargates . . . they were also working on that there. But some of the Montauk reports on the net are unconvincing to me. I've seen some of the photos of the equipment they're supposed to have used, and it's junk, just a pile of junk.

[Bill] I've always had a problem with the idea of time portals because I don't see how or why they'd stay with the planet at a certain location as it moves through space. If a portal was created in spacetime, you'd expect it to be left behind somewhere very quickly as the earth rotates, and moves round in its orbit, and the solar system itself is orbiting the galaxy in a huge cycle. I mean, everything's in motion, all the time, and this is well known. Can you explain this?

No, I can't . . . but I know what you mean, and the portals do stay in specific locations, kind of anchored to this planet. That does happen that way. Why they don't get left behind or just kind of float off somewhere, I have no idea. Maybe they're gravitationally anchored in some way. Your guess is as good as mine.

One of the portals connects to Mars, and it's a stable connection, no matter where Earth and Mars are in their orbits. We have a base there established in the early sixties. Actually, we have a number of bases.

So we've explored Mars already.

Sure, a long time ago. Have you seen *Alternative Three*?

Yes.

That had some truth to it. The Mars landing video was all a spoof, and other parts of it were as well, but there's truth there.

What else do you know as a physicist working on these projects?

OK. This may interest you if you have a physics background. You know what signal non-locality is, right? When two particles in different parts of the universe can apparently communicate with each other simultaneously, no matter what the distance. Communications devices have been made for communicating across vast distances and also locally using a methodology that's impossible to eavesdrop on, because there's nothing traveling between the two devices that can be intercepted. It's impossible to crack or codebreak or eavesdrop because no signal travels anywhere, so there's no signal to be intercepted or decoded. It just doesn't work like that.

The beauty of it is that the devices are actually so simple to build. You can create two chaotic circuits, on a couple of small breadboards using cheap components which anyone can buy, and they communicate with each other in this way. You can build these if you know how.

Are there any other applications besides communication?

[pause]
Yes.

What else can you tell us about this?

That's it. Oh, I should say that I didn't realize at first that you were also the guy who created the Serpo website. Let me just say that it wasn't called by that name. And I doubt that the travel took nine months. That's not how they traveled there.

Oh, you mean the travel was instantaneous?

[pause]
I don't think they traveled the way they say they did on the Serpo website. Maybe there were other programs. There may have been many. But travel across large distances is best done using portals. Anything else is really . . . it's just inefficient.

You mean they used Stargates?

I guess you could call them that, yes.
I also suspect the system isn't Zeta Reticuli. It sounds to me like Alpha Centauri. I think you mentioned this on the site.

Do you have a reason for saying that?

Well, Zeta 1 and 2 are a long way apart from one another. Alpha Centauri and Promixa Centauri are close together. Alpha Centauri has a solar system very much like ours, but it's older. The planets are in stable orbits. There are three inhabited planets, the second, third and fourth. No, wait, the fifth, I think. Second, third and fifth.

That's astonishing . . . you knew this professionally? I mean, you came across this in the course of your work?

Yes. This is known. It's comparatively easy to get there, less than five light years away, and that's, you know, it's right next door to us. The . . . people . . . there are very human-like. They're not Grays, they're like us. The human form is very common in the universe.

[Bill] Is one of the planets desert-like? That's what I saw in the photo I described. Two setting suns, over a desert landscape. It really blew me away. *[See this article on the Serpo website]*

Yes, it is. A desert planet.

Wow.

Are you familiar with Project Looking Glass?

That sounds kind of familiar . . .

It was a kind of technology that Dan Burisch told us about that involved seeing into the future. Were you involved with it?

OK, that technology wasn't developed by us. We were given it, or it was taken from a craft we acquired. I didn't work on that.

We heard they have a man-made Stargate as Los Alamos. Are you familiar with that?

[looks at us without answering, slight enigmatic smile]

What can you tell us about Los Alamos?

There's a Los Alamos website I'll send you, and then you can search there under "gravity shielding" and things like that. It's all there.

[Note: the website is http://lanl.arxiv.org.]

Now, it may have been an error that it's in the public domain. You might want to advise people to archive the pages they find there before they're taken off the web once this gets out, if it does. But right now you can see it with your own eyes.

It's hard to know what else to say.

What can you tell us about the ET presence?

Look up the movie *Wavelength*. It's based on a totally true story. Have you seen it? It's based on an incident that took place at Hunter Liggett. This is a hot one.

No. Where's Hunter Liggett?

90 miles south-south-east of Monterey, California. My primary station at the time was Fort Ord.

I was working there back in the early 70s, when I was in the military, and I was working under CDCEC, which is Combat Developments Command Experimentation Command. You can go look that up.

We were doing testing of all kinds of devices, and we lived out in the field there. We wore laser protection goggles a lot of the time and we had our eyes dialated routinely to check our retinas for burns. Some of the cattle in the fields even wore modified goggles! This was the most bizarre sight you could ever imagine.

Well, one day something happened while we were testing. A disk came into the area and it was hovering, it hovered right directly in front of us, out in a field. So [pause] we shot the ****ing thing down.

You shot down a disk?

[shaking head] We should never have done it. It wasn't me personally, but the group did. Between us we had all this gizmo weaponry and I guess they panicked and thought they were in a movie or something. The disk was disabled and it was captured, and so were the occupants, and I saw these very briefly. They were small child-like humanoids, with no hair. And they had small eyes, not large almond-shaped eyes. I don't think anyone knows about this. As far as I know it's not on the internet.

This is incredible. I've never heard of this incident.

Most of the other witnesses ended up in Vietnam and many were killed. I may be the only living witness to what happened . . . I don't know.

The rest of the story is in a sci-fi movie called *Wavelength*, which was released in the early '80s. I'd never heard of it until I ran into it years later, in Arizona. Did I just say this? [laughs, for the first time]

When I saw the video, I was expecting some, you know, light entertainment with a beer or two, but I mean, my mouth just hung wide open. The beginning of the film just completely clearly and accurately describes the incident, and the film is very close to the rest of the story, including the use of an abandoned Nike base in Southern California to store them.

Go find it. It's all basically true. I was just amazed when I saw it. The person who wrote it must have been there, or knew someone who was there. But I don't know who.

I had a genuine alien photo once. I showed it to someone, a woman, a very talented woman, who was a microbiologist working for one of the agencies. It scared the s*** out of her. I couldn't believe it. She just didn't want to deal with it at all. And I'd say that just suggests that the public, even scientists aren't ready for this information to be released. And this person was really smart. It didn't stop her from freaking out, just not wanting to know. She was just, you know, totally spooked.

Do you still have the photo? Can we see it?

I don't know. I may still have it somewhere, and if I can find it, I'll forward it to you.

Can you describe it?

It showed a small being with dark skin, kind of black and wrinkled. He was a sole survivor of an incident. But he died shortly afterwards. He had a suit that was self-healing, ah . . . self-repairing. It was a kind of fabric, or something, that would actually repair itself. And he had an artifact with him that was some kind of remote control device, and that was taken away from him.

He was the survivor of a crash?

[pause]
    No.

A time traveler?

You know everything, don't you?

No, but you're giving us verification.

I mean, it's just so incredibly complicated. It's so complex it's possible that no one person has all the information. Most of the agencies don't know what the other agencies know and everything is heavily compartmentalized right up the wazoo. No-one talks to anyone else about this stuff. Sometimes entire projects are duplicated at the cost of God knows how many billions because the existence of the other project is unknown, it's kept from them. I mean, I'm a scientist, and scientists sometimes have one arm tied behind their backs because they can't communicate freely. In fact, they can't communicate at all [laughs]. And there are dozens, hundreds of classified projects, I mean major ones. It's just a total mess.

Look, there are many groups of ETs, and besides our own *ancestors* are mixed in there. There are time loops upon time loops, and it's all a mess. You'd need an IQ of 190 to figure it all out.

Tell us about the time loops. By the way, can we ask you again . . . you've not heard of Dan Burisch?

Not that I remember. It's not familiar to me.

We interviewed him last month. He was next to John Lear on the web page.

I did see your interview with John Lear, talking about the moon photos and the way they're airbrushed. NASA does that all the time. He's quite a character, by the way. I'd like to meet him one day.

What few people know is that radar reports for the National Weather Service are also airbrushed, so that certain radar images aren't released. I don't mean airbrushed as in by hand. The radar images are electronically filtered using software. Some of these radar traces are huge. In addition, the weather radar won't record traces that are moving faster than a certain high speed, a couple of thousand miles an hour. But there are still traces which need to be removed.

UFOs?

Sure. They're often optically invisible, but usually show up on radar. They're also visible in ultraviolet . . . I don't think this is generally known by people.

So what can you tell us about the time loops?

Right. [long pause]

The situation with time loops is that there are a large number of parallel timelines, lots of branches. There are no paradoxes. [draws a diagram] If you go back in time and kill your grandfather, that's the grandfather paradox everyone talks about, there's no paradox. When you go back and change the past, it creates a different timeline, which is a new branch of the original one. On that timeline, you'd not be born and wouldn't exist, so that aspect of the paradox is true. Do you see? But on this timeline, which you're on here and now, you do exist, and continue to do so. There's no paradox. It's simple . . . do you see? You're dealing with different branches of a kind of time tree. No principles get violated. All future events are possibilities, not certainties. That's kind of pretty important, an important . . . distinction. That's really all I can say about that.

Do you know anything about chemtrails?

OK. Chemtrails were developed by Edward Teller and are basically the seeding of thousands of tons of microparticles of aluminum on the upper atmosphere to try to increase the albedo of the planet, the reflectivity of the planet, because of global warming. Now, *gold* microparticles, real gold, were used once in a similar situation on another planet, but I guess they had lots of gold, and we used aluminum instead. Global warming is partly because of the greenhouse effect, and that certainly makes things worse, but most of it is because of increased solar activity. Solar activity is the real problem.

Why isn't this information in the public domain? It seems like people should know and would like to know, and there's no security risk if what you say is true.

Scientifically, it's just a total gamble. Not nearly enough is understood. It may work, or maybe it won't. It could easily make things worse. There may also be health side-effects, weather side-effects, God knows what. It affects the whole planet and here you have a unilateral, non-democratic decision, unconnected with the political or democratic process, to launch a huge technological special project that affects everyone on earth. If that's not controversial, I don't know what is. The solution is to keep it secret. It's the usual kneejerk solution, too.

Will it work?

I don't know.

Is this also connected with weather wars?

[pause]

Yes, there are weather wars. The Air Force will own the weather within two years.

What else can you tell us?

Read *The Report from Iron Mountain.* Much of that is true. I was working with a group down in [_____]. They called us in and passed out a report. The weird thing is that it wasn't even connected with what we were working on, and it came just right out of the blue, out of nowhere, and none of us were expecting it. The guy said, and I'll never forget it because it struck me as just wrong: "There are the wolves and there are the sheep, and we are the wolves." Then they told us to go and read the report, and that was that.

There wasn't any choice, and there still isn't. The way they see it is there are too many people, and, you know, they're right. That's true. So they figure they need to eliminate them and they're planning solutions to this. I happen to think it doesn't have to be that way. Apart from what I've mentioned so far about the spacetime problems, the problem is overpopulation. It's as simple as that. There are programs to reduce global population for everyone's benefit. Believe it or not, the intent there is positive. It was put together by Kennedy way back then. The RAND Corporation was involved, and one of the Rockefellers, I forget which one, probably Laurance, I think.

By killing people off?

Basically, yes. Artificial viruses that have been deployed using a number of means and are hard to detect or identify and nearly impossible to cure. Medical people in the public domain can't identify what's happening.

How do you feel about this personally?

Very mixed. [pause]

As an individual flesh-and-blood human being, I'm appalled. And as a scientist trained to look at things from a high vantage point, a high overview, I have to say that I can understand the thinking.

You have to understand that I'm not defending or condoning this. It's just a comment from an abstract scientific perspective. But the problems we face on this planet are so huge that very few people have the training or experience to view it all, to see it all in the same field of vision.

My situation was different, and I got a chance to see a lot of things because of the nature of my work. Most people don't see it all. But I've worked with many agencies, and I have the big picture.

Do you know that it's legal to test biological and chemical agents against US citizens? It's *legal*. You know, all that has to be done is to get the approval of the mayor of the city, or his equivalent in any area. Or some representative official. No-one knows this, but it can be checked out. Go look it up. It's all carefully hidden away in the law somewhere, but it's all in the public domain. It's all there.

You've revealed a whole lot of extraordinary material here in our conversation. What's the most important message you'd like to leave people with?

Look, I don't want to shock anyone, but I'm not optimistic. The problems facing us as a race on this planet are *huge*. I don't believe most civilians are ready and able to comprehend and deal with the sheer scale and complexity of it all. They have enough trouble managing their everyday lives, and these problems are on a completely different level. Overpopulation is really the biggest issue. Everything else facing us is connected with this.

You see, I can understand the military taking matters into their own hands. If there was full disclosure of all the problems, and all the proposed solutions, do you really think it would help any of us? I suggest the answer is probably not. It would just complicate matters further.

But deep down I *do* feel that everyone should know these things, or else I would not be talking with you. The essential message I want to leave with is that I do actually hope and want to believe that we as a people can handle all this, but sometimes I wake up in the morning and doubt it, but deep down I want people to know the important things that have been kept from us all. But sometimes I do wonder. You don't know what I haven't told you.

On 27 September, three weeks after the initial meetings, and after we had strongly urged him to view the three part Dan Burisch video interview on the Project Camelot website, we received the following e-mail. It is quoted verbatim and in its entirety.

Dan Burisch is telling the whole truth.
I confirm this.
timelines and all
best wishes

6 October 2006

We've had ongoing continued communication from 'Henry Deacon' (see below). For Henry's background and for a transcript of our first interview, please *click here*. It should ideally be read first before studying the information which follows.

What we present here was not recorded in any interview. This is our own compilation of what we consider important from a number of conversations and written exchanges since our first meeting.

We know Henry very well now. He's a values-driven man of high intelligence, acutely aware of the import of the information he shares at considerable risk to himself. He's softly-spoken, kind, and has a delightful sense of humor. He does not seek any limelight. He is extremely concerned about the state of the world . . . and the direction in which it seems to be heading.

Because of the professional posts he's held, he's in a privileged position to be able to see everything that's happening—which he describes as a bewildering complexity, all the details of which are known or understood by very few people—from an overview vantage point. The picture he presents is complex, challenging, and significant.

Dan Burisch's testimony

In our *first interview*, Henry told us he had never heard of Dan Burisch. Because parts of Henry's story seemed to overlap with Dan's, we strongly urged him to view our three part *Dan Burisch video interview*.

On 27 September, three weeks later, we received the following e-mail. It's quoted verbatim and in its entirety.

Dan Burisch is telling the whole truth.

I confirm this.

timelines and all

best wishes

This was of extreme importance, and we checked this with Henry in person as soon as we could. It appears that Dan's claims—as extraordinary and incredible as they may seem—are true.

Henry did not comment on J-Rod, the treaties, or Lotus, as he has no experience or knowledge of those. But he did confirm that the greatest secret in the classified world—about which many insiders themselves have *not* been briefed—is that there is a complex problem concerning alternative timelines, that some of the visitors are indeed humans from the distant future, and that there are significant issues with a possible future event which may seriously affect the earth and its population.

The last of these is the reason that all this has been kept so highly classified, and why disclosure is so problematic.

Roswell

The Roswell visitors were future humans—as Dan Burisch has stated. They were not from another planet, but from a future Earth—stepping (which is a better word than "traveling") back in time to 1947 to attempt to deal with the problems which had occurred in their history. Apparently, Dan was also correct in that the Roswell visitors were from earlier in the future than some other visitors who arrived subsequently. But Henry didn't give details or time frames.

The Roswell visitors were on a purely altruistic mission. They did not have to do this, but chose to . . . out of compassion. But the mission went disastrously wrong—not just because they crashed (an accident caused by high-powered radar—later the military realized this and adapted the radar as a weapon), but because they had a device with them which was their only means, as an orientation device in time and space, to get them home and back to their own time.

The device was a little box, far smaller than the "Looking Glass" that Dan Burisch and Bill Hamilton describe as being subsequently utilized by military scientists in various experiments. When the box was acquired and investigated by the military, this became a catastrophe in itself. It made the timeline problem many times worse, because this both introduced the time-portal technology to us at the wrong time . . . and also told the military what lay ahead.

Henry could not stress too strongly how totally calamitous for us all the Roswell incident was. It was a major, major setback, right at the start of the future humans' project to help fix the problem.

Since then, there have been continued attempts by the future humans to remedy the exacerbated situation. This is what has caused an overlay of timelines, creating a tangled complexity which apparently is challenging for even the most brilliant present-day minds to understand fully.

We asked Henry why the disks kept on crashing, almost year after year. He said that it seemed strange that the visitors had not evaluated the risk from radar before they returned in time, but he explained it was just very dangerous for them to be here for a variety of reasons, despite their advanced technology. Crashes have been caused by many factors, including offensive action. Importantly, Henry emphasized that the Roswell visitors were *not* the Grays. (He did not mention who the Grays were.)

NOAA, the Dark Star, and global warming

Henry at one point did some work with NOAA (the *National Oceanic and Atmospheric Administration*) and it was there that he learned about what he called the "second sun". This is a massive astronomical object which is on a long elliptical orbit around our own sun, on an inclined plane to the rest of the planets. It's now approaching, and is causing resonance effects on our sun in various ways. A small organization within NOAA is aware that this is a cause of the warming of *all* the planets, not just the Earth. This information is classified, but has been known for a number of years.

We told him about Andy Lloyd's excellent *'Dark Star' website*, which he had not seen before. We also offered to send him Lloyd's *book* 'The Dark Star'; but he declined, saying that there was a risk that it might 'front-load' him too much before he had the chance to recall more information.

This issue is connected with the Roswell catastrophe described above. The problems the future humans were attempting to address were multiple, but principally—as Dan Burisch has stated on many occasions—featured a possible event triggered by a massive 'spike' of solar activity at some point in our currently near future.

Henry, like Dan, is at pains to emphasize strongly that this event is only possible (having been observed in Looking Glass devices in a possible future) . . . and currently is evaluated to be unlikely.

Henry explains that the increase in solar activity is caused only in part by the 'Dark Star', multiple factors being at play. These are complex. Some of them are on a galactic scale, and are associated with natural, periodic events which the Earth has suffered through a number of times previously. What makes this particular time completely unique for our planet is that there is a convergence of serious factors—such as carbon emissions, overpopulation, and our propensity for choreographing war—all of which combine with these major, cyclic and solar events to simultaneously threaten the well-being of ourselves and the biosphere.

Mars

Henry confirmed the existence of a large manned base on Mars, supplied through an alternative space fleet and also through stargates.

Signal non-locality

Henry told us that he had personal experience of interfacing with a team which effectively conducted Alain Aspect's pivotal and conclusive

1981 experiment to prove Bell's Theorem, under the auspices of a classified project at Livermore in the late 1970s. The results obtained were never published in journals—as is usual in the case of "black budget" research.

The mathematical physics is complex, but the technology developed is now regularly used to communicate instantaneously across vast distances.

The disk shot down at Hunter Liggett

We asked Henry to tell us more about this event. He gave the location of the incident: within a mile of the center of *this map*. It occurred in late 1972 or early 1973. His team were testing experimental laser weapons, targeting various materials in the field environment. Suddenly a disk-shaped craft appeared out of nowhere—measuring about 100 feet diameter by 25 feet high. It was about 150-200 yards away. Someone shot at it with Air Force experimental laser which was under test at the time. The primary laser system was contained and operated from the back of a slightly modified M-35 2.5 ton truck called a "Deuce and a Half". There was no visible external damage from what Henry could see or remember, but the craft was disabled. Three small child-like aliens (not classic "Grays") were captured, alive and in good health, and were transferred to a Nike base located in the hills near Tilden Park, directly east of Kensington, CA. All this happened very quickly and was quite a shock to all concerned.

17 February, 2007

Click here for a further update (May 2007)

We have heard nothing from Henry Deacon for the last five weeks. Prior to our last communication, in the last week of March, he had told us he was being 'coerced'. Up to that time he had been in very regular communication with us. We now feel certain he has been forced into silence.

While this newest update contains fascinating further information on a number of topics (including more information about Mars and some details of the inside story of 9/11), we continue to believe that the most important information he has shared is to be found on this page.

2 May, 2007

———————————

Click here for an important new statement (December 2007)

Henry has broken his silence after six months to tell us of his pessimism about a number of current serious global problems. He again references *The Report from Iron Mountain*, and urges us to pay attention to *Dr. Bill Deagle* and to Alex Jones' *Endgame*. He then talks more about Mars and the solar system, the secret space program, Arthur C. Clarke, the Apollo moon landings, and more.

17 December, 2007

A further update from 'Henry Deacon',

A Black Operations Whistleblower

For Henry's background and for a transcript of our first interview, please *click here*.

A few months later, we published *this update*, which was a compilation of further information from ongoing communications we had received.

Both should ideally be read carefully before studying the further compilation which follows here.

———————————

our last communication from Henry was on 30 March, 2007. Since then he has been totally silent, despite all our efforts to re-establish contact. Prior to his last communication, he had told us that he was being 'coerced', and that he feared he might be obliged to cease communication with us.

Although he was very careful (and sometimes enigmatically so) with his messages—which reached us through a variety of sometimes very creative means—he was also under no illusions about the capacities of some agencies to monitor information; see below for details.

We continue to regard Henry as a good friend whom we'd grown to like, admire and trust. We miss his wry messages, his sparkling humor, and his strong principles and values. Wherever he is and whatever he is doing now, we hope he is safe, and wish him well. We assume that if he's still alive he'll be reading this page. Henry, we miss the lassi! It just wouldn't be the same without you.

What follows is a further compilation of information of interest. Although some of it is literally incredible, we believe he has always told us the truth.

Monitoring

Henry warned us that there exists advanced technology that can monitor conversations even in the outdoors. He told us that satellite lasers now have the capacity to pick up vibrations *on a person's clothing*. Monitoring speech vibrations from a glass window pane is elementary, and is older technology.

This is important for us all to understand: bugs no longer need to be physically planted in someone's apartment. Cellphones can also be activated to relay conversations, even when switched off; the only true safeguard is to remove the battery. Our conversations can be heard almost anywhere, at any time . . . if the agencies choose to listen in.

9/11

Henry told us that besides knowing way beforehand that something like this was planned (see his first interview), *he'd been briefed about it in detail a few hours before it happened in his place of work at the time*. The briefing took place with a group of colleagues. He reported being shocked not only at the low-key way it was announced, but at the comparative apparent lack of reaction of many of his colleagues. The purpose of the announcement was so that when the employees heard about it on the news later that day, they would not be alarmed: "When you hear this in the news later today, don't panic, because this is what's going on."

We are of course under no illusions about the importance of this. Five years after the event, Henry may be the first insider to come forward with confirmation that 9/11 was an inside job.

Further details:

- The planes that hit the twin towers were remotely controlled (over-riding the pilots, and with software modifications that would permit tight turns that would normally be impossible due to autopilot default limit settings). The remote control was enacted from thousands of miles away.
- The plane that hit the Pentagon was a small, remotely piloted Navy jet. A full-sized Boeing 757 could not have been used because the aerodynamic ground effects would have prevented it coming in so low at full speed without it having to slow right down as if to land.
- The well-publicized cell-phone calls had been fabricated.
- He also thought that control over Flight 93, which either crashed or was shot down over Pennsylvania, had been lost, and that that part of the plan had gone wrong (there had been another targeted building, but he did not know which one).
- When we asked what had happened to the passengers and the plane of Flight 77 (the plane that supposedly hit the Pentagon), he replied that he did not know.
- Osama bin Laden had nothing to do with any of it apart from the fact that he was a USG asset.

He told us that that was all he knew.

Multiple timelines

We received this e-mail message from Henry in February 2007. We reproduce it in its entirety.

Do we exist on multiple timelines? Your reality, in one way, depends upon "your" perceptions, awareness, selections . . . from a vast ocean of "probabilities" . . . but this is only an extremely limited model to work with in an attempt to conceptualize an answer. The English language alone does not accommodate for communicating an answer to this question. Most humans in today's Earth-bound world cannot grasp an answer to this question due to lack of exposure to other aspects of "reality". The question itself is not complete, again, due to the constraints inherent in most world languages (the languages being at best, related to four dimensions) (or on the other hand, utilizing "time" as a part of the language(s)) and possible non-exposure to other reality sets, by the person attempting to ask and to understand the nature of so-called timelines, etc. The concept of timelines is only a model attempting to explain what cannot be explained in "words" here. There ARE other communication modalities available to us here, which can aid in understanding certain concepts. Explain colors to a person who has never had sight. Explain this world, as you know it, to an unborn child, who has active senses, but yet to put them to use in our "world". Remember the story of the Flatlanders—or even Sphereland? At this moment, the focus should be on waking the people . . . preparing them for a "reality" shockwave . . . to minimize suffering. Best Wishes.

Stargates

Regarding Montauk, Henry said most of Al Bielek's information is correct. There are apparently several kinds of star-gate, notably: (a), The kind where you step through a portal and leave the device behind, and (b) the kind where you take the device with you.

The latter he said was a bit like "Think about where you want to go, and you're there." The mental interface is significant, apparently. He confirmed that as far as he knew Dan Burisch's information about star-gates was 95% correct, but the missing 5% was that he had no knowledge of the large-scale *Looking Glasses* Dan described. (We showed him *Dan's diagrams* and also *Bill Hamilton's source's text*.) Henry emphasized that this didn't mean they didn't exist, because there was so much compartmentalization—but simply that he had no knowledge or experience of those devices.

We showed him Dan's diagrams in person. We watched him while he examined them carefully. Then he suddenly said: "Did he tell you about the one in Iraq?" We asked him whether had he not heard *that part of Dan's interview*. Henry told us that he'd not finished watching the videos. We asked him what he knew. He said the Iraq star-gate was what the Iraq war was really all about, that its location was one of the biggest secrets, and that the war was at least partially about control of it. We asked him how he knew all this—did he read it in a briefing document? No, he said, not a briefing document. The only thing he would say was that it was "first-hand knowledge".

The distant future

Henry told us that in approximately 6,000 years time the Earth will be practically barren, and there will be an attempt to repopulate it. He said that large numbers of children have been abducted from the present and taken to the future Earth, because their genome is undamaged. (In future history the imminent catastrophe significantly damages the human genome.)

He also confirmed that somewhat later than 52,000 years time, the Looking Glass data seems to go blank and no further information is accessible. This is is exactly what was stated by Dan Burisch. (This information was given in a meeting. Henry finished the sentence for us as we were asking the question.)

(Note: Henry clearly confirmed the existence of "Looking Glass" devices which could "see" into the future, or into possible futures. It was the details of the type of technology described by Dan Burisch (see *this page*) which Henry was unable to confirm personally.)

Henry told us that he had thought that there was some kind of barrier in place which meant that future humans were only able to visit us at certain intervals of about 6,000 years. He stressed that he didn't know much about this. Interestingly, a quick calculation shows that concerning the two principal dates frequently referenced by Dan Burisch, $45,000 = 7 \times 6,500$ and $52,000 = 8 \times 6,500$. This struck us as potentially significant. We had wondered about why there had not been visiting future humans from the intermediate periods of 46,000, 47,000, 48,000 (and so on) years in the future. Henry seemed to be offering one explanation.

Incredibly and significantly, it seems that the Mayans (whose famously accurate calendar ends in the year 2012) apparently had access to information left to them by visiting time travelers.

An environmental threat

Henry very much wanted to visit Egypt. He was quite anxious about it, and told us it was a problem. When we asked why, he said that there was very little remaining time in which to travel there. We pressed him for a reason, and he responded that it was not connected with war or politics. We pressed him further, and he eventually said simply: "an environmental threat". He refused to elaborate, nor would he reveal how it was he had access to this information.

Underground and undersea bases

Henry confirmed that many undersea bases exist.

Chaotic resonating circuits

Henry elaborated on the chaotic resonating circuits he mentioned in our *first interview*. He confirmed that they were relatively cheap and easy to make, and that the information, in segmented pieces, had been fairly widely available in a certain academic community in the 1970s. He had retained no records of the circuit diagrams, but intriguingly he said he was 80% confident that he would be able to locate them if he spent time searching public domain records in a particular university library. We know the location but for understandable reasons are not revealing it at this time.

Important contacts

Henry, on several occasions, referenced and strongly recommended the work of *Bernard Pietsch, Stan Tenen*, and *Richard Hoagland*. He maintained that Pietsch knew everything there was to know about the Great Pyramid; that Tenen was an inspired genius; and that Hoagland knew a very great deal of accurate information about the solar system.

Mars

The story appears to be extremely complex, and that may be an understatement. Henry fed us snippets on an intermittent basis. These were sometimes very enigmatic. What we were able to gather together was this:

- Henry told us that the Mars base has a large population—670,000 as of a few years ago. This seemed a huge number to us. We asked if these were all human. "It depends what you mean by human", came the reply.

The base has been in existence for an extremely long time ("tens of thousands of years"), and its population has waxed and waned over the centuries. It lies "at the bottom of an ancient seabed". It is "not far" from the location of *this NASA photo*, taken by the 1976 Viking 2 lander on Mars' expansive *Utopia Planitia* ("Nowhere Plain", sometimes referred to as "Utopian Plain").

- Henry stated that the *most recent NASA images*, discrediting the notion of a "Face on Mars", have been doctored, as have the *colors of the Martian sky* on most officially released NASA photos. (It is apparently more blue than we are permitted to believe.)
- The Anunnaki are one part of the mix. Henry referred to them a number of times as being operational in present time. He said that they originated from another star system, but he could not remember where. The Anunnaki is our name for them, and they are the same race referenced and described in Zecharia Sitchin's work.

The Anunnaki themselves are split into a number of factions, some of which are friendly, and others not. The most sinister thing he alluded to was that one faction of the Anunnaki sometimes preyed on human flesh, having acquired a taste for it. Other factions of the Anunnaki sought to prevent this. Such was the extreme nature of this piece of information that (in conversation) Henry was obliged to make repeated oblique references to it before we finally grasped what he was trying to tell us.

Click on the image on the right to read a screen shot of an e-mail we received which conveyed a very small part of this story. Note Henry's emphasis on the word "*appears*" (typical of his cautious and exact approach), and the typo in "Sumerian". The e-mail is presented exactly as received.

- Transport is by two means: stargates for personnel and small items, spacecraft for larger items of freight. The alternative fleet is codenamed SOLAR WARDEN.

We had first heard of this from another source, and queried Henry about the codename. We sent him two separate messages, each simply of one word: SOLAR, and then WARDEN. We offered no context, or reason for our communication.

The reply came immediately in three e-mails, each from a different address. The first said MARS, the second said ALTERNATIVE, and the third had as its subject "Not listed here" and gave *this URL* as its only content. We were impressed.

• Most controversially, Henry gave several hints, on separate occasions, that he had been to Mars himself. The first occasion was in conversation, when he was talking about the base. We impulsively asked him if he had been there.

There was a very long pause. Eventually he smiled and said: "I played a lot of ping-pong and watched a lot of TV". He repeated this obscure allusion later on two further occasions. Every way he referred to the base was consistent with his having visited it personally.

Later, in separate conversation, he said that a journey through a star-gate was "instantaneous", and he gave the impression that the experience of the star-gate transition was at once disorienting and exhilarating. He described the appearance of a manmade star-gate as a featureless gray surface. Natural star-gates, he said, have a different appearance which is much harder to detect.

He appeared to agree with us strongly when we remarked that the prime protection of claims of this nature was their sheer unbelievability.

We have withheld this part of Henry's story thus far because of the obvious risk that this may appear to discredit him in the eyes of some. Intellectual honesty compels us to report this, now that Henry appears to have been silenced.

2 May, 2007

Click here for an important new statement (December 2007)

Henry has broken his silence after six months to tell us of his pessimism about a number of current serious global problems. He again references *The Report from Iron Mountain*, and urges us to pay attention to *Dr. Bill Deagle* and to Alex Jones' *Endgame*. He then talks more about Mars and the solar system, the secret space program, Arthur C. Clarke, the Apollo moon landings, and more.

17 December, 2007

 -FAMILY BRIEFING DOCUMENT-

---

Earlier I told you about the "Philadelphia Experiment" and how the NAVY scrapped the project after the unforeseen problems with it occurred, and how after that, they sunk money into Einstein and the H-bomb to win WWII.

After WWII the NAVY went back to the Philadelphia experiment to see just where they went wrong and tried to figure it all out. This began with the "Mauntak Project" also associated with "Project Rainbow" and one other project that I can't recall, but all the same project. What these projects did was ultimately develop time travel. I won't go into how it works but will brief you on its operations and some of its experiments.

Now remember I told you that in WWI the Germans found a UFO. To the best of my knowledge it was dug up. Well the molecular composition of this craft had some very extraordinary capabilities. Scientists found that the pilot's chair of this craft was somehow connected as an interface to the ship. In other words when you sat in the pilots chair you could somehow control the craft with your thoughts or it sensed and understood your thoughts and commands. So this chair, was of huge interest in more ways than could be explained.

What they found was that the molecular structure or the material that the chair was made of, could feel, sense, and respond to the most subtle electromagnetic biorhythmic telekinetic thought wave activities created by the human brain. The metal could resonate in sympathy with, and respond to, the thought waves of someone sitting in the chair.

Experiments were conducted on the chair with a human in it who's brain wave activity was monitored by supercomputers through electrodes stuck all over his head as well as electrodes stuck all over the chair. What took place was beyond belief. While in the chair, a human thought was sent through the chair and out to a series of huge amplifiers to raise the thought waves energy output. With enough power they could manifest anything into existence. The human would think of a can of beer in front of him on the table and a can of beer would appear. It would actually appear out of thin air and you could drink it.

The unexpected thing was that the beer would not instantly appear. Usually it would appear within 72 hours or less. After stopping to figure out what was going on they realized they had a time machine on their hands. So now they told the human in the chair to imagine a time portal to 1943 and within a few seconds a portal opened up. With a few experiments they could send someone through and back. Not long after time travel was perfected. Soon after that they realized they no longer needed the chair or the human. All they needed was just the computer recording of the supercharged thought pattern that was created by the chair and they could open the portals at will.

These experiments took place in the late 1940's and were perfected by the NAVY and the AIR-FORCE. Today it is now referred to as "Jump" technology.

Prior to these experiments the military was also out at Groom Lake at the Nevada test site developing and detonating atomic weapons. At this point we attracted the attention of ETs. In 1947 we accidentally downed a UFO with a high powered radar dish. It seems that the craft are electromagnetic and work in the opposite as static electricity. Basically the craft is made of two plates connected in the center by a series of poles that run from the top of the craft to the bottom. The bottom of the craft is frozen somehow and then a very high voltage charge is sent down the poles which creates a radioactive coronal discharge initiating anti-gravity and the craft is lifted off of the ground creating an electromagnetic bubble or force shield around the craft known as a bow-wave. This electromagnetic bow-wave is what allows the craft to enter and exit water without creating a splash. In this hyper energetic state and encapsulated in an electromagnetic bow-wave this ship has no resistance from the laws of physics as we know them. The ship simply resonates to the frequency of the desired dimension within the time space continuum and slips into hyperspace to instantaneously shift itself to that location in time-space. Bob Lazar has explained this in his interview far better than I have here and I'm sure I do it no justice in comparison at all. But I did read somewhere that due to our planet being mineral rich a UFO was found over 2 miles deep in the earth stuck in, if I recall correctly, a gypsum deposit. The craft was unharmed but the occupants were dead. It seems gypsum has the ability to short out or ground out the ships electromagnetic hyperdimensional propulsion system stopping it cold and rendering it stranded and stuck in the deposit if it encounters one large enough as it did on this occasion. These craft have the ability to slip in and out of dimensional travel at will and can therefore move through mountains or any other solid object as long as it does not encounter a molecular structure that is the magnetic polar opposite of its own.

Here is the schematic along with the mathematical formula explaining the physics behind its capabilities. The letter that follows that is a communication from Tom Bearden. He is Lt. Col. T.E. Bearden (retd.), 1990. Colonel Bearden is a nuclear engineer, war-games analyst, and military tactician with over 26 years experience in air defense systems, tactics and operations, technical intelligence, anti-radiation missile countermeasures, nuclear weapons employment, computerized war-games and military systems requirements. The guy is a genius and has probably been involved with back engineering all this equipment for every agency on the planet. I'll let Tom explain it, apparently NASA is all over this stuff as well and has a patent on one of the "lifters". I pick back up again after that.—(Sabre)

The Gravitational Spacecraft from Fran De Aquino

Warping to the deep space . . .

Courtesy of Fran De Aquino

*created on March 21th, 2000—**JLN Labs**—Last update April 20th, 2000*

It is known that photons have *null* inertial mass (m_i = 0) and that they do not absorb others photons (U = 0). So, if we put m_i = 0 and U = 0 in Eq.(1.04),

$$(1.04)\ m_g = m_1 - 2\left\{\left[\,1 + \left[\,U/m_1 c^2\left[\,\tfrac{1}{2}e_r m_r\left[\,(1+(s/we\,)^2)^{1/2}+1\right]\,\right]^{1/2}\right]^2\right]^{1/2} - 1\right\}m_1$$

the result is mg = 0. Therefore photons have *null* gravitational mass. Let us consider a point source of radiation with power P, frequency f and radiation density at distance r given by D = P /4???r^2

Due to the *null* gravitational mass of the photons, it must be possible to build a *shield* of photons around the source, which will impede the exchange of gravitons between the particles inside the *shield* and the rest of the Universe. The *shield* begins at distance rs from the source where the radiation density is such that there will be a photon in opposition to each incident graviton. This critical situation occurs when $D = hf^2 / S_g$, where S_g is the geometric cross section of the *graviton*. Thus r_s is given by the relation, $rs = (rg / f)(P/h)^{1/2}$

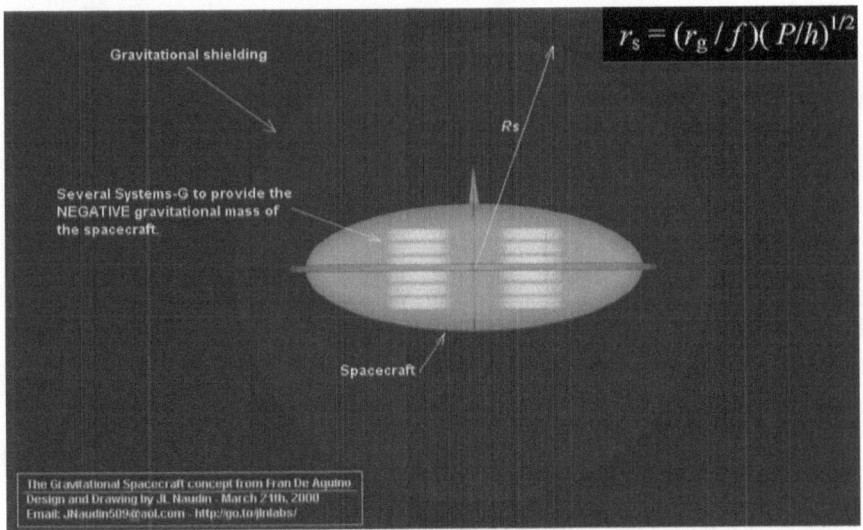

We then see that the ELF radiation are the most appropriate to produce the *shield*. It can be easily shown that, if f???1mHz, the radiation will traverse any particle. It is not difficult to see that in this case, there will be "clouds" of photons around the particles inside the *shield*. Due to the null gravitational mass of the photons, these "clouds" will impede the exchange of gravitons between the particle inside the "cloud" and the rest of the Universe. Thus, we can say that the gravitational mass of the particle will be null with respect to the Universe, and that the space-time *inside the shield* (out of the particles) becomes *flat* or *euclidean*. It is clear that the space-time which the particles occupies remains non-euclidean.

In an *euclidean* space-time the maximum speed of propagation of the interactions is *infinite* $[c \to \infty]$ because, as we know, the metrics becomes from *Galilei*.

Therefore, the interactions are *instantaneous*. Thus, in this space-time the speed of *photons* must be *infinite*, simply because they are the *quanta* of the *electromagnetic* interaction. So, the speed of photons will be infinite *inside the shield*.

On the other hand, the new relativistic expression for mass, Eq.(2.06),

$$(2.06) \quad |M_g| = |m_g| \left[1 - V^2/c^2 \right]^{-1/2}$$

shows that a particle with null gravitational mass isn't submitted to the increase of *relativistic mass*, because under these circumstances its gravitational mass doesn't increase with increasing velocity. i.e., it remains null independently of the particle's velocity. In addition, the gravitational potential $\varphi = GM_g/r$ for the particle will be null and, consequently, the component $g_{00} = -1 - 2\varphi/c^2$ of the metric tensor will be equal to ???.

Thus, we will have $ds^2 = g_{00}(dx_0)^2 = g_{00}(icdt')^2 = c^2(dt')^2$ where t' is the time in a clock moving with the particle, and $ds^2 = c^2\,dt^2$ where t is the time indicated by a clock at rest ($dx = dy = dz = 0$).

From the combination of these two equations we conclude that $t' = t$. This means that the particle will be not more submitted to the relativistic effects predicted in Einstein's theory. So, it can reach and even surpass the speed of light. We can imagine a spacecraft with *positive* gravitational mass qual to (m) kg, and *negative* gravitational mass (see System-G in appendix A) equal to ???(m ??? 0.001) kg. It has a *shield* of photons, as above mentioned. If the photons, which produce the *shield*, radiate from the *surface* of the spacecraft, then the space-time that it occupies remains *non-euclidean*, and consequently, for an observer in this space-time, the *total* gravitational mass of the spacecraft, will be $|M_g| = 0.001\,\text{kg}$. Therefore, if its propulsion system produces F=10N (only) the spacecraft acquires acceleration $a = F/|M_g| = 10^4\,\text{m/s}$ (see Eq.(2.05)).

$$(2.05) \quad F = |M_g|\,a$$

Furthermore, due to the "cloud" of photons around the spacecraft its gravitational interaction with the Universe will be null, and therefore, we can say that its gravitational mass will be null with respect to the Universe. Consequently, the inertial forces upon the spacecraft will also be null, in agreement with Eq.2.05 (Mach's principle). This means that the spacecraft will lose its *inertial properties*. In addition, the spacecraft will can reach and even surpass the speed of light because, as we have seen, a particle with null gravitational mass will be not submitted to the relativistic effects.

Fran De Aquino, April 20th, 2000

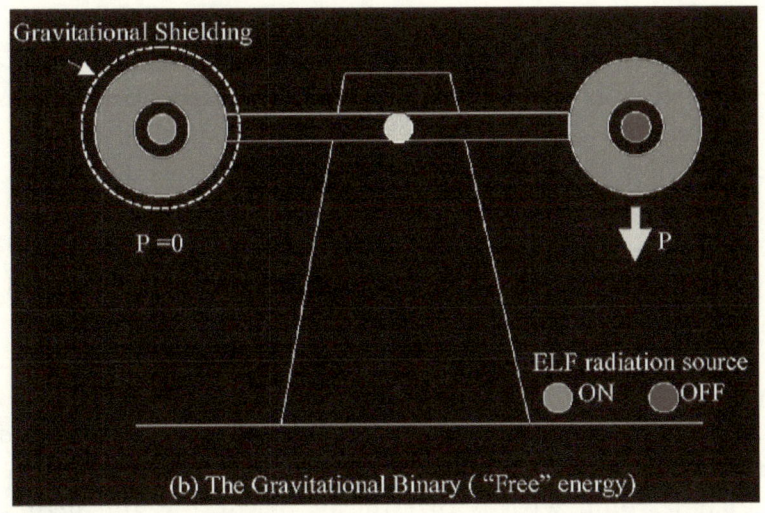

See also:

- The *System-G Experimental setup* tested by Fran De Aquino on January 27th, 2000
- *Engineering the System-G device by JL Naudin and Steve Burns*

Reference documents:

The Gravitational Spacecraft by Fran De Aquino (physics/9904018)

Gravitation and Electromagnetism: Correlation and Grand Unification by Fran De Aquino (gr-qc/9910036)

Email: *JNaudin509@aol.com* or mail to the *JLN Lab's eGroup* at: *jlnlabs@egroups.com* if you are a *team member.*

Return to the *System-G* home page

Dear Tim,

Happy to receive your nice letter, and by visiting your website at *http://www.americananantigravity.com* I was able to see the remarkable experimental work you have been doing. I'm particularly impressed when researchers such as yourself also publish detailed information on exactly how to build the devices and the circuits and repeat the experiments. So researchers wishing to test this area themselves, can in fact built a fairly inexpensive lifter and experiment with it. They can prove for themselves that it works, and that it really doesn't fit the electrodynamics they were taught in university. Congratulations also on your videos taken of actual tests. I feel that this particular research with "lifter" technology is long overdue, and the conventional scientific community has been much remiss for decades in not vigorously funding research in this area.

What I would wish is that the hard-working and dedicated experimental researchers such as yourself could be funded by the scientific community, DoE, or large nonprofit organizations, so that a large group of determined researchers keeps digging into the phenomenology. In any new area where things are not yet understood, it is the phenomenology and its detailed exploration that eventually leads to a breakthrough understanding of the field. Once that happens, then good theoretical models—and technology and engineering—follow apace. So in my view, what you are doing is of extraordinary importance to the development of science and particularly to the further extension of physics. It also is the forerunner to developing actual usable technology. If ever we really wish to explore space, we are certainly going to have to find and develop better propulsion and lifting systems than rockets! If the nation can afford to spend 50 years and billions of dollars attempting to conquer hot fusion, and still seem to be another 50 years from it, then surely we can afford to spend 20 years and a billion or two dollars on this vital area of research with such great potential.

Also it was good to see so many fine researchers in this field! Just to mention a few, there is Tim Ventura (yourself), Jeff Cameron, Hal Puthoff, Jean-Louis Naudin (and quite a few other fellows corresponding on his website), Woodward, Rueda, Haisch, Campbell, Ning Li (now returned to China), Podkletnov, Serrano, Kulikov, Corum, Cox, Black, and others too numerous to mention. Also, an appreciable patenting activity has been occurring, with many patents being issued (perhaps more than 100).

Even NASA has a lifter patent assigned to it that seems to be a variation of the T. T. Brown capacitor effect.

Aside from these researchers, some distinguished theoreticians and academicians are also working in directly related areas. There are many theories or branch theories of gravity, of course. Sachs's unified field theory—which is engineerable by higher group symmetry electrodynamics, including SU(2)XSU(2) advocated by Barrett and O(3) advocated by Evans and Vigier—is a case in point. The Alpha Foundation's Institute for Advanced Study (AIAS), spearheaded by Evans, has in fact published a paper on antigravity: it is M. W. Evans et al., "Anti-Gravity Effects in the Sachs Theory of Electrodynamics," Foundations of Physics Letters, 14(6), Dec. 2001, p. 601-605. Many other fine papers struggling with the problem of positive and negative gravity are also being published.

The real problem, I feel, is the hopelessness of conventional classical electrodynamics and electrical engineering with respect to this work. E.g., the standard EE model erroneously assumes an inert vacuum and a flat local spacetime. The inane EM model used in every electrical engineering department actually excludes every charge in the universe as an acceptable Maxwellian system. Instead, it is unable to model or solve the "source charge problem", the fact that the charge sits there and continuously pours out real, measurable EM energy in all directions in 3-space, with absolutely no observable EM energy input. In short, the classical EM model and electrical engineering assume that every charge in the universe is a perpetual motion machine, freely creating energy out of nothing, continuously, and pouring it out. The solution to that problem has been called the "most difficult problem" in quantal and classical electrodynamics (Sen, Fields and/or Particles, Academic Press, London and New York, 1968, p. viii.). Yet the basis for its solution has been in particle physics for 45 years, with the award of the Nobel prize to Lee and Yang in 1957. One of the things proved by Wu et al. in early 1957, when they proved Lee and Yang's strong 1957 prediction of broken symmetry, is the broken symmetry of opposite charges—such as are on the opposite ends of any dipole. Take an "isolated charged particle", for example. As is well-known, it is clustered around by virtual charges of opposite charges in the vacuum—the well-known polarization of the vacuum. That effect has to be accounted for, since it shields part of the charge and the magnitude of the charge that is observed is dramatically different from the magnitude of the "bare" charge if there were no such shielding. Now take a differential piece of the observable charge, and pair it with any of those virtual charges of opposite sign. Voila! The "isolated charge" is a set of composite dipoles, so it is a set of broken symmetries. This means rigorously that it continuously absorbs virtual photon energy from the vacuum, transduced it into real observable photons, and pours them out at the speed of light in all directions in 3-space, creating the associated fields and potentials and their energy, eventually reaching across all space. That is the solution to this "most difficult problem" in electrodynamics. We published that solution in 2000. (Bearden, "Giant Negentropy from the Common Dipole," Proceedings of Congress 2000, St. Petersburg, Russia, Vol. 1, July 2000, p. 86-98. Also published in Journal of New Energy, 5(1), Summer 2000, p. 11-23. Also carried on DoE restricted website *http://www. ott.doe.gov/electromagnetic/* and *www.cheniere.org*.). Later we uncovered very powerful support of that proposed solution, from quantum field theory and a slight reinterpretation (slight correction) of Whittaker's 1903 decomposition of the scalar potential. The charge is therefore a special kind of system continuously extracting real, usable EM energy from the vacuum. So is every dipole, including the source dipole formed between the dipoles of a battery or generator.

Now we can understand how every generator and battery already pours out enormously more energy than the shaft energy input to the generator or the chemical energy available to the battery, as Heaviside discovered, Poynting never considered, and Lorentz arbitrarily discarded.

It also means that there is not now, and there never has been, a single electrical engineering department, professor, or textbook that even teaches what powers and electrical circuit or the power grid. It isn't the tranducing the shaft energy input to the generator or the chemical energy available to the battery. All that burning of hydrocarbons, use of nuclear fuel cells, building of dams and windmills, of itself does not directly place a single joule of energy on the power line. Instead, all that mess just makes the dipole—that the standard closed current loop circuit destroys faster than the circuit can power its external load. So we have to keep destroying and polluting the biosphere, ruining the planet, etc. just to keep restoring the dipoles in our primary power generators etc., while the engineers happily design the systems to keep destroying those dipoles faster than they can use some of the extracted vacuum energy to power their loads.

Also, by assuming a flat local spacetime, the EE model assumes there can be no change in the energy density of the vacuum—falsified by every EM wave, potential, and field. If rigorously applied to itself, the model—

with Lorentz symmetrical regauging—"eats itself" and is an oxymoron. The easiest thing in all the world is to extract EM energy—enormous amounts in a continuous great flow—anywhere in the universe. Just make a little dipole or charge up something. That's it. The only problem is to intercept some of the freely flowing energy, collect it in a circuit, and then dissipate it in a load without using half the collected energy to destroy the source dipole that is gushing forth the EM energy extracted from the vacuum.

The electrical engineer's Lorentz-regauged model forbids any open EM system far from equilibrium with an active environment (the local active vacuum and the local curvatures of spacetime). Consequently, it forbids every electrical charge and magnetic pole. But it also assumes that all the EM energy processed comes from those very source charges. I.e., it is therefore an oxymoron. This is also part of the problem that has prevented practical electrogravitation. The huge extra Heaviside nondiverged energy flow component is not accounted and not used, but just wasted.

For example, Laithwaite published a paper pointing out the implications of the extra energy flow term in Heaviside's energy flow theory. As you know, Heaviside and Poynting independently and simultaneously discovered EM energy flow, in the 1880s after Maxwell was deceased. It is known that, from the terminals of a generator (from the source dipolarity, once created), there pours out a continuous stream of EM energy flow, filling all space around the conductors of the external circuit. Poynting's theory considered only that component of this external energy flow that is intercepted by the external circuit and diverged into the conductors to power the Drude electrons. Heaviside, on the other hand, included not only that "caught" component, but also included the remaining component that is not intercepted by the external circuit, but misses it and is wasted. The wasted energy flow is orders of magnitude greater than the energy flow caught. However, this meant that from every generator there already pours out far more EM energy than the amount of mechanical shaft energy input to the generator—and that is indeed true. Since no one could explain what could possibly be furnishing such a torrent of excess energy, obviously there was a bit of a problem with the law of energy conservation as it was understood at the time. Since then, the broken symmetry of opposite charges—such as the opposite charges on the ends of the source dipole, once formed between the generator terminals—has been proven in particle physics. In short, now it is known (in particle physics, not electrical engineering) that the source dipole, once formed, continually absorbs virtual photons from the seething vacuum, transduces (coherently integrates) it into real observable photons, and pours out that torrent of real, observable photon energy streaming from those generator terminals. This is the solution to the problem that so puzzled Heaviside and so vexed Lorentz, goading him into creating a neat little trick to get rid of the problem itself.

Unable to solve the problem of the source of that enormous EM energy flow from the terminals of every generator or battery, H.A. Lorentz, who understood the work of both Heaviside and Poynting, reasoned that the excess nondiverged Heaviside energy flow component had "no physical significance" since it did not power anything. So Lorentz integrated the energy flow vector around a closed surface assumed around any volume element of interest. This little trick arbitrarily and neatly disposed of all accountability of the bothersome Heaviside component, while retaining and accounting the Poynting energy flow component. All EM textbooks and electrical engineering to this day repeat Lorentz's integration trick, and dutifully (and arbitrarily) dismiss that Heaviside component. It is still present in every field/charge reaction and outside every electrical circuit, since the Bohren experiment proves its existence, and is readily replicated at any proper university laboratory. So the irony is that electrodynamics and electrical engineering—as they are still being taught in university—arbitrarily dismiss this very large nondiverged, nonreacting component of EM energy surrounding every field/particle reaction. Even the EM fields are misdefined in terms of what is diverged or wrenched out of them—a gross non sequitur. My point is that every EM interaction involves far greater EM energy than is presently accounted for, since Lorentz discarded that huge Heaviside non-diverged component.

In honor of Heaviside, I have nominated that very large unaccounted Heaviside component as what is responsible for the excess gravity holding the arms of the spiral galaxies together—as a solution to the "dark matter and then dark energy" problem. (Bearden, "Dark Matter or Dark Energy?", Journal of New Energy, 4(4), Spring 2000, p. 4-11.)

Heaviside eventually realized in his latter hermit years—spent in a little garret apartment—that his extra energy flow component (which flowed in closed loops, in his theory) had gravitational significance. After his death, thieves ransacked his little garret apartment. Later, beneath the floorboards where he had stowed his draft notes, there were found handwritten papers by Heaviside, developing his theory of unified electrogravitation, with that extra component of energy now converted to a gravitational component. See E. R. Laithwaite, "Oliver Heaviside—establishment shaker," Electrical Review, 211(16), Nov. 12, 1982, p. 44-45. Laithwaite felt that Heaviside's postulation that a flux of electrogravitational energy combines with the (E′H) electromagnetic energy flux, could shake the foundations of physics.

This is interesting and possibly of great significance, because of the tremendous magnitude of that long-neglected excess energy flow component.

In addition to publishing a paper on the potential significance of Heaviside's gravitational work, Laithwaite even suggested that Newton's laws of motion might be in trouble. A presentation of this work to the Royal Institution in 1973 and a demonstration using a heavy gyroscope to prove it to the assembled Royal Institution members, led to the rather abrupt curtailment of Laithwaite's rising career. During the lecture he simply showed them that a very heavy gyroscope, difficult to lift when not turning, could be lifted easily with one hand when turning at speed. Anyone could try it for himself. For the first time in its 200 year history, the Royal Institution did not publish a proceedings of an invited discourse—that one by Laithwaite in 1973. Laithwaite's rise toward grander things was ended.

In 1970 Laithwaite had also completed and delivered a working model of a device that continuously moved itself with "indefinite motion", using a linear motor primary rolled into a cylinder to form the stator of a motor. Laithwaite showed that, under the proper circumstances, a steel washer (a little over an inch in diameter) could be made to roll continuously in a vertical plane around the inside of the stator. Somehow, a combination of centrifugal force and magnetic attraction (and the ever-present force of gravity) maintained the washer in contact with the stator at all times. The little washer would roll indefinitely and continuously. This working model was delivered by Laithwaite to the Centennial Center of Science & Technology in Ontario. So far as I am aware, no one ever tried to analyze Laithwaite's successful experiment in terms of a unified field theory. We know that Laithwaite worked on sophisticated gyroscopic systems for the latter years of his life, finally achieving a mass transfer effect of some kind. He and William Dawson obtained a patent in 1995, with a U.S. patent following in 1999. Sadly, Professor Laithwaite died in 1997. We recall that Laithwaite was for some years a professor at the Imperial College in London, one of the pioneers of the linear electric motor, and also pioneered portions of the MagLev (magnetic levitation) train concept. I had the pleasure of meeting him once, many years ago, at the Imperial College.

A while back, I also visited Transdimensional Technologies here in Huntsville, where I spoke to the Chief Scientist Jeff Cameron and his team. One can see their website at *http://www.tdimension.com/*. Jeff kindly came in from a day of vacation, and he and his team gave Ken Moore and I some very good demonstrations of the lifter technology and their rotor technology as well. I was able to examine the equipment, etc. and can personally vouch that this experimentation is for real. Pictures of the rotor device and the simple lifter are posted on the website. The rotor was tested in vacuum, to prove it is not an ion wind effect. Jeff made me acquainted with NASA's Breakthrough Propulsion Program (BPP), established in 1996, which has had very limited funding for some research in this area and really should be given greater funding and greater priority. I believe that program, or what is left of it, is still managed by the Glenn Research Center, sponsored by the Advanced Space Transportation Program, with its overall management by NASA's Marshall Space Flight Center here in Huntsville, Alabama. The BPP has sponsored some important research, and for example the myth that the lifters could work by ion wind effects has been disproved. Apparently two or more conferences have been held under the auspices of the BPP each year, with papers given and experimental results presented. The real problem in the area seems to be that no one yet has a truly viable theory, although several have been advanced, at least tentatively. Cameron and team are now working on what I would call "second generation" equipment and techniques, have filed several more patents, and expect to be into practical lift vehicles in about five years.

To finish things off, I visited Jean-Louis Naudin's website, where a remarkable collection of photos, videos, information, etc. on lifter technology is given at *http://jnaudin.free.fr/html/lifters.htm*. All in all, I spent quite some time on the web, visiting some other sites as well, in "catching up" to what has been going on in this field.

All this and my visit to your website vividly brought back memories of the antigravity experiment I designed and convinced Floyd Sweet to perform back in 1984, following a theory I had had since Georgia Tech in 1971. That experiment worked beautifully, but it absolutely depended upon access to a COP>>1.0 EM power system. The COP of Sweet's device was 1,500,000 and it had to be pushed to nearly double that. But the experiment did reduce the weight of an object (the power device) on the bench by 90%, at a power level of 1,000 watts. In my view, it proved my theory of antigravity, but of course that still remains to be seen. Eventually we published a paper on the device that included that experiment, which paper is Floyd Sweet and T. E. Bearden, "Utilizing Scalar Electromagnetics to Tap Vacuum Energy," Proceedings of the 26th Intersociety Energy Conversion Engineering Conference (IECEC '91), Boston, Massachusetts, 1991, p. 370-375. I wrote the paper, but placed Sweet's name first, which was appropriate since he invented the VTA (vacuum triode amplifier) being used to perform the experiment with a new output section I convinced him to make. Unfortunately, much of the secret of how Sweet activated his barium ferrite magnets into such powerful self-oscillation was lost when Sweet later died.

So I was delighted to hear from you and receive the photos. I'll ask Tony to post this correspondence on the website in the correspondence section, and also post the photos for all to see. Those persons interested in further information can visit your website, that of Transdimensional Technologies, and Naudin's website and find reams of additional important information, experimental results, ongoing work and investigations, etc.

It is my hope that philanthropic wealthy persons and well-heeled non-profit institutions will recognize the importance of such research, and that funding will be made available to you fellows to continue this vital work.

Very best wishes,
Tom Bearden

This is more witness testimony from a Black Budget Operator compliments of Project Camelot at www.projectcamelot.org—(Sabre)

Jake Simpson: The biggest secret

On return to Europe from the 2008 NEXUS Conference in Australia, flying out of Sydney, we stopped off in Thailand to visit a close friend who lives on the island of Koh Samui. Samui is well known as a travelers' international crossroads, and is also a place where a number of expats of all nationalities have settled.

There we had the good fortune to meet Jake Simpson [an agreed pseudonym]. We spent several days with him and got to know him, and his family, very well. His story, which we heard in great detail, was one of the most important and interesting we have ever heard.

None of the many long conversations we had were recorded, and we hope to capture an audio interview with Jake soon. The following concise summary was compiled upon return to the US, and has been checked for accuracy by Jake himself.

For reasons which will become obvious, we were at first quite unsure whether to release this at all. After a great deal of thought, and further consultation with Jake, we decided to make this information available. Assuming it is true—and we believe that it is—it could hardly be more significant. It dovetails with everything else we know. And in many ways, we wish that it did not.

Much of what follows is barely believable, so this may be its own best defense. Those who choose not to believe this information at all can relax—just a little, and maybe for a little while—with the knowledge that none of it can currently be proved. For the benefit of those whose jobs it is to monitor this information, we do not have any documentation of any kind. We are delighted, however, to consider Jake a close friend. We are absolutely certain that he is exactly who he says he is, and have talked with a number of people who have known him for many years.

In the report that follows, exact written quotes from Jake himself are presented indented and in italics, as in this closing paragraph.

Enhanced abilities

Jake Simpson was in a 'specialist field' [his preferred term] working for a nation friendly to the US. His early training, which is still with him, featured an enhancement of his ability to absorb written information. Incredibly, he has a reading speed of between 80,000 and 100,000 words per minute. To graduate from that class, he was given a copy of George Orwell's *Animal Farm* and had three minutes to read it before being tested on every detail. He scored 90%. This was adequate for a pass, but was not the highest score . . . another student scored 100%.

Jake is a very able psychic intuitive to this day. He is able to perceive when the AI [artificial intelligence] information gathering system was 'focused' in his direction and would periodically pick his exact moment to relate something to us. We observed this again and again. While it at all times seemed the same to us, sometimes Jake told us it was safe to talk, while at other times it was not. It took a little while for us to understand exactly what was going on here.

See below for more on this. We have not heard this information anywhere else in the literature, on the internet, or mentioned by any other witness. But I [Bill] had one experience that showed me directly that this was very real indeed.

Approaching Project Camelot

Jake had written to us by way of introduction:

I have spent many years previously working in abstract areas of national security on behalf of the various parties concerned. I have been very impressed by some of your interviewees' comments. Keep up the outstanding work.

Jake approached Project Camelot after our work had been brought to his attention by a friend. At first, he wasn't convinced that we were 'real'. However, his connections enabled him to do the requisite background checks, and he told us that it soon became clear that we passed muster.

He told us he knew everything about us, but that we had "nothing to worry about". Everyone who had ever tried to do something like we were doing before, apparently, had been killed. He told us he'd watched every one of our videos, admired what we did, and that we had a lot of courage. He assured us that some of what we have reported is very close to the truth, and that we had his respect.

Jake is one of many in military/intelligence circles who are the 'white hats'. Idealistically motivated, he made his career choice when young, wanting to work for mankind and play his part in helping Planet Earth become a better place. Despite discovering the true, bewildering complexity of the world he had entered, he retained both his idealism and his job . . . and many years later decided to approach Project Camelot with a portion of what he now knew.

He told us that we had many of the correct puzzle pieces—and that, furthermore, it was understood that we work with integrity, and that we're not trying to breach any legitimate national security. He stressed that we were quite liked by a number of the 'white hats' who were monitoring us closely, despite our being on a number of 'watch lists' of every kind.

Jake helped us understand that if we kept our information general and didn't try to *prove* anything (with documentation or by any other means), we would remain safe. He stressed that it was very important not to get too specific on certain sensitive issues, and to be *very* wary of ever getting hold of any definitive documentation.

Classified technology and the secret space program

Jake emphasized to us that the current state of classified technology was something like 10,000 [ten thousand] years ahead of public sector technology—and was accelerating away from public sector technology at a current rate of 1,000 years per calendar year.

This got our attention.

Jake did not blink when we mentioned time travel, the Mars base, or the advanced fleet of craft which we had been told by *Henry Deacon* serviced it. Jake told us that some of the advanced craft were capable of traveling from geostationary orbit (22,300 miles) to treetop height in five seconds. (Work it out: that's about 16 million miles per hour—although Jake made it clear that the craft would not actually be moving through space in the normal sense . . . and would also never be seen unless this was intended.) Some of the craft were "larger on the inside than outside".

Had they traveled to the outer reaches of the solar system? *Yes.* Beyond our solar system? *Yes.* Are some of them superluminal (i.e. capable of faster-than-light travel)? *Yes.* Were some of them very large? *Yes.* By this time, we were no longer surprised by Jake's answers. The significance of the superluminal craft would be stressed in a subsequent conversation.

The human race had had contact with extraterrestrials since before World War II. Jake told us that it was very probable that Eisenhower's 1955 heart attack was at least partially induced by the stress of some of the information he had learned from the extraterrestrials who he had personally met a short time before (after several previous set-up meetings with senior military officials).

Taken all together, Jake told us, the ET visitors came from various races, systems and times, and that human DNA "was compatible" with hundreds of different races. All these ET races, in some meaningful sense, could be said to be "human or human-like".

AI surveillance and access to knowledge

The AI surveillance system, Jake told us, was literally "out of this world". It operates hyperdimensionally, based on a highly advanced quantum computing model that is basically our development based on acquired alien technology. This system is so advanced that the ETs themselves are unhappy that we have it.

Not only does it enable access to what any given person is saying, or even thinking—if targeted for investigation—it can also transcend time itself and thereby access information about the thoughts and words of historical figures. Whether this system can look into the future—the Tom Cruise movie *Minority Report*, based on a story by the prolific author Philip K. Dick, comes to mind—we omitted to ask.

Jake's actions in being sensitive to this device (if device is the right word)—by waiting for exactly the right 'window' of opportunity to tell us certain things—were not fully understood by us until I (Bill Ryan) had the following experience.

On our last night together, sitting out in the open after a barbecue, at about 2 am, Jake decided to tell me some things he had not previously revealed, surveillance or no surveillance. As he began to speak, he immediately encountered problems, as if trying to force himself through a barrier. Simultaneously, I found I was being put to sleep and could hardly keep my eyes open. We both spotted what was happening, and remarked on it to one another.

Jake forced himself to keep talking, and I made myself keep listening through a spell of overwhelming dopiness. This episode lasted half an hour or maybe more. We were being forcibly stopped, in real time, from communicating effectively, as a direct and immediate response to our intentions.

It's very important to understand that this was unconnected with electromagnetics, hidden microphones, targeted beams, or anything else of that nature. My own reference point for what happened was a kind of negative *radionics* (which also works hyperdimensionally, but as a positive health modality).

In the end we concluded our conversation, now pretty tired. The next morning I simply could not recall what Jake had told me—and still can't. Upon meeting him again and reporting that I couldn't remember a thing about our conversation except for the weird effects we had both experienced, he replied wryly:

Maybe it's just as well.

The biggest secret

The international network of deep underground bases, Jake confirmed, had been built in a continuing program since soon after the end of World War II costing trillions of dollars. The issue here was that military leaders had learned through ET contact that a potential catastrophe of huge magnitude, occurring early in the 21st century, was possible. This information was certainly known to Eisenhower, Jake said (and may have been partially responsible for his heart attack), and was very possibly known as early as World War II.

Just as we had presented in our important summary article *The Big Picture*, the problem is one which involves massive potential Earth changes that could, *in extremis,* threaten our civilization. The situation had been extensively studied and evaluated and the conclusion had been reached that the public could not be told.

Jake described the threat—*metaphorically*—as a wave that was heading our way. It was unclear whether this 'wave' is a product of an area of space which the solar system is entering—or whether it is the result of a close fly-past by a large rogue celestial body, or even a combination of two or three simultaneous situations or other unusual and impending cosmological events. But when I asked how this is all known, the answer came back that the superluminal craft have gone out to take a good look at what is around, and have returned with the information.

Jake stressed that IT IS NOT KNOWN what the effects of this situation will be, nor precisely when this may occur. The military are preparing for worst case scenarios, which is what they do best. Readers familiar with our work will note the connection with *the report from the Norwegian Politician*, and also *Dan Burisch's information* which culminates in the report on *Timeline 1, variant 83* [T1v83].

Of particular interest is the contradiction with the T1v83 information, in which Dan told us that in the latter half of 2007 a highly classified time portal intelligence retrieval project (for lack of any better phrase) had analyzed a number of possible alternative future timelines and concluded that variant 83—the most probable of the many that had been investigated—demonstrated that the 'Timeline 2' catastrophe had been averted and that while civilization would not be under threat, the next few years would bring major problems. These included nuclear exchanges in a prolonged period of global conflict, under an administration in which Hillary Clinton had been elected President of the US.

Since that information was researched—a year ago at the time of writing this article—it has become very clear that that timeline variant has been 'broken', and that we are now instead hurtling along on another, uncharted timeline. In Dan's words to us a few months ago, "All bets are off"—and when we put Dan's phrase to Jake, he responded:

That's about right. I wouldn't disagree with that at all.

Jake's information is that it had *never* been certain that the catastrophe had been averted, and he confirmed that the governments of many first-world nations were continuing to make their detailed and extensive preparations. Australia, we were told, was the "Ark of the World", and had been designated as such many years previously.

Jake confirmed that he had personally seen some of the classified maps showing dramatically altered future coastlines, and also confirmed the possibility of a very advanced high-speed 'shuttle-like' system that connected many places, like the US and Australia under the Pacific Ocean—a longstanding but always uncorroborated rumor within the UFO community that had acquired semi-mythological status since the startling reports of John Lear and William Cooper in the late 1980s. Jake told us that

. . . the acceleration presses you back into your seat for a very long time.

This has all happened before

One of the most startling snippets of information Jake revealed was that in some locations the base construction engineers had broken through into much older facilities that had been there for thousands of years prior—apparently built for an identical defensive purpose. All this, Jake had told us, had happened before: the catastrophic events are cyclic.

Because of what had been learned through breaking into older facilities built by a prior Earth culture, in some locations decisions had been made to increase the depth of the new facilities to as much as 30,000 feet [9000 meters].

The great classified libraries of the world, in the Vatican and elsewhere, all contained detailed accounts of the destruction of prior civilizations. The Flood Myth, as many anthropologists have described, is evident in many dozens of different cultures all over the world. All this is described in our article *The Big Picture*.

The threat of stealth viruses

These catastrophic events, Jake told us, would happen *not* in 2012 but several years after that, though the dates were not precisely known. When we put *Bob Dean's date of 2017* to him (in the context of the coming of *Nibiru*), Jake's response is that that would be close, as best he knew.

More immediate, said Jake, was the threat of the deliberate release of viruses followed by

. . . the hideous effects of spontaneous eruptions of new generations of opportunistic bacteria like Necrotizing Fasciitis and more advanced versions of golden staphylococci which would further reduce the world's population after the initial first line of worldwide disasters had occurred.

This would trigger worldwide infrastructure breakdown, cause chaos, and make populations easier to control.

Jake predicted that it was quite possible that sometime before the end of 2009—or possibly early 2010—there would be a sudden and rapid escalation of international reported outbreaks of extremely dangerous viruses (whether manmade or otherwise). He emphasized that announcements of a global pandemic could suddenly emerge from nowhere "within hours", and that it would be smart to be prepared: he stressed that some countries could quickly become quarantined, or choose to quarantine themselves, with major implications for international travel and port or airport controls.

Coupled with these outbreaks there will be a very real possibility of food shortages. Even more importantly, there will very likely be shortages of quality foods containing all of the necessary levels of active and absorbable vitamins and minerals of the necessary types, and in the right or proportional quantities, to allow for the human body to properly and adequately nourish itself.

These shortfalls and omissions of some of the fundamental basic substances of various foods will prevent people's immune systems from operating at optimum levels, thus leaving them exposed to these new virulent types of diseases. This will leave literally hundreds of millions of people exposed to disease vectors through 2011-2012 onwards.

Technological access to other dimensions

Finally, Jake told us of research that had unlocked technology surrounding access to other dimensional states of existence.

In these other alternate states of reality sometimes it turns out they can very briefly and spontaneously manifest, very occasionally quite naturally, here on Earth and in any other part of this universe. In very special circumstances, these can spontaneously manifest across not only this universe but indeed into alternative universal realities.

There is a massive amount of research funding being applied to this very obscure part of the broad spectrum of the special access programs of the world's budgetary allowance for these types of programs. These funds are managed and funneled from every imaginable area of the majority of the world's countries through an amazing array of abstract instrumentalities and public funding projects.

The research involved with this subject is at the top of the world's power elite's priorities. This is why there are very selective sightings of some of the largest scientific programs, currently ongoing across the world.

The geographical location (in three dimensions) is absolutely critical in some special cases. Jake wrote:

The Earth's specific rotation is also of consideration and its relationship to the sun and the other planetary bodies contained with in our solar system—especially the larger ones. There are very many localized effects experienced subtly, here on Earth in the course of a year, in relation to the specific location to other localized celestial bodies, apart from our sun, contained within and without the heliosphere.

On a larger scale, the position of our solar system in relation to our galaxy is also of vital importance. Jake's words again:

Our galaxy's position, speed and direction relative to several neighboring galaxies is also important. Beyond this is an understanding of a technology that not only allows for absolute universal travel and capitalization of the fantastic power that entails, but provides the ability to move or travel dimensionally.

Jake explained that this knowledge is essential to the next phase in 'Earth human' development . . .

. . . if we Earth humans are ever to be freed from this particular part of existence and universal / multiversal constraints.

Jake stated that as in every aspect of human history here on Earth, there have been supporters, detractors and outright enemies . . .

. . . both on Earth and not from Earth. Nothing much has changed over all these eons of time. The play is still largely the same with all of the same motives and allegiances.

The pursuit by some at the expense of the many seems to always end up to be the ultimate price, along with the loss of the Individual and Eternal Grand Self/ Soul/ Spirit, particularly when the individual sells out all that is sacred to the ongoing existence of our race.

As Jake explained, some of these power elite's motives constituted

. . . a viable and sustainable level of a heavily manicured and vastly reduced human population, under the pretence of saving first the planet, and next the vast majority of all of the other diverse range of different species here on Earth.

Jake told us that it had been explained to him:

"You have to break a few eggs in order to create a really great meal. The Destroyer of Worlds [sic] *brings with it the promise of massively renewed and clean prolific growth for yet another new direction in Earth Human evolution. Roaches will always be roaches. Someone's got to keep them under control."*

The quote reminded us of *Henry Deacon's reported briefing* in which he had been ordered to read *The Report From Iron Mountain*, and had been told: *"There are the wolves and there are the sheep, and we are the wolves."*

Jake said that what he had been told was highly disturbing to him, and stressed just how at odds he ultimately was

. . . with the paramount agenda of a significant element of the absolute power elite.

Jake's strong personal ethics and morality always prevented him from receiving the rich benefits that he told us he knew could have been his. He told us he was

. . . happy to remain obscure while still contributing where appropriate to assist humankind in meaningful and positive ways.

We gained the impression that the power elite have largely left Jake alone. It was evident to us that he has an intimate knowledge of their culture and knows how to ensure his and his family's safety.

We close here with this remarkable man's own words:

This suggests an even more powerful, off-world group or association that I may have a very special relationship with, that is lending a very discreet helping hand . . . where possible.

We are hopeful that we might record a voice-disguised audio interview with Jake, and although this has been discussed it has not yet been arranged or agreed. We will post more information as soon as we can.

We thank our friend here for his courage in providing this remarkable testimony. Kerry and I are as certain as we can be that this man is both well-informed and has the highest integrity.

I've added this schematical picture from www.eaglesdisobey.com. in reference to the preceding conversation with Jake Simpson concerning, (I'm assuming) "Star-Gates". It is a diagram of the Earth revolving around the Sun in correlation to its tilt that also references the magnetic fields of these two bodies. Do the experiment!—(Sabre)

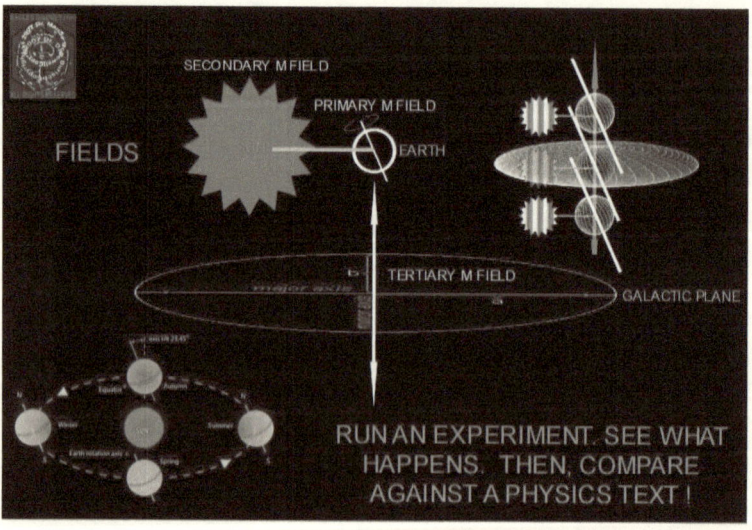

 -FAMILY BRIEFING CONTINUED-

What you are about to read is a thread that was started by an individual under the pseudonym "Astralwalker". It is called (2012 NEXUS EVENT-"Unknown form of energy comes our way.")

This is an attempt to explain the "energy" heading this way that is responsible for increasing the planetary rotation we were warned about by "Captive" at Area-51.

This thread was located on project Avalon (*www.projectavalon.net*) and this site was also started by Project Camelot, (*www.projectcamelot.org*) as a place to share information and for groups to converge to physically locate one another for safety and survival purposes.

The information laid out by Astralwalker was insightful and informative. Much was backed with web links to the scientific data in support of his theories. This informative thread took literally months to release and was followed by a very large group of meditators, spiritual healers, and teachers who had just as much to add to the thread and the overall body of information contained there-in as did Astralwalker himself.

I have added the first book to this list of works and information that was transcribed and re-released as a free PDF in hopes that it will be even more widely distributed for more to read and to add to. There is a second book that follows this book, both are free, fantastic and highly recommended to be read.

The ultimate resolve of this thread by this individual was to get everyone reading it to meditate at the same time, on the same days, with the same thought in mind, to gather the necessary energy required to effect the critical state of mass to induce the change in the morphic field of this planet by assisting it to resonate in sympathy for that which was being called forward to manifest itself into this 3 dimensional reality, which were love, light, peace, and harmony. But the real trick was that in order to really effect the unified field of this planets conciousness the meditations had to be done on, or at these sacred sites, and that would have actually been the cu-de' gra of Astralwalker's ultimate resolve. I think this may have actually happened with a small contingency of people who had the monitary ability to travel to the sites closest to them.

This was very much indicative of the American Indians in which they gathered for the sacred dance with the rest of the tribe in a quantum culmination incorporating physical motion and mind consciousness energies, surrounded and assisted by, the four creative forces of the universe, to include fire in the center, to bring forth that which was being called forward into this 3 dimensional reality, rain.

Like the American Indians, Astralwalker knows that human beings' can actively create and manipulate morphic fields through causality and be co-creators' of their own reality around them. These are the physics that were taught to man-kind in the cradle of civilization and only reconstituted once it was finally discerned again that the universe itself was in motion around the galaxy, the galaxy in motion in the universe around the solar system, the solar system in motion around the planets, the respective moons in motion around the planets, and finally the planets around their respective sun's. All working together in harmony to create, promote and sustain life. This is the way of the one true living GOD and the spirit that is extended from him that lives within all of us. They are NOT working against one another in discord to create chaos. This should FINALLY be a lesson to us all, if not the first lesson on the road to understanding this ability.

The second lesson is harmony promotes prosperity which creates abundance giving birth to love for the continuance of all life that is known as amazing grace.

"Out of Chaos comes order." And where does that all end? And who decides where that ends? And then where does that all end? Do you see where I'm going with this?—(Sabre)

The following statement is courtesy of Project Camelot

- We have a newly refined announcement from *Astralwalker* regarding the ongoing *Planetary Meditations* being orchestrated by a group that has formed calling itself the Rainbow Warriors for world peace and healing of the planet.

Although we support these meditations—from what we understand via testimony from our witnesses such as Jake Simpson and Dan Burisch, our solar system has *already* entered this area of the galaxy encountering waves of energy i.e. a galactic superwave based on the research of *Paul LaViolette*: *"(1981-82): LaViolette was the first to measure the extraterrestrial material content of prehistoric polar ice. Using the neutron activation analysis technique, he found high levels of iridium and nickel in 6 out of the 8 polar ice dust samples (35k to 73k yrs BP), an indication that they contain high levels of cosmic dust. This showed that Galactic superwaves may have affected our solar system in the recent past."* . . . and are now affecting our planet.

Some of the proposed effects are thought to act upon our consciousness as well as our DNA. As we move into alignment with the galactic center these effects are expected to increase and to reach a sort of peak around the years 2012-2013.

- **Nexus 2012 Series of Meditations, aka *The Gathering*:** As many of you are aware, we are assisting Astralwalker and *www. HealingExperiment.com* by announcing a series of meditations for peace and oneness, for healing of the Earth and creating a new paradigm . . . featuring what they have termed *Rainbow Warriors* around the world; this includes Ground Crew members, indigos and all other gifted children (all children are gifted), indigenous people and anyone else who wants to participate.

These meditations are happening starting on February 8th and will continue every Saturday, **culminating in a serioes of global sacred site meditations that will begin on May 9, 2009.**

2012 Nexus Event—Unknown Form Of Energy Comes Our Way

Author—"Astralwalker"

This 332 page document comes from a Project Camelot/Project Avalon forum at the following link http://www.projectavalon.net/forum/showthread.php?t=8441

Copies can be obtained free of charge at Notes by compiling editor Michael Knight.

What you have here is the far-reaching work of an individual whose user name on the forum is "Astralwalker." This document has been compiled and made available with his approval. Project Camelot is one of the few Internet sites dedicated to freedom of information,

- Astralwalker Thread: MUST READ

We have a major and interesting new thread materializing on the Project Camelot section of our Avalon Forum. It's freely available to non-subscribers on a read-only basis. This researcher is laying out in great detail the potential astronomical dynamics behind our move into alignment with the galactic plane and the effects on our solar system. Highly recommended.

—Kerry (Project Camelot).

This document has been edited into this Portable Document Format (.pdf) in order to distribute the information it contains as widely as possible around the world. Please do so.

Much of this knowledge has been suppressed, changed, or rewritten and "spun" out of context by a group sometimes known as "the elite"—in order to keep the bulk of humanity unaware of what is really occurring as we approach the year 2012. ASTRALWALKER has done a remarkable job of presenting information about

- Earth Changes
- 2012
- Crop Circles
- The Mayan Long Count calendar
- Astral travel (with instructions on how to)
- Remote viewing (with instructions)
- Seeing while blindfolded (with instructions for the student)
- Hollow Earth (and hollow planets—with multiple pictures)
- Consciousness and its effect on the planet
- The Illuminati or "Elite" and their control agenda
- Possible safe places during Earth Changes
- How we can influence and change the future—for the better.

The many included links worked at the time they were first posted.

In the copy and paste process, questions by other forum members were either included or excluded according to whether or not they appeared relevant or essential.

This document is provided free of charge and without obligation. It may be passed on free of charge but includes absolutely no resale rights.

It is the reader's choice and responsibility as to how he/she perceives and/or acts upon this material.

Michael Knight—www.buycontacthasbegun.com

Editor Earth Change Report (a free International Newsletter about Abrupt earth changes) Director "Contact Has Begun" Author of the ebook "Earth Changes—Mind Matters" Author of the ebook "How To Survive Earth Changes, Economic Depression and Martial Law."

Hello everyone!

I would like to share few thoughts and conclusions with all of you who want to listen about my research on the **2012 scenario**. They will be simplified to the point where some scholars and scientists could argue about a part of which will be presented, but the main facts are true.

I will start with the ancient Maya knowledge . . .

Deep in the center of the Milky Way there is a Black Hole which scientific community called Sgr.A. (Sagittarius A).

The **Mayans** called it **The tree of life** and it was passed to them that in the center of the galaxy a point known as "**Hunab Ku**" exist and it's a location from where consciousness comes. According to Ancient Maya records, it's a place from which "**The Supreme Creator**" crates and destroys, and that most of all that is created in this region of Space comes from it. However, in the Mayan culture there are no images about "**Hunab Ku—Supreme Creator**" because it was considered as **Ultimate Force of Creation** which in fact does not truly posses visible form and cyclically influences the life in the Galaxy through galactic core.

There is a high possibility that the black hole in the center of the Milky Way represents a portal into a higher Universe which possesses completely different reality. From the cosmic aspect, the black holes are commonly located in the centers of the galaxies and from the latest research are in fact penetration points in and out of this Universe. And they are all connected as grid of cosmic portals. Inside them there are **Points of Singularity**, where all **known laws of physics** fall apart and does not work anymore.

The "**Ones**" that left the knowledge to Mayas, insisted that "Hunab Ku" is the Consciousness that organizes all mater and antimatter. In fact, fallowing the same logic, all black holes in the centers of the Galaxies can be referred as Hunab Ku—a places from where the Will of the Ultimate Creator is manifested. Hunab Ku is so called "**Mother Womb**" which constantly gives birth of new stars and which also gave birth to our Sun and our planetary system.

The "ancients" insisted that we understand the message that the Hunab Ku runs everything that happens in the Galaxies and acts through periodical extreme powerful explosions of Consciousness Energy from the galactic center.

Today, the astronomers confirms that the center of the Galaxy which we call "Milky Way" has form of flatten disk with the black hole in the middle which swallows stars that exists and in the same time it gives birth to new ones.

However, the major part of the "mainstream science", are still considering the **Time** as **Linear** and that it flows in just one direction.

On the other hand, the "**Ones**" that left the knowledge to the Mayas clearly pointed out that the **Time** actually flows in a circular manner or in **cycles**.

"**They**" clearly showed that there is a "**beginning and an end**" of all things, but also a restart after the end of every Time Cycle. The data that was left indicate the importance of understanding the "**periodical nature**" of all manifested phenomena on which the **Mayan Long Count Calendar** is actually based. Those highly sophisticated teachers left high knowledge about the Extreme Time measurements and instructed the Mayas that the Time itself originates from **Hunab Ku** and it is controlled by it in the same time. In fact if we draw a parallel to all other "Ancient Cultures" like **Egypt**, a culture that supposedly worshiped the Sun this picture comes forward:

"**They didn't worship the Sun as a Sun, but as the intelligent Creative Force behind it which brings life**". "**Ahau**" is term which refers the **Sun** and "**Ajpu**" is the term which refers the **Light** from which we all had came and where we all will return. Both terms are used in abundance to show us "**The End**" of our familiar physical existence in **2012** and "**The Beginning of new awareness**" or existence in a **higher dimension** of our Universe. The "**thing**" which is very important for all of us, currently present on Earth in this time which according to Gregorian Calendar is November 2008, is that the "ancients" had delivered a message through the Mayan Long Count Calendar that on a date in the future which matches the Gregorian Calendar on 21 Dec 2012 there will be a **cosmic alignment** between the Earth, Sun, The star cluster Pleiades and the center of the Milky Way Galaxy—Hunab Ku.

On this day, according to ancient Maya and the later Meso-American Civilizations will happen the return of Quetzalcoatl. Is this is going to take place, its still unknown. So, from the astronomical point of view, in 2012 our star will be in alignment with the center galactic point. This was very, very significant to the ancient people. Besides that, we hear about the hidden Astrology and the various aspects of the 12 signs of Zodiac, than about the effect which comes from the alignment of the Pole of our planet and the galactic center and etc.

On **21 Dec 2012**, 11:11 AM **Universal Time** (also known as **Zero Time Point**), our Sun and our planet will be at perfect alignment with the Galactic center.

On the same day, if we consider that everything stays "by default" or on the same track, our Sun will be at its solar maximum, and the ecliptic of our solar system will intersect with the Galactic plane, which also is called the "**Galactic Equator**" of the Milky Way! Here is a link to a interesting website which has lots of important data: *http://www.viewzone. com/endtime.html* Let's go on.

This also is called the planetary alignment and by the Ancients was considered to be from a highest importance for the life on this planet. According to Carlos Barrios and some other Mayan Day-Keeper's, the "Ancients" considered the date 21 Dec 2012 as a date of rebirth . . . beginning of a New Era which is result of crossing the solar meridian through the galactic equator and the alignment of Earth with the center of the Milky Way.

This **Cosmic Cross** is considered as embodiment of the Sacred Tree, (**The Tree of Life**), a **tree** which is remembered in most of the ancient spiritual traditions as alignment with the cosmic heart of the Galaxy, when some kind of opening or a **channel of cosmic energy** occurs and passes though **Earth**, in the same time **cleaning** the whole dirt of the life forms that live on this planet, and **lifting** them on **higher level of vibrational**

existence. Even the greatest sceptics (low consciousness evolution beings) are having difficult time rejecting the so called "**Mayan Cosmology**" because those time keepers has left such **Time Keeping Methodologies** that are so precise that makes you spin in your head, because they are **more accurate** then the ones we use today.

Their **pyramids** like the one in **Chichen-Itza** in **Mexico** are so precise build, that are accurately showing the precise moments of **Solstices** and **Equinoxes** during the year, which are caused by the **Earth's precession** (known as very slow wobble or movement of the planet around its axis). I have to mention the other pyramids in (**Teotihuacan**) Mexico, where we see an extremely precise depiction of all the orbits of the planets in our solar system. **They have a Sun pyramid, Mercury pyramid, Venus pyramid . . . a pyramid for all planets.** The mathematics used in the construction and correlation with each other is so precise that blows one's mind. This extreme precision could not come from primitive people. This extreme precision could come only from a society which had acquired knowledge that is far beyond the one we have today, at least in the public arena. The evidence that such civilizations had exist before are all over the world.

For example in Lebanon there is a location 44 miles from Beirut with fascination ruins. This place is called Baalbek and it is also mentioned in the Old Testament Book of Kings. Those stones called Trilithon are officially the most massive stones in the world used for construction purposes. **Despite the fact that some of the scientists consider that they weight around 800 tons, the truth is that no one with 100% certainty can not say how much actually they weight.** It could be double or at least much more then 800 tons.

They are simply **so big** that we do not know for sure how much they weight. Those stones were somehow cut, shaped, and moved to Baalbek from a location some **five miles** away. **Then they were lifted in the air, perfectly erected and connect that you can not put a paper or riser in between.**

How in the world, they managed to do that, it's still unclear. Besides there some rumours about the use of **antigravity technology**, there is a **legend** that speaks about a race of giants who were here immediately after the great Flood and that those giants had actually build the Baalbek. In any case, one thing is for sure—**today** "officially" we do not posses a **crane** which is capable of lifting those weights. **Not even close.** Even for weights few times smaller then the Trilithon stone weights, we have to

use few cranes with combination of helicopters. If we today can not lift those stones, how was this achieved by the builders of Baalbek thousands of years ago? And not to mention "again" that they were transported 5 miles. There is no tree that we know of, that can hold such weights, and even the trees with the hardest wood are very fast smashed by the weight of the heavy Trilithon stone so the Discovery Channel lies for building Pyramids in Giza, does not work here. The official stories presented by leading media about how the Giza Pyramids were build are ridicules and absurd. Not to mention Yonaguni, the ancient underwater structure that probably is older then 12,000 years.
http://www.altarcheologie.nl/index.h . . . troduction.htm

I can go on and on, but let's summarize this for a moment:

—Despite our impotence do admit that we are far away from the peak of Evolution of the life that once existed on this planet, there is huge amount of evidence all around the world. First there were, so called "**Men in Black**" who were responsible for retrieving, hiding or destroying if necessary everything that can serve as evidence of what exactly happened here on Earth.

If you do not know what happened before, you do not understand why the things are like they are; you do not have a clue what you are and why you have only this double helix awareness. Today after they destroyed this "Men in Black enigma" with movies which reflects SF comic nonsense and with absolutely not even close story to the real events that happened, it's probably a group with a different name but with a same purpose. Anyway, that is why we must not be so ignorant about the knowledge and messages that ancients had left for us. That is why it's extremely important that we study that ancient knowledge which is speaking about the beginning and an end of cycles that are from crucial importance for the life on this planet.

The people, who do not know their history, are lost and can be easily manipulated. While few ones from so called "**Elite**", are in possession of the real information what exactly happened here in the past, are in great advantage that gives them a huge space for easy ruling. However, the ancients were clear that all this is only a part of a cycly, and every cycle has a beginning and an end. This time the cycle ends on **21 Dec 2012**, the last step in the dollar pyramid, so it seems that they will not enjoy in their luxury and wealth gained on the misery of billions other hungry, not educated, false guided on purpose though the mainstream media, political agendas, so called masters, gurus, and spiritual teachers. In any case, lets start to analize from astronomical aspect what will possibly happen around the date that Mayas determined as a last in their Long Count Calendar.

Despite the fact that are **1845 days** left, our world is starting to experience unseen, extreme and rapid climate changes. The number and intensity of those "natural" catastrophes increases rapidly. The scientists had recorded extreme unusual behaviour of our Sun which directly influences all the planets in our solar system. Did you notice the extreme brightness on the all the planets in our solar system?

Does strange **energetic emanations** from the **center of the Milky Way** detected from the **Dr. Scott Hyman** are the cause of so visible changes occurring in our system?—**The Mayas would clearly say: Yes!**

They will also say that colossal emissions of "unknown form of energy" will arrive from the center of the Milky Way, which will change the fundamentals of the physics of our world, new material and immaterial conditions for life which will last till the end of the next cycle. On 21.12.2012 according to the ancient Maya records the humanity will enter into **New Era of Higher Consciousness**. It is interesting that at this moment all that are present on the planet are experiencing more or less a Shift in Consciousness. More and more human individuals are starting to awake, asking who are they, what life is really about, can not eat flesh anymore and simply experience a clear disability to live in old ways.

In 2012 we all have a chance to transcend the old ways and to learn to live in peace and harmony with all the rest that exist in the Cosmos. That process of the rebirth to half-etheric beings, according to the input that was generated from the pictograms in the crops all over the world but mostly in UK, will start somewhere between **13-21 Dec 2012 (The end of the Fifth Sun)** and it will be complete on **28 March 2013** when in fact is **the end of the Sun-Venus calendar** and in the same time **the start of Sixth Sun**.

During this process from the direction of the center of the Milky Way a cosmic rays will arrive which will bring the final **DNA reprogramming** of life that exists in this solar system. That is why it is **extremely important** that we all pay attention to the messages that are coming in the crops. Huge amount of input will come from there, and what has to be done about 2012 event.

—**end of part one**—

Part 2

Earth Bombarded with Gamma Rays

When this alignment happens our planet will be bombarded with the most deadly gamma rays ever found emitting from the Milky Way galaxy, 3.5 trillion electron volts.

From the data that arrived in the **crop circles** we will exposed on this **gamma ray stream** at least few months when the reprogramming of our **DNA** will take place. The other planets in our solar system are being affected by **Global Warming** too. **Possibly because our solar system is getting closer to the galactic equator where trillions of electron volts are spewing out of the dark rift. Wikipedia: The electron volt** is a unit of energy, it is e times V, so it is written eV. It is the amount of energy gained by a single unbound electron when it accelerates through an electrostatic potential difference of one volt.

Physicists find evidence for **highest energy photons** ever detected from Milky Way's equator. Physicist at nearly a dozen research institution including **New York University**, have discovered evidence for very high energy gamma rays emitting from the Milky Way marking the highest energies ever detected from the **galactic equator**.

Previous satellite experiments have seen gamma-ray emissions along the galactic equator reaching up to energies of only 30 billion electron-volts.

What does that mean? Well imagine this, our suns particle energy is around 500 thousand electron volts. What they found being

**emitting from the galactic core is up to 7 million times the energy of
our sun and travelling near the speed of light. Wow, up to 7 million
times the energy output of our Sun.** This is the **reason** the temperature
on each planet in our solar system is increasing.

**Conclusion: We are going to align with the galactic equator and be
able to view the galactic core entirely through the dark rift opening
on December 21, 2012 as astronomically predicted by the Maya.**

This alignment will indeed cause increased temperatures on every
object in our solar system including our sun.
    **That much radiation directly is believed to be extremely lethal
and at that strength will destroy all organic matter.**
    So yes, it is possible that when this energy engulfs our planet, we will
all die December 2012. Most of the **Galaxies** require a massive compact
energy source of energy of enormous straight to stabilize the orbits of
stars within them.
    The **engine** which drives the rotation and shape of each galaxy consist
of what astrophysicists refer to as a super massive black hole located in
the direct center.
    The **black hole** located in the center of Milky Way Galaxy contains
an **estimated mass** of over **1 trillion stars** and it is believed to be over **2
trillion miles** in diameter. Many leading astro and quantum physicists
also believe that super massive black holes are spinning in unbelievable
high rate due to their overwhelming mass. At this high rate of spin the
black holes projected gravitational field is no longer spherical but rather
flattens out to form a massive yet extremely thin spinning disk. Milky
Way's galactic plane can easy be identified by even an untrained eye as
one views the shape and the characteristic of the galaxy. From images
provided by the **Hubble telescope**, and projected computer simulations
we can easily see that all mater moves and it is formed around this flatten
gravitational influence. This also explains why almost all galaxies are flat
and circular. If you look in Milky Way galaxy you can see a dark bend
which shows you where this gravitational plane is located. **It also shows
you where galactic dust and mass is been collecting since the birth
of the galaxy.**
    This is where we can easily seen the location of the galactic plane,
which is in fact what the modern science calls a galactic equinox. **But
what this has to do with humanity future?—In the Milky Way which**

is an active galaxy, our solar system cyclically moves above and bellow the galactic plane. As stars and planetary systems including our own, approach this galactic plane, the gravitational influence increases, which disturbs the stability of each planet including Earth.

The passage to the densest portion of the gravitational plane is the direct cause of the devastating cycles and pole shifts that we see recorded throughout history. This cyclic nature of our solar system as we move through the Milky Way is precisely how many ancient civilizations based their calendar systems.

The **Mayans** themselves described what they referred to as "**The Dark Rift**" or the "**The Galactic Plane**". **The Mayans stated that on the end of each Age, which brings world wide devastations is defined by world sitting on a Dark Rift.** Even that the Mayans don't clearly present the science behind why the Galactic plane causes severe implications upon Earth's stability on modern terminology, it is very clear that we are talking about the same event.

An event when Earth passes through the Galactic equinox, the Dark Rift or the Central plane. **It all refers to the cyclic and destructive gravitational influence created by the super massive black hole at the center of the Milky Way.**

Researchers and scientists agree that we are indeed approaching the end of what the Mayans defines as the current Age.

Even that many scientist groups are still debating the full implications on what passing the Galactic plane will have on Earth, we already began to see the early stages that the increased gravitational influence is having upon our world and other planets in our Solar system.

As we approach the **gravitational plane** we will continue to **experience** severe whether, geological effects such as earthquakes, tsunamis, hurricanes and volcanic activity, with increased frequency and intensity.

As we penetrate the **densest portion** of the **galactic plane** and experience the full gravitational effects we may witness unprecedented solar flares, unexpected meteor showers and unfortunately geographical **Pole Shift**.

Subsequently celestial objects may pass through the solar system as they too are influenced by the gravitational plane.

This may count for the ancient records describing several world wide catastrophes accompanied by passing bodies, comets or what some researchers refers to as **Planet X. Computer simulations** utilizing the collection of knowledge, we have decades of galactic models and satellite data, tells us that our solar system will definitely began passing through the galactic plane in the very near future. The most severe effect that will cause world wide devastations, a Pole Shift are most likely to occur beginning some time between **2008** and **2014**.

—end of part 2—

Part 3

Super Wave

Dr. Paul LaViolette

Perhaps if we want to understand things that are coming, it should be useful if we look at the work of scientific genius Dr. Paul LaViolette.

He is the author of number of famous books as **"The Talk of the Galaxy"**, **"Earth Under Fire"**, **"Genesis of the Cosmos (Beyond the Big Bang)"**, **"Subquantum Kinetics"** and others.

Dr. Paul LaViolette has **9 degree's** in physics from **Johns Hopkins**, **MBA** from the **University of Chicago**, and **PhD from Portland State University**.

He has also published many original papers in physics, astronomy, climatology, systems theory, and psychology. He has served as a solar energy consultant for the UN, Greek government, and Club of Rome Goals for Mankind Project and has also consulted Fortune 500 companies on ways of stimulating innovation.

Research he conducted at **Harvard School of Public Health** led him to invent an improved pulsation dampener for air sampling pumps.

Related work led him to **develop** an improved life-**support rebreather apparatus** for protection against **hazardous environments** and for which he received two patents. Recognized in the Marquis **Who's Who in Science and Engineering**, Dr. LaViolette is the **first** to predict that **high intensity volleys of cosmic ray particles** travel directly to our planet from distant sources in our **Galaxy**, a phenomenon now is confirmed by scientific data. He is also the first to discover high concentrations of **cosmic dust** in **Ice Age polar sheet**, indicating the occurrence of a **global cosmic catastrophe** in ancient times. **Based on this work, he made predictions about the entry of interstellar dust into the** solar system **ten years before its confirmation in 1993 by data from the Ulysses spacecraft and by radar observations from New Zealand.**

He is currently president of the **Starburst Foundation** interdisciplinary scientific research institute.

For him, people say that he is a genius a **brain sharp as a razor**, and a unique person. One key area of **Starburst** research is concerned with the investigation of **Galactic superwaves**, intense cosmic ray particle barrages that travel to us from the center of the Milky Way and last for periods of up to a few thousand years. One thing which is relevant for the **2012-2013 Nexus Event** is that the **Starburst Foundation** discovered that at least **two superwaves** with the streinght to generate **New Ice Age** are travelling our way from their place of origing distant around 26,**000 Light Years—Galactic Center of the Milky Way**.

Attention: Some scientists have measured the **distance between our Sun and the galactic center** as 24 Light years; others consider that, it is probably something closer then 26,000 years. I personally consider that the second distance is correct. Starburst researcher **Dr. Paul LaViolette** began alerting the scientific community to the existence of superwaves in **1983** through his published papers and scientific conference presentations.

He also raised the public awareness about the superwave phenomenon through his book **Earth Under Fire** as well as through various magazine articles.

Many aspects of **Dr. LaViolette's superwave theory** have since been verified by recent observations.

The **primary goal** of Dr. Paul LaViolette's research has to do with the **galactic core explosions**—violent explosions occurring in the centre of the galaxies.

The research in early 80's led him to **ice samples** from **Greenland** and **Antarctica**. Particularly in **Greenland** he found high levels of **cosmic dust** in the **Ice Age part of the core**. Which was confirming a hypothesis which was that there wood been an arrival of cosmic rays from Galactic centre around that time.

That arrival had pushed a galactic dust into our solar system causing an extreme climatic change on all the planets in our system.

Those cosmic rays filed the area with dust included the Sun and the radiation that reached the Earth was coming with a different spectrum.

In the physical view, the doctor is speaking about more reddish and dusty sky, difficulty of perceiving the stars in the night, but most of it, about the radiation in the **infrared spectrum** and created what Paul LaViolette calls a *interplanetary hot house effect*.

Like **Earth**, our entire **solar system** has its own **atmosphere**, called the **Heliopause**. This "**bubble**" surrounds the Sun and planets as it travels through galactic space. Like our earth's **magnetosphere**, the movement of the **heliopause** creates a rounded "**head**" and a narrowing "**tail**."

Actually, it's more egg shaped. Until recently, astronomers believed that our solar system was a region relatively free from cosmic dust. **The cosmic dust and frozen material of space were kept outside this protective bubble.**

This was confirmed when the **IRAS** and **Ulysses** spacecrafts showed infrared images of the solar system, surrounded by whispy clouds of cosmic dust that increase in density just beyond **Saturn**.

So if the **cosmic dust** is surrounding the heliopause, what would make it suddenly enter the heliopause and how would this coincide with huge solar flares? **LaViolette** envisioned something disrupting the heliopause from the outside, impacting it and drawing cosmic dust inside with it and energizing the Sun.

The energy of such an impact would be immense. The **most logical place** to look for such enormous energy was the **Milky Way Galaxy**.

Perhaps related to this is the puzzling fact that, even though we have witnessed no Galactic explosions or "**bursts**", the measurements of cosmic dust streaming inside the heliopause as been steadily increasing to almost three times since the **last solar maximum in 2001**.

During the solar maximum of each **11 year cycle**, the polarity of the Sun shifts—**North** becomes **South** and **visa versa**. This brief period of **magnetic instability** allows some **cosmic dust** to enter the **heliopause** because the **Sun's "shields"** are reduced. But once the **new polarity** is established, the Sun usually quickly blocks the dust. **This time it didn't happen.** Cosmic dust has been streaming in from the Glactic center and astronomers are at a loss to explain why.

This is the first time something like that happens, at least in the time frame of our Sun monitoring. Not to mention so many anomalies and new phenomena that appeared this year together with climate change on all other planets. **And those changes are far from over.**

It's likely that our solar system is already experiencing the **invasive energy** from the **Galactic equator** as we move into position and align with it on **2012**.

The recent data shows that dramatic and potentially deadly effects can result from solar flares and coronal mass ejections. Substantial data suggests that an event, similar to the one anticipated in the 2012 "doomsday" scenario, occurred about 14,950 years ago and was recorded by ancient humans.

This event appears to have lasted for several years in duration and was responsible for the abrupt end of the last ice age as well as a substantial culling of the human population.

The surprising findings of LaViolette, supported by other research, suggests that the extreme solar event corresponded to powerful radiation coming from the center of the Milky Way Galaxy and was associated with gamma rays and cosmic dust. Recent observations have shown a dramatic **increase** in gamma ray energy in the Glaxy's equator which will be in **maximum** while the alignment with our solar system on December 12, 2012.

The past records in ice cores (strata from 13,880 to 13,785 BCE) suggest that intense radiation from this last event could have lasted many years.

It seems highly likely that this alignment will cause another extreme solar event since other factors precipitating a "**solar maximum**" (i.e. the opposition of major planetary barycenters) also converge on this exact date.

The fact that galactic centers routinely radiate lethal gamma rays makes it unlikely that life, at least as we understand it, can survive in the universe. Sooner or later it is destined to be zapped.

A new genetic study of **Y-chromosome** variation by **Dr. Marcus Feldman** of **Stanford University** shows that the population from which the world's present population is derived consisted of about **2,000 individuals**. Somehow, humans, flora and fauna did survive the past doomsday and some may yet survive **2012**.

Anyway, let's get back on the findings of LaViolette:

I already mentioned that the **primary goal** of LaViolette's expeditions was **search for evidence** of **increased levels of cosmic dust in the Ice Core Records which matches the last Ice Age**.

—How can you tell if there is a presence of cosmic dust in the ice? Is there a presence of a particular elements that he searched?

Yap, he was looking for the examples of iridium and **nickel** which are present in **high levels** in the cosmic dust but on **Earth** they are much lower.

Particularly the precious metal—**Iridium** is about one hundred times more abundant that the rest cosmic materials in the cosmic dust. **Gold** was another one, and he also found the presence of elevated levels of that element in the polar ice.

What about **beryllium**?

No, to do that, Dr. Paul points out that you need a different technique of analysis. **Beryllium can be found with mass spectrometers**.

Paul was using neutrino activation which was when you bombarded your sample in reactor, make it radioactive and than you measure gamma rays.

From that you can tell what is in the Ice. With the **Beryllium10** the **radioactive isotope of Beryllium** that's created in the atmosphere when the atmosphere is bombarded with cosmic rays and to separate that particular isotope from the other Beryllium isotopes you need an accelerator. It's a totally different approach but other people had done that work. And they **found** high levels of **Beryllium10** in the ice. **So it did happen before.**

Dr. Paul LaViolette testing was conducted from different depths or to be more precise—the ice samples that match the periods between **12-16.000 years ago** which was the end of the last Ice Age. Also those samples were matching the times of most known animal extinction on Earth. **The time of dinosaur mass extinction.** It is **my belief** that at the end of the **2012** we can expect a hit of at lest **five superwaves** with power to cause a **New Ice Age** and they are travelling this way with speed higher then the speed of Light, from their starting point, distant **26,000** Light Years away—**The Black Hole in the center of the Milky Way Galaxy**.

What is **not clear**, did the **galactic core** of the **Milky Way** exploded 26,000 years ago, because if the beams are travelling with the speed of light it takes the same time period to reach here, or **sooner** which suggest that they are travelling high above the speed of light. What on the other hand is not imposible because our **telescopes** had recorded speeds of **galactic jets** moving at least five times speed of light.

—**end of part 3**—

Part 4

How Did Dr. Paul

CONNECT THE DOTS?

The discovery of the Dr. Paul LaViolette had echoed deep in the scientific community. Despite the fact that it causes huge amount of controversy and disagreements, he succeed to reject all the attacks to his claims because there were based on pure scientific facts with a huge backup of material evidence.

To have different opinions from the one in the mainstream science and in the same time to stay highly respected in the scientific elite, can work only in one case—if it a case of a genius who has courage and will to look and analyze where the mainstream had stopped looking for a long time or that is already researched thermally and no other scanning on checking is necessary.

But the genius of this man is extraordinary and he often founds more and more parts of the puzzle that are missing so he can see the big picture.

When he was asked how he did it, he said that simple he knew where to look. When they asked him to explain that better, he responded that in the same time he was conducting his scientific research he also conducted a thermal and detail study and analysis of ancient myths, traditions, manuscripts, maps, drawings and coded messages on the walls of most ancient temples.

From the extract of useful and confirmed data that he made from the research, he had gain a conclusion that on our planet there was once a very advance science. **He found not breakable evidence of the existence of Space Age Science advanced at least as our or far more advance. Vedas** themselves were speaking about the technologies familiar and far advanced to ones that we use today, so nothing that we see today around us is discovered for the first time but everything in the process of rediscovery. This **stands** for every **direction of science and aspect of existence**. Everything we see today, existed before and much more. While **Dr. Paul LaViolette** was searching for some extremely important lead in the ancient records and myths, as I already stated, he also conducted complicated research and measurements in the field of **Astronomy**, worked on new theories in astrophysics, and he spend much time on studding **Pulsars, Evolution Cosmology** etc.

While doing this two parallel research, looking for some connection between them, one day he noticed that in the **Western Zoodiac**, is **Sagittarius** pointing towards **Scorpion** and it become clear to him that this is **extremely important** because it contains a **hidden message** that is thousands years old.

Then he draw the **trajectory** and come to a point which is very close to the **galactic center**. It is than that he discovered that the arrow of the **Sagittarius** is placed there to point directly at the **center of the Milky Way**. All the stars had moved since that time, thousands of years passed, and that was little unclear to the **Dr. Paul** because there are several of Zodiacs.

We have Zodiac which is used in the western astrology; the Mesopotamians on other hand had 18 constellations, while the Chinese had other number of constellations. But the message that he is talking about can be found only in Western Zodiac. In any case, whoever created this message he used certain cryptographic techniques. Also in the message were included some **constellations** that are out of the ecliptic as **Saggita** and **Aquila**.

Just for comparison, **Aquila** is the **eagle that holds the arrow (Sagitta)** and actually represent a clear **Illuminati symbol** on the dollar bill.

On the other side (**on the southern hemisphere**) in the message are implemented other constellations known as **Centaurus Constelation** and **Crux**.

Now, if we move our look for a distance of **1 radiant point** we see directly into the **Sagitta Constellation**.

On the other hand, if we simulate the same on the **southern hemisphere** and move our look for a distance of **1 radiant point** we are looking directly on **Centaurus Constelation** and **Southern Cross.???**

For everyone with little more knowledge of astrophysics and astronomy, it becomes clear in an instant, that, this could not be left by chance or that is simply a coincidence. If we take in the consideration the both sides we come up with a simple conclusion that in the **Western Zodiac** there are **two arrows**.

One of the arrow is located in the **northern hemisphere—Sagitta** and points or shoots **out of the galactic center** while the other one is located in the **southern hemisphere—Sagittarius** and it points or is shooting directly **into the galactic center**!

Way . . . this is heavy information, if you are aware what is coming in 2012.

When you analyze what is illustrating you can clearly see that the arrow of the **Sagittarius** is indicating an explosion day and the arrow of the **Sagitta** is the **result of that explosion. In other words, the cosmic rays.**

Lets get back to the explosions itself. **Can core samples give as a solid evidence if similar explosion or explosion happened before?—Yes.**

In the **Beryllium10 graphs** that indicate **cosmic rays density**, the **strikes in the Earth's atmosphere** we can clearly see peaks.

There is **one major peak** right at the time when the message is indicating that something was happening and previous times. There is one very large peak around 37000 ago which corresponds according to what we know, to the extinction of the Neanderthals. If you look in the ancient myths of Hopi Indians, Incas, Mayas, Vedas and others, you can see that there talking about Ages and races which once existed and become extinct at certain times.

So this could be for example reference to one of the species that once existed on the planet, and they also they are talking about suns . . . forth sun, fifth sun and so on.

And when you look at the **ice core records** you see that the times of the beginning and the termination of the **Ice Ages**, not only the previous Ice Age but the one before that, **corresponds** with these huge peaks of cosmic rays.

So we can conclude that there was **inner glacial period** of something like 9000-11000 years between those two last Ice Ages. And suddenly there was another peak of cosmic rays and glaciations began. **So this cosmic rays peak seems to be climate triggers even for Ice Age even for an extreme wormer Age.**

Dr. Paul LaViolette points out that this had to do with how long the superwave is lasting.

Anyway, today it is scientifically accepted that these galactic core explosions do happen despite the fact that the galactic core is not directly visible to telescopes because its heavenly secured by the interstellar dust. We know that **cosmic rays** are **released** when such galactic event happens. **Also we are seeing it happening in other galaxies.** Quasars for example are example for very intense explosions. **From what is discovered, this happens in cycles. In basic, the large explosions appear in every 13000-26000 years.**

From the recording of similar events happening in other galaxies, we can calculate that those cosmic rays are travelling around the speed of light, but this is questionable, the speed could be faster in some major explosions.

That is why you have **frozen mamuts** and other animals with food in their mouth like something unexpected happened on a global scale. **This is not impossible when we know that in the case of the Milky Way center there is a Supper Massive Black Hole which with its extreme gravitational power is pulling a huge amount of cosmic dust and gas towards the central point and when the process reaches its culmination, boom . . . an explosion so powerful that releases a visible beams "Jets" that consist mostly of ionized gas and plasma. Cosmic rays** that were travelling at such high speed can to a certain point explain the phenomena of **frozen mamuts** and other animals with a **food** in their mouths just as **something unpredicted and suddenly happened on a global scale** without any warning and caused an **Ice Age**. However, if we **summarise** this, the **message in the Western Zodiac** indicates that **on this trajectory date** a **galactic center energy creation event becomes visible!**

In our case it will be right after the appearance of the **New Moon** on **December 13th 2012**, what also can mean a start of extremely huge catastrophic effects for our planet. **Scientists found that the core of the active galaxy can shine brighter then the galaxy itself. That is cosmic rays electrons braking away from the galactic core, close to the speed of light produce a bluish bright light, so bright that it masks the life of galaxy spiral arms and surrounding mass of stars.**

Since **our solar system** is currently at the location of joining point **between Sagittarius Dwarf galaxy and the Milky Way** it will certainly affect us. Somehow the ones who left the knowledge to the Mayas and other cultures precisely knew about these events which are happening in cycles. **Strangely, the ancient Hopi Prophecy says:**

[b] There will come a time, when a **blue star** will appear in the sky . . . Its light will shatter the darkness of the night . . . This blue light will bring a wind . . . A wind like that has not been seen on Earth for a long time . . . The blue star will bring a fire . . . This fire will be so bright and hot that will transform the matter of the Universe . . . The blue light from this star is a signal that forth world is ending . . . The blue star will cause the oceans to rise and top up towards the land, flooding the world . . . **Almost all living things will physically perish in this great catastrophe**

Strangely, but this ancient "**Hopi Prophecy**" does not look so ridicules any more, but I contrary. As we were saying earlier, since we ca not see the **galactic core** because its **intervening dust clouds**, most of the **blue light** from the **galactic core explosion** will come from the cosmic ray electrons emitted from the core.

During that **26 000 light years journey to us**, those **powerful cosmic ray electrons** would continuously generate and beaming forward a **bluish light**.

Soon after they become visible for us on **13 Dec 2012**, the **superwave cosmic rays** will start to penetrate the **protective magnetic field** of our solar system.

This will cause a network to form, faintly luminous cob web like filaments stretching out forth from the sky. Perhaps the most frightening phenomenon to occur in this early stage will be the prompt arrival of the electromagnetic pulse and perhaps shortly afterwards the arrival of a giant gravity wave.

It will **impact** our **Sun** and **all the planets** in our solar system, causing a severe bad weather and natural disasters, a shift of the magnetic and geological Poles which we all know where all that leads.

The **gravity** which holds the **water** in **oceans** in the moment of flip-over (**Pole Shift**) will no longer able to sustain **equilibrium** (balance) and the water from them no longer controlled by stable gravitation will **escape** on one side which means nothing else but a **New Global Flood**! That will trigger Tsunamis few miles high, tearing apart of the continents, ash from the volcanoes and wind like no other seen so far.

This will happen when the **precession** of the Earth ends on **21.12.2012**, (it takes **25,765 years** to complete) the rotation will come to a stop and **when that happens** on the surface the ones that survive till then will experience what ancient records describes as the **"heaven and the earth switch places and the next day the Sun raise on a different side"**.

If you consider how much scientific and esoteric data streams down to that exact day, it makes you spin in your head, because different from everything else before all the riversare flowing to this date in space in time . . . **zero time point.**

The **Maya Calendar** is primary based on a point in **Pleiades.**

Is this spot will no longer be on the same location in the sky on Dec 21 2012, and that is why their calendar does not function anymore?

Or they just run out of paper?—Common people. Think. It's coming this way, and its coming soon!

The ancients left clear warning about this event, thousands years ago. We have to stop being ignorant! But what about the Sun?

We are still not clear 100% what this exactly means for our **Sun**, but from the data that is confirmed or passed through the pictograms in the crops our star will **drastically increase its mass.**

The frendly **"ETs"** or the **"Visitors"**, if we can call them like that, despite the fact that there is a high probability that they are here longer then our race, clearly had draw on **22 July 2008** in a huge crop field in **Avebury Manor / UK** that our **Sun** will go through expansion in the mass and that will swallow the **Mercury** and **Venus**. I'm **aware** that is sound outrages and fantastic, but you better wake up and see what they are desperately trying to tell us, because the mainstream media is corrupted and the property of Illuminati. **Wake up!**

Despite you are aware or not, we are in the middle of a Spiritual War!

There is not sitting on two chairs. Even you support what Shadow Government is doing by the influence of negative ETs or other hand you risk all you have and stand on a side of Light, to protect this planet and life that exist on it.

There is no two sides, you have to wake up, to choose a side, to understand that you are spiritual being inside the body, that there is heavy manipulation going on here on Earth, to understand your responsibilities and to act according what your heart tells you is right.

Look at this links and make up you own conclusions:
http://www.cropcircleconnector.com/2 . . . anor2008a.html
http://www.cropcircleconnector.com/2 . . . anor2008b.html

We know that **extreme solar storm** with unseen **solar flares** is coming before the **end of 2012** and we know how all this will culminate and have affect on the life on this planet even if we are lucky to have the **most soft 2012 scenario** . . .

—**end of part 4**—

Part 5

Basic Input About Solar Storms

It is **well known** that the surface of the **Sun** pumps up huge amounts of **heat** and **light** in the space. From time to time, those huge **explosions** become **solar storms** and travel to our planet.

When those solar storms hit into the Earth, in most cases they do not have fatal effects on the living forms but the dense energetic grid which makes our lives and our existence simpler, can easily serve as conductor or channel for the destructive solar storm energy and it can completely neutralize our electrical systems.

Day or **two** without **electrical power** is torture but months or years is simple not comprehendible. Despite the fact that the **Shadow Government** had foreseen this option long time ago, there are high chances that this will happen on the surface around the end of **2012**, even sooner. We have to take in consideration that even before the 21 Dec 2012 when the **planetary alignment** will happen, our **Sun** is already exposed on the influence of the **strongest gravitational force** that exist in the center of the Milky Way. That is why, you can not look for long at the Sun, because is starting to burn like never before. There are huge changes on the Sun. People who **meditate** in the nature for at least 20-30 years waiting for the **Sunrise** knows what I'm talking about and that the Sun is not the same as before. Despite there is a data, that Shadow Government is using some kind of **plasma weapons** to **bombard the Sun**, using **HAARP** and **Tesla Technology**, I personally think that it has to do most with the alignment with the central galactic point with the strongest gravitational pull.

It's like there is some king of increased **Aether exchange** going on, and in more intensive and powerful way, that was not the case when we were little away from the central point of the galactic equator. As we start to understand this new dynamic of new astro-physics it becomes obvious that the old model of our understanding of Universe is totally false in some areas, and not complete in most areas. So we have to change it. Importunely, NASA and Shadow Government are hiding information from us, they have technologies centuries ahead, tested in secret underground military bases around the world **but we have go on with what we have and to try to understand this before its too late.**

Matter of fact, is were we are heading, even without **2012** in the next 10 years there will **NWO** on this planet, and it will not be a nice place to be anymore . . .

Anyway, the thing will go worse and worse, the **Sun** will burn hotter and hotter like never before, because the gravity from the central belt of the galaxy its already doing its "**magic**".

This **gravitational pull** and exchange of energy is also coming through some minor cosmic rays that are already hitting our solar system **But make no mistake**, those minor cosmic rays are not comparable with the proportion of the Nexus—A Storm of Superwaves that is coming. Those minor cosmic rays that are arriving everyday are only overture for what is coming. But even minor they bring a change to whole solar system, including to our Sun.

Also, make no mistake, the fact that our **Sun was spotless** for more then a month (it happens first time like this (Not just in the Solar **Cycle 24**)—since we are monitoring the Sun) and even now the Sun Spots are very rare.

Officially we did 400 years of tracking, but we all know that only in the last 50-60 years we had devices and technology for advanced and more precise measurements and observations.

Make no mistake—This is like a nasty silence before the storm comes.

Put in other words, when the right conditions will create inside the Sun at the end of **2012**, and we can see that they are slowly forming, the solar storm that hit **Quebec** will look like a chilled game. **Please** don't understand all this I'm presenting one way or another. The things that

I'm presenting are just like that, **parts of natural cycles** and it is better that they are understood in that way.

Without fear, without panic, and without prejudices. The Nature is not like this or like that . . . Its simple is. It must become clear to everyone that the Universe is bigger then us and our lives.

We are only transit phenomena, although a part of whole cosmic reality.

This is neither good, neither bad. Its simple is. From what we know, the **Sun** is a **magnetic variable star** that passes through a **magnetic cycle** that usually lasts **22 years**, with the intense **sun spots** every **11 years**. During the time of the solar maximum there are higher chances that Sun can produce massive solar storms.

During such a storm, the Earth would be under the influence of the Sun's bombardment, and the highest layer of the atmosphere would become such electrically dense that every equipment or technology that orbits our planet would be exposed to danger. This **hitting** of the **Earth's atmosphere** will become bigger and bigger as those currents circulate above our planet.

The **hardware** and the rest electronic equipment will be burned and they will start to fall down. From bellow, they will be seen as **meteors** which burn on their way down through the atmosphere, but some of them will pass and eventually will hit the ground cause more or less damage.

That is why besides **Hubble Telescope** and other advanced telescopes in orbit, they started to build Telescopes like the one close to the South Pole (**btw** . . . did you know that there is a **hole** in the exact **South Pole** and our **planet is hollow**!) and elsewhere. The new ones are new are built with cover over them or underground with the seal that is opening and closing. This new Telescopes has many objectives, but few of them is tracking the **Nexus** or **The Superwave Event** that is arriving from the galactic center, than tracking in infrared huge object **Planet X** which will intercept our solar system close to Pluto.

They are changing their strategy because they know that all the Telescopes and Satellites will be burn from the Sun, or destroyed by someone else . . .

Anyway, **it does not look very nice for us, ordinary people**, who in the eyes of the Shadow Government are **expendable. But we are not expendable and we are going to prove it! Let's go further.**

The **normal** Sun's activity (if we can use that term because everything that we see around us everything but normal) is hardly detected by the surface people, because the **Earth** protect us with its **magnetic fields** called **magnetosphere**.

But this magnetic field has limits. During those intensive magnetic storms that comes from the Sun, our planet is hit like hummer from the magnetic waves, which results with magnetosphere stretchening and weakening of its defence.

The polarity of the **solar wind** is very important, and one polarity reflects from the magnetosphere like from a mirror, but the other polarity penetrates directly through **Earth's magnetic shield**. This means trouble for us.

That lives us with low defence. As we drifting slowly to the point of alignment with the central galactic point, we are clearly seeing that the conditions for this are developing inside the Sun.

If this continues further with nominal space rhythm, the sun's surface will release a billions tons cloud of electrified gas known as **Coronal Mass Ejection** or **CME.**

In **normal** conditions, it takes 8-10 hours of travelling before the massive solar flare hits our magnetosphere. But many things will not normal in space around the end of **2012**. If the conditions are changed than the strength and the speed of the **CME** are different according to that change. Today, we know that in a case where huge amounts of **cosmic dust** and **gas** start to fall into the **Sun's surface**, in the moment of impact produces **conversion** of the gained **kinetic energy** into **heat. In the same manner**, in case of penetration of huge amounts of cosmic dust and gas into our solar system **during** a **super wave passage**, this cosmic dust and gas will certainly reach our Sun, they will penetrate under its surface and cause irritation and make the **Sun** to burn much **hotter**.

Besides that, **cosmic dust cocoons** will surely surround the **Sun**. That would generate **increasing** of the Sun's surface **temperature** as a result of the interception of **sun's energetic flux output**, and immediately after, a **reflection** of the same, **back** to the **sun's surface**.

Those **two effects** would surely produce **extreme solar wind**, appearance of extreme big and dangerous **solar flares** for life on Earth. In other words, the **cosmic dust** and **gas** that will arrive at the end of **2012** into our solar system with the coming of the **Nexus—Superwave Event**, will cause our **Sun** to start to behave like **T Tauri Star**. T Tauri **stars** are stars that are surrounded by thin cloud of **interstellar gas** and

dust which induces that they become **highly bright** because of starting of **continual coronal activity. In the end they become strong emitters of cosmic rays!** In case of our **Sun**, where the **solar flares** (used to) occur **1-10 times** in the **year** period, in **Superwave Event** they will start to appear **frequently** on the Sun's surface . . .

—end of part 5—

Part 6

Messages From Above

But let's see what the **friendly "ETs"** have to say about the **2012 event** through their **complex crop designs** around the world.

The **detail analysis** that was conducted by "experts" clearly indicates that high number of pictograms has "**Mayan symbolism**" and "**Quetzalcoatl connection**".

Conclusion was made that the crop circle makers, are measuring the Time according to the "**Long Count Calendar**" and "**Sun-Venus Calendar**" exactly the same as **ancient Mayas** and **Aztecs** in **Central America** did long time ago.

Both Calendars are clearly depicted in the crop pictograms that appeared in UK, in **Silbury Hill** in **2004** and in **Woolstone Hill** in **2005**.

As we already know, **The Long Count Calendar** ends on **December 21, 2012**, which will mark the **End of the Fifth Sun** which started exactly on **13 August 3114 BC**. Different from the first calendar, the **Mayan Sun-Venus Calendar** ends on **28 March 2013**, which marks the end of the current **52-year Venus cycle** that began on **April 10, 1961** and more important it also mark the start of new **5000-year cycle** of **Sixth Sun**.

The messages were perfectly clear, like in the pictogram that appeared 2006 in **Wayland's Smithy**. This pictogram clearly depicts that at the end of the **Sun-Venus Calendar** periodically emitted cosmic rays from the galactic core will arrive on Earth.

The preliminary analysis of the pictogram that appeared on **Wayland's Smithy 2006**, had shown series of "astronomical rays" emerging from some central source in all three spatial directions **x**, **y** or **z**: Those rays have been drawn as "**square**" along their lengths, because the intensity of any **astronomical wave** or ray becomes weaker as it proceeds further from the source by a factor of distance-squared. What could be the central source of energies shown there?

The answer is completely clear—**Ultra Massive Black Hole** in the center of the **Milky Way** known as **Sgr A***.

A **ring** of **12 stars** surrounds **Sgr A*** in space, just like a ring of **12 points** surrounds the empty centre of that **crop picture**. http://www.cropcircleconnector.com/a . . . time2007m.html

Interesting is the every of the three longest rays seems to contain a **7 x 8 = 56** grid of **minicircles** within any open square end. **Could that be some kind of time code, telling us when the longest ray will reach Earth?**

For someone to understand what "**they**" are trying to tell us, it is better that we firstly understand that every **Sun-Venus conjunction** lasts for **292 days**, whether inferior (**Venus between Earth and Sun**) or superior (**Sun between Earth and Venus**). Ever since the pictogram appeared on **July 8, 2006**, two years ago, when there were still "**9 left**", they constantly remind us that the countdown to **2013** has continued! Then on **August 12, 2007**, the Mayan number "**six**" appeared at Stanton St. Bernard. It was intended to tell us that "**only 6**" **Sun-Venus conjunctions** remained until **Venus** transit on **June 6, 2012**, or "**only 7**" until an end to their **Sun-Venus** calendar on **March 28, 2013**.

In summary, the crop pictograms clearly are showing astronomical event that will happen in **late 2012** or **early 2013**, close to the end of the current **Sun-Venus calendar** or between the lines—the arrival of some kind of cosmic rays from the **Galactic Center**.

Yet one other pictogram from **West Kennett (July 13, 1996)** did seem to describe the same near-future astronomical event in a slightly more precise way, so we will study that one next. **West Kennett 1996**, suggests that a new Moon and comet will align in Earth's sky with the galactic centre, when we first see rays emerging (**possibly on December 13, 2012**). It is clear depiction of the **Ancient Mayan symbol** for the Center of the Milky Way with "**waves emerging in all directions**" from it.

There is an ancient Mayan prophecy which says that "**the Moon will be eight days old**" on **December 21, 2012** when their **Long Count calendar ends.**

Indeed, December 13 plus eight days equals December 21.

???—Do you see the connection? They transmitted everything loud and clear. What more do we want? Lets go further . . .

What about the comet?

Some scientists had **determined** to found out (at least closely) on which comet clearly depicted in the crop design they refer to? After a search for all known comets that will arrive at **perihelion** (the closest point to the **Sun**) between **2008-2012**, they come up with possible solution—**Comet 152P Helin—Lawrence**.

This comet was the closest candidate and it will be near the new Moon and in alignment with the galactic center in the middle of December 2012.

However, what the comet is concern, we are not 100% sure that it is **Comet 152P Helin—Lawrence**. Perhaps they refer to some other comet!

But, one thing is for sure!—the people of Earth for the first time in a long time period will have an opportunity to visually see in the sky the coming of cosmic rays from the galactic center exactly on **Dec 13th 2012** when **New Moon** and some **Comet** will be aligned with the galactic center. It becomes, obvious that those rays from the galactic center will arrive periodically according to their size and strength. The last and the biggest cosmic ray of all from the "**Nexus—Superwave Event (Unknown Form of energy that travels our way)**", will finish the reprogramming of the DNA of the organic life forms in this system. It will arrive exactly on **28 March 2008**, when as we mentioned earlier, officially is the "**End of the Sun-Venus Calendar**", and the beginning of the "**Sixth Sun**"! **What this means?**

At the heart of the transformation of humans on **Earth** is an evolutionary movement brought into play by galactic forces and celestial movement. This involves the bathing of our planet and the biology upon it in gamma ray energies projected from the **Galactic Center**. This **bio-cosmic** event has been anticipated throughout humanity's presence on this Planet and appears in the cultural records of many peoples as the Perennial Wisdom. The perennial wisdom describes the triumph of compassion and intelligence over violence and ignorance.

This is to be achieved by the actual transmutation of human genetic material, our **DNA**, when it is bathed in galactic light emanating from our own Sun, acting as transducer of the Galactic Centre itself, passing directly through the Earth and along the axis to the Pleiades Star System.

This event has been foretold repeatedly in the cultures of Earth. It is the current emergence and new information and its convergence with traditional wisdom that gives us the language and tools to negotiate the extraordinary and unprecedented psychophysical metaphysical adventure evolution is delivering up for humanity.

The emergence point of this event in time is recorded as **December 2012**.

Ultimately, it is love in its personal application as compassion and its universal manifestation as creation, which both guides and characterizes the advent of the noosphere. The ordained celestial movements are transforming homo sapiens DNA to the DNA of homonoosphericus. That is the desire of love and the intelligently designed outcome of biology. Our collective success in Universe is assured by love's intention for the success of humans—its intended embodiment of divine intention in time—in all dimensions and universes! But, what exactly, from geological aspect, this means, for our planetary system and our Sun?

Hold your pants . . . you gona like it! Let's put some light to the pictogram that appeared in **Avebury Manor** on **July 15, 2008**. The pictogram is clear depiction of our planetary system on **23-24 Dec 2012**, two or three days after the End of the **Long Count Calendar—December 21, 2012**. **It is a clear selection of two groups:**

One group consist of thin circles and depicts the orbits of the Mercury, Venus, Earth, Mars and Pluto. The **other one** with **bold** circles consists from Jupiter, Saturn, Uranus and Neptun. "**They**" insist that we pay attention on two things:

First: the first anomaly is in the Pluto's orbit, which clearly depicts that this heavenly body will be influenced by outer gravitation from some passing body, planet or comet.

Second: the second anomaly is splitting on two groups. Can this be understood as separation of our solar system on two groups, as result of extreme Sun's activity in combination with some extreme powerful gravitation of passing body or comet?

The message is perfectly clear—**it's our solar system on 23-24 Dec 2012.**

All this would not be so significant if 7 days after, on **22 July** an **update** didn't appear. **And what was on it, says absolutely everything we need to know!**

It indicates huge **geological changes** and movements in our **solar system** as a result of **Sun's expansion in mass** and strong **gravitational pull** from outside.

Have a look at the **update** and see for yourself:

The mass of the Sun is bigger for more then 25% and the planet Mercury and Venus are completely burned. **Attention:** The person who depicted this diagram had made an error in the orbits of the Earth and Mars. They are both put in the same orbit, and its only a technical error. **Let's proceed with the analysis:**

The **Earth** with its companion (if the dot in it represents the **Moon**) together with the **Mars** are staying in their orbits but dangerously close to the **Sun**.

It does not take to be a genius what this **CME** activity and Sun's close gravity pull mean for **Earth's balance**, magnetic and geological Poles.

According to this, we are facing new extinction of the human kind!

This is enough, but let's see what else reveals the update of the 15 July pictogram. **After short analysis, it becomes clear that the second pictogram (the update of the first one) is not depicting our solar system on 23, 24 Dec 2012 (it was only used to bring our attention to this time frame of 2012) but 8 days before the End of the Long Count Calendar—13 Dec 2012.**

Everything becomes **clear** from this point . . . and there is no doubt anymore what they are trying to say. **From the left side we see the approaching of huge object (number 10) that is probably Planet X which has strong gravitation which is influencing the Pluto's orbit and then a New Moon and Bright Comet from the right side.**

Expansion of the Sun is obvious, together with the swallowing of its two closes planets. **Mercury and Venus are completely burned by the Sun.**

If this pictogram is correct, our solar system will never be the same!

What is not completely clear to me is the huge circle outside of the planetary orbits. **Mr. Harold Stryderight** considers that they are depicting an expanded view of the lunar orbit.

With that, that in the planetary depiction of the Earth there is a dot inside, and on the bold circle outside the planetary orbits similar symbol is shown, there is a high possibility that it is the zoomed orbit of the Moon around the Earth through the year. About the other smaller figures close to the huge bold circle I can not comment because I honestly don't know what they are trying to say with them.

Maybe if everything stays as it is, and the Shadow Government does not implement NWO at least on summer season, new input will come and people will surely decoded them. Also surely there is someone overthere that probably has the answers but we have already received the **basic download.**

I'm not sure about this but it is not bad if we see the pictures and the analysis of **Mr. Harold Stryderight**. However, this is a **confirmation** from the **1996 pictogram** where it is **clearly shown** that **from the Earth for the first time some kind of cosmic rays from the galactic center will become visible.**

What **effect** they will have on our solar system and on the life forms that live here, we have already explained.

We still need more input, but even from this little that we got, its 80% clear what is coming in 2012 . . .

Alignment that is very important for us and also was for the ancients is on 21 Dec 2012. This pictogram depicts our **Solar System** on **23-24 Dec 2012.**

Here is the link to **Solar System Live:**

http://www.fourmilab.ch/cgi-bin/Solar
Please type: **2012-12-21** 11:11:11 Compare the pictogram of the 22 July 2008 and it becomes clear.

The final part

2012 Nexus Event—Unknown Form of Energy comes our way!

Useful Leads:

There are **high indications** that **climate change** will increase, to **extreme**, as we approaching the point of the **perfect alignment** with the **galactic center** on **21 Dec 2012**.

This will be a natural result from the exposure of the **maxim gravitational pull** which comes from the **Sgr A*** and from the **passing celestial body** that is close to our **solar system** and which as presented in the crop data, will **intercept** in **2012** in area close to **Pluto**. If the **galactic core** which is in the center of the **flatten disk** has gravitational power to hold stars and planetary systems distant to **50,000 light-years** in each direction, you better believe that has **power** to do much more.

The stellar disk of the **Milky Way** is around **100,000 light-years** in diameter and to hold all this together and to maintain the spin it takes extreme power. As we stated before, this engine is the **Super Massive Black Hole**, and **Hubble Telescope** discovered that is the case also with most of galaxies. **In fact they are starting to find them everywhere but . . . it's a huge Space, everything is possible.**

Anyway, it will soon become obvious that the **climate** is out of control and more and more "**surprises**" will start to appear. **Tornados, Cyclones, earthquakes, volcanoes, shifting in the planetary magnetic balance, tsunamis and you name it what else, that will eventually lead to Shift** in the Magnetic polarity of the Earths energetic grid and finally to a Shift in Geological Poles.

It starts to become clear to more and more people what **Pole Shift** in the geological aspect means.

It means new Global Flood, tearing apart of tectonic plates, sinking parts of continents, rising of new land, deadly CMEs and etc. Not to mention the **Sun**, which by "**coincidence**", will go through **Pole Shift** again, which so far, as far as we know, was doing after **every 11 years**. The last Pole Shift was in 2001 the next one is in 2012.

Interesting, the more you dig, more and more it's coming back to you, and its saying: "**What more do you want, are you stupid or something . . .**"

In magnetism **North** rejects North, but attracts **South** and **vice versa.**

Imagine if the **Sun** comes as close as it is depicted in the pictogram, and **switch its polarity**, in the moment of **flip over** its magnetic pull will certainly **interfere** our magnetic and geological Poles.

There are already drastic changes in the magnetism of Poles and to mention changes in geological Poles. It's all connected with planetary spinning, then on outer influence from the Sun, then directly Sgr A* and so on and on.

Imagine the Sun grow in mass! From our perspective it's hard to comprehend all this and it looks as SF, but somehow the creators of the crop formations are not laughing on this possibility, not to mention that our Governments had build deep underground facilities. So in reality, the only people that someone is laughing at, is us. They consider us **stupid**; they give us **Aspartame** and **Fluoride** in everything we consume, because it affects the function of the brain, so we can not think **clearly**. If we can not think clearly, we don't see clear and we don't understand what is going on.

That's how they rule. But it has to stop. They are killing us one by one, and we are allowing all this to happen. So it is time that we wake up from a long dream to stop consuming flesh and other poisons and after a while and with the help of meditation and Dharma practice our consciousness will become clear and we will be able to see what's coming. It's about time. In fact, everything has its order and timing, so everything is happening in the rhythm and speed as it should be. **Wow!**

Small group of powerful people, that they proclaimed themselves as "**Elite**" with a strong **unmoral accent** control the huge population through their world leading media, economic agendas, political games, military structures and etc.

This self pronounced "**Elite**" had take advantage of the confusion and the habit of the human kind to be comfortable with someone else thinking for it, has build vast underground world before our sleepy eyes.

We all have to pay big Gratitude to this man **Phil Schneider**, for all his effort to bring what is going on underground to the public. Your stay here in this physical plane of existence was not for nothing and many people appreciate what you have done. **Deepest respect my friend . . .**

This "**Elite**" knows what's coming, but it will not inform people. The analysis from the world's political and military strategies has shown that vast chambers are built as well as huge prison camps. That means that there is a strategy that major parts of the human population to be wipe out by the events that are coming. Also the actions of some ongoing programs in the background are indicating that a selective part of 1/6 of the human population will be taken to underground world. If you fallow the same logic, many things becomes visible. There is a high probability that there exist completed lists of names that will be taken down immediately after they pronounce a Marshal Law. It has to do with genetic compatibility, slave and pleasure purposes, flesh and organ supply and who knows what else.

There are many in the highest military and political world, that are fond off that they gained a place in the underground world. Not to speak about the polluted rich people, businessmen, doctors, lawyers, pharmacists, and the rest of "scum" . . . believed or not—**going down is huge mistake**.

Everyone that will be taken down, even if it is by force or by free will, will have to go through **Microchipping. By the Microchip injection, you have ended your choice.** Once the Microchip is in the body they can wirelessly control the subject, his health status, behaviour and eliminate him if they consider nessesary.

Even those on the **Top** of the political world who think that they are going to be saved, are also manipulated and they too will have to go through the process of Microchipping. It comes to this: **there is no bargain with the Dark Side. You always lose at the end!**

So for all of you, who are among those groups that I mentioned, and I know you monitor everything and that there is no privacy, please consider your actions. Choose wisely because your superiors are laying to you. This is a critical time for all of us and we all have to choose a side. Despite ridiculed by whole world this **1977 Vrillon message** is authentic. You can not expect the same Shadow Government to announce the authenticy of the messages when it has been conducted extreme cleaver **Cover Up projects** for at least 60 years.

The message is authentic, and its your choice to believe is it genuine or not. There are many scientific and technical data that can be brought out to prove that hijacking five major UK transmitters in fashion that was done, is extremely hard technological achievement in 1977 and that was not done by us but from beings that have multidimensional flying crafts. It was real but they Cover it up very well using media. Normally the people fall for it. They swallow everything, every time, so why should be this time any different. Anyway, here it is once more: http://www.youtube.com/watch?v=axxa0wcMCnw

Full 1977 message transcript:

This is the voice of **Vrillon**, a representative of the **Ashtar Galactic Command**, speaking to you.

For many years you have seen us as lights in the skies. We speak to you now in peace and wisdom as we have done to your brothers and sisters all over this, your planet Earth. We come to **warn you** of the **destiny** of your race and your world so that you may communicate to your fellow beings the course you must take to avoid the disaster which threatens your world, and the beings on our worlds around you. This is in order that you may share in the great awakening, as the planet passes into the **New Age of Aquarius.**

The **New Age** can be a time of **great peace and evolution** for your race, but only if your **rulers** are made aware of the **evil forces** that can overshadow their **judgments**. Be still now and listen, **for your chance may not come again.**

All your weapons of evil must be removed. The **time for conflict** is now **past** and the **race of which you are a part may proceed to the higher stages of its evolution** if you show yourselves **worthy** to do this.

You have but a **short time** to **learn** to live together in peace and goodwill.

Small groups all over the planet are learning this, and exist to pass on the light of the dawning New Age to you all.

You are free to accept or reject their teachings, but only those who learn to live in peace will pass to the higher realms of spiritual evolution.

Hear now the voice of Vrillon, a representative of the Ashtar Galactic Command, speaking to you. **Be aware** also that there are many false prophets and guides operating in your world. **They will suck your energy from you—the energy you call money and will put it to evil ends and give you worthless dross in return.**

Your **inner divine self** will protect you from this. You must learn to be sensitive to the voice within that can tell you what is truth, and what is confusion, chaos and untruth. **Learn** to **listen** to the **voice of truth** which is within you and you will lead yourselves onto the **path of evolution. This is our message to our dear friends.** We have watched you growing for many years as you too have watched our lights in your skies. You know now that we are here, and that there are more beings on and around your Earth than your scientists admit.

We are deeply concerned about you and your path towards the light and will do all we can to help you. **Have no fear, seek only to know yourselves, and live in harmony with the ways of your planet Earth.** We of the **Ashtar Galactic Command** thank you for your attention. We are now leaving the plane of your existence. **May you be blessed by the Supreme Love and Truth of the Cosmos . . .**

>>**Btw,** this was **1977**, and do you know the **correct date** that referred **Age of Aquarius** starts?

—28 March 2013 (**The Start of the Sixth Sun**).

—**Its starting to make sense. Does it?**

Anyway, there is something that this "**Elite**" is afraid. They are not afraid of any human or alien factor. In fact they build their underground world above one that allready existed that hosts ETs that are here for a long time. Many of them are not friendly. So the ultimate enemy is not human factor. **It goes deeper then that.**

I will not go into details about this subject, and most of you that are searching for something important (you don't know exactly what you are looking for, but you feel that you are waiting for something, and that once you will see it you will know that it was that—the right stuff you were looking)! It comes to **red pill-blue pill** choice. For all of you who are reading this, in high military installations, underground, everywhere, and you can do difference, please make a wise choice before its too late. Look deep into your heart and see what is telling you.

The rule "**Need to Know**" is deliberately put there, so you don't know what other projects are doing, so you do not see the big picture and what exactly this **Elite** is doing. You have children too, think of them. Do you think they will be safe down. **DON'T count on it**. It's a mistake and not just one in a life time but one in thousands of years. I know you are scared and you don't know who to trust, and you feel that something is terribly wrong. Fear Not, but act responsible and do what you can. We all risk a lot for bringing this out, but if we as humans do not do what it takes and I don't know if someone overthere will intervene in our behalf.

Reject the system, find similar people and do what is right. Its not enough if you are good in heart but you are not doing what is moral and right, you are equal as them if you do not accept your responsibility. You are all in a places where you were suppose to be and to do your part in this complicated **Matrix Scenario**.

So wake up. Don't be afraid to do something because if we do not do, we will lost not just our bodies, but our planet and our souls.

We are also in the middle of a spiritual conflict, the **Dark Side** is deceptive and powerful but its wrong option. First thing is to reject the system. Video files like **Zeitgeist 1** and **2** and similar provides a lot of clue what is going on.

As I was saying earlier, this ruling **Elite** or **Shadow Government** is afraid of what is coming this way. With the last wave that arrives on **28 March 2013** they will all pay for they did because against the Will of the **Creator** no one can.

They have flying technologies that can go to Moon, Mars and far beyond that but they **can not escape** the mirror of their actions done in the past. **SO, WHAT IS SMART TO DO? Many things.** But mostly, **to stay alive** till the arrival of the first few beams. They will become visible on **Dec 13 2012**, and the first one will arrive **after 8 days** exactly on **11:11** when the alignment will take place.

—**After the arrival of the first few beams from the Galactic center, we are safe. They will be strong enough to clean the astral plane, and afterwards it doesn't matter even if we lose our physical bodies. We are safe. The Nexus will clean all the dirt in it. Advice 1:** if the things went as they will, don't go to the **underground facilities** that will be offered to you. It's better to stay outside then to hide underground.

The Way Down leads to **New World Order**—distorted idea that stretches all the way back to the **ancient Atlantis. Besides that**, there is a big question with tectonic anomalies like previously presented asgravitational pull from the Super Massive Black Hole in the Center of the Milky Way, our Sun and the huge celestial passing body close to our solar system, **will something be left** of what is build under the Earth's surface. **At first look**, people will say, this is not the first time its happening, underground structures, basses and tunnels will survive again. **Wrong, this time is different.** This time we can **free ourselves** or we can lose much more than our physical bodies.

Advice 2: Around the World a certain percentage of people are more or less informed about **2012 scenario**, and they constantly check the new info about 2012. They connect with each other through what is called "**survival groups**".

This and similar Forums on the Web are doing great Job in informing the people. It's the only thing we can use to communicate. But they will introduce some **major internet restrictions soon**, so I do not know how long we can surf on this huge library of knowledge and information.

Perhaps you join one of those groups and start preparing. Its not early it's just around the corner. Despite, those groups are preparing solidly, both technically and financially and they look for location high in the mountain ranges, choose wisely. What **Europe** is concern, whatever scenario takes place, it will be **under deep water**. There will be **tearing apart** of the **tectonic plates** in the region of **Ex Yugoslavia, Adriatic Sea, Italy, and down to the middle Africa.**

In fact two tectonic plates will brake on the same line. **Euro-Asian and African Plato.** It will be a huge opening and you can imagine the earthquakes of **10.0+ on the Richter scale** in the surrounding area.

There was some channelling that **New York City** will have nuclear blast in **2012**. It could be related to the 2012 June 06 (**Transits of Venus**).

Will this really happen, remains to be seen. We really **don't relay** on **channelling** but there are people outhere that are really **gifted**. Far more gifted and skilled that the 14 Oct 2008 Event.

When those people are trained with special programs, you get some extraordinary results.

Advice number 3: fallow the input that comes from the **crop circles**. Through them all the **safe locations** will be shown. It is up to every each of you to choose will you believe those messages or not. Whatever you choose it's your choice and the benefit or consequences is yours.

The **Divine Creation** will make sure that the right info reaches every one who is worth it, but the choice is yours. There are many **websites** with good update of the new crop formations and as far as I know there are some good engineers working on decoding those messages. If you want to learn to read crop circles, start to learn **Sacred Geometry**. It has to do much with fractal geometry, free energy source and even building a space craft. **From my point of view**, if someone has bad intentions it is not in his/her/its benefit to give you a knowledge, how to build technology, to alert you about the cosmic events and etc, but it is in his/her/its benefit **to keep you** in the **dark**. The more you are a sleep and you do not understand anything the more easily you can be manipulated. **So, makes no sense**. They are trying to help. So learn the Seed of Life, Flower of Life, learn Sacred Geometry, learn the mathematical language and read what they are trying to pass to us. Anyway if you want the see if the **crop diagram** is real or not, one of the ways is to use **software to spin them**. The creators make their signature—it's usually **pentagram, cube or some 3D shape** if the input is concern microwave or instructions for building **technology**.

http://www.youtube.com/watch?v=nKy358q8RCk
http://www.youtube.com/watch?v=q6PjF . . . eature=related
http://www.youtube.com/watch?v=xAhwi . . . eature=related
http://www.youtube.com/watch?v=KwGaZDeAWqU
http://video.google.com/videoplay?do . . . 40261221&hl=en
http://video.google.com/videoplay?do . . . 26989180928496

Did you notice that they are **shooting them down** and most times immediately the craft appear above the crops, **black helicopters** with no marks are released to shoot them down or to chase them away. Here is a link to one of the biggest crop circle archives: *http://www.x-cosmos.it/cropcircles/* However, the communication with the "**friendly ETs**" becomes more frequent and there is a huge **UFO** activity. There are many **Orbs**, which are living **multidimensional entities** who are trying to help. Is in it pathetic, that we are getting help from above and our own human government are laying to us?

Summarized, there are is a heavy positive and negative **ET presence** at the moment. **Both sides** are doing their own influence on this world so the human kind has to be aware of this fact.

Advice 4: In the past times they also were facing similar events. They build huge Arcs to be safe them selves: Here is a link where they are: *http://www.youtube.com/watch?v=6cvlzXz17A4* perhaps those locations will be of some use to us, but the one in Australia does not count anymore, because (according to the channelled data) in the moment of the heaviest gravitational pull from the Sun, the most of the continent will break and sink under the water.

We are hopping that the channelling is wrong . . . but. The Queensland's coastal cities will be flooded and only the central regions of Queensland along the inner area of the mountain range together with a part of New South Wales will remain.

The breaking line of the continent is between **Adelaide (Port Augusta)** and **Normanton** on the other side of the continent. **Central** and **Western part** of the **continent** will **sink down**, the **coastline of Eastern part will be flooded**, the **mountain range** will **redirect** the coming **water** towards **Indonesia**, just enough a part of **land** to stay in one peace and intact. **For the first time in my life, I hope the channelling is wrong.** Anyway I passed to you so what you make of it, its your own choice.

There are other places that are safe, but most of others will probably not sustain suitable climate, because after this **Global Warming process** we are facing **Ice Age** in **most part** of our planet. **That is why they build the Underground World.** It's already **90% operational** but if the **Norwegian politician** is correct it has to be finished by the **start** of **2011.**

But they are all wrong. Besides it is build with the "**other**" technologies and with the help of the renegade aliens, **this cycle is different.** The **Shadow Government** is deceived by those entities. **Isn't ironic?** They had deceived all **human kind** (their own kind) but soon they will face that they also were deceived by the **negative ETs** and have not been told the truth and the real purpose why underground world is build.

As I was saying earlier: **YOU CAN NOT BARGAIN WITH THE DARK SIDE. AT THE END YOU ALWAYS LOSE. IT IS IMPORTANT THAT YOU REMEMBER THESE WORDS WHEN TIME COMES!** So, for all of you who will read this. Wake up! And do what you heart is telling that is moral and right to do. Can you see? I'm not an enemy. I just want to help! But I can not do it alone. We all have to unite to one rhythm of consciousness and to step out of the system and to say NO . . . This has to Stop! We are not playing your game anymore. We don't need you. We don't need politicians to tell us what to do. We do not need police and army to defend us. Defend us from what? The elite groups in the military and police institutions are the worst criminals.

See this link of **Dr. Bill Deagle** lecture: *http://video.google.com/videoplay?do . . . 52945040630461* As **Phil Schneider, David Icke Michael Tsarion, Maxwell Jordan, Alex Jones** and others . . . he is determined to take his responsibility and to step on the other side of the circle.

And how more and more of us will awake, more and more they will be stepping to the other side of the cirle. And it's about time.

Advice 5: Do not ever allow you or your child to be injected with vaccine. Not ever and not at any cause. I will not comment further on this.

Advice 6: Under "**certain circumstances**" I **had left** the **physical body** before the **beams** had arrived? **What now?**—In this moment the **astral space** of this planet is not safe "if I may say that" and not many beings can go safely out of it.

There is high chance that (we have to be open to all possibilities) that the most of the population with the coming events, after leaving the physical bodies will come directly to the **astral dimension**.

The adjustment of the **consciousness** has to be done as soon as possible. I would **not recommend** entering to **dark tunnels** with the light at the end. This could be easily one more manipulation from the multidimensional entities that are not friendly to us. If you are not sure, do not get "**friendly**" with any entity before the **beams** from the **galactic center** arrives. **There is a possibility**, that on the **astral plane**, **huge ships** will be parked, with a plan to use their advanced technology to trap as many **consciousness** they can. The same consciousness that were previously in human bodies, and then to cut their awareness, to put them in dream like awareness so they will not know what is going on, (**believe that they have technology and the skills to do that**) and then to put them inside the energetic containers, then transported to other systems and finally to put them in other bodies so the manipulation and the explanation of the generated energy field called aura which is result of the fusing of our consciousness with the energies in the surrounding space, can continue.

We also are **multidimensional beings**, but something happened that lefts us with this double helix awareness, which is enough for reading, testing and perceiving this physical reality (**although limited only on five senses**) and partial awareness of the astral plane in a form of dreams, lucid dreaming, and partly conscious out of the body experience. That is why it is advisable **physical dislocation** to **zones** that have multidimensional portals above the ground, places with strong and dense positive energy. Put in simple words—**places on the physical plane where the astral dimension is save and secure.**

I know that all this is probably **hard to swallow** for most of you, but its happening. For few of you, still open minded, the **Shield** that is keeping our **consciousness** attached to this planet, its weakening and with the right determination can be **penetrated** and consciousness can **escape**. **Leaving in any direction**, means **infinitive choices**, but I recommend redirecting your astral body directly towards the **center of the Milky Way**, high speed astral flight in that direction and intercepting the beams that are coming this way.

It the only place, that I'm sure it's safe!

Please remember all this, before they remove this from the net!

If **your astral skills** are not so good, try to stay awake in the astral plane and avoid any tunnels, vortexes and etc., in the astral plane. Resist any outer pressure from anyone overthere to enter into craft, portal and etc., and wait for the **Nexus** to arrive.

If you stay while in the **astral dimension**, you will soon use to this new density and if that happens don't wait, but leave this planet and go directly to the directions of center of the galaxy. The moment one of the beams of the Nexus touches you, you are safe!

Will the aliens land officially on Earth before 2012?

—I have no clear answer for that question. **Many of those races are here all the time, many penetrate our plane of existence from time to time, doing their own agendas and they leave.** Will some of those races will decide to come to open before the **2012**, remains to be seen.—**In any case**, many of this is still an **open book**, and like in a chess game, depends on the moves of the **Malevolent races, Neutral and Benevolent races** present in this solar system, **Illuminati, Shadow Government, Increase of the level of the awareness of the global population etc.**

—Is there is a possibility that **nothing** of this happens and that all this 2012 scenario is only very good loaded New Age Scum?

—**From the best that I know, no**. Not just that is going to happen, but its already happening. **Just look at the evidence around you**. Climate change, Pole Shifts of other planets, increased of brightness of the planets (search Google for increased brightness of Venus and Jupiter, and tell me that this is normal and has happened before), and I can go on and on for hours just to make a list of evidence.

—**There are some indications that Shadow Government is conducting some kind of experiment using the "other technology" to move through time with the intention to stop this 2012 event?**

—Waste of time. No one can stop what is coming.

—**Should we be afraid?**

—**No.** The **Cosmic reality** is bigger that we are. We are on the start of the change which is natural culmination of things. We should fear ignorance and prejudice. Only the **true knowledge** will set you free and it will be there for you when you will need it the most.

—**If some entity is trying to communicate with me, how should I know if he/she/it is positive or negative?**

—Try to feel what you feel deep inside of you. Try to sense the vibration that your **heart** is sending to you, and that is your answer.

—**Are we on a door of something extraordinary?**

—**I believe Yes.** All our **troubles** will soon be over and soon we will exist fully conscious on higher density and become aware of our true nature and the true nature of all things in the Universe.

—**What else is smart that we do?**

—**Look back** to your lives and see where obstructs and where you find unpleasant mark. **Fix** the mistakes that you have done, be brave to **apologize** where you need to and to ask for **forgiveness** where you need to. From every **enemy** in your life try to make friend or it that isn't possible make effort to gain a neutral relationship with that person which will automatically lead to **disintegration** of long time generated mental energy.

    Remember, we are all **connected**. We just don see it, because of this five sense limitation we have. We are **one giant conscious organism, one being** but because of the DNA degradation we don't feel it and we don't see it. The **beams** will reprogram our **DNA** and you will see what I'm talking about. Do the **things** that you wanted to do for a long time. Give yourself a **time-out.** Spend time with your family and bring joy and happy moments. **Clean your life. Try to avoid doing evil in any form. Turn to your cosmic nature. Stop to eating flesh, because by consuming it you approve what is doing to our race.** Make a change. Turn to the nature, meditate, expend you perception. **That is smart to do.** The **rest** will be done by the **beams.**

—**Should we be joyful that we are present here in this moment in time, just before the end of a long cycle?**

Yes. If you look with the eyes of your **common sense** and intellect, you see **Pole Shifts**, unavoidable catastrophes as result of natural cosmic conditions and laws. **But** if you look with the **eyes of you heart,** the picture looks different and deep inside yourself you know that in the end, everything will be **Ok.**

—**Something about the future?**

—**Stay awake.** Follow the path of Love and the Universe will open its paths that will take to **safety**. Just open your heart and listen where it points.

—How do I picture the form of the Creator and how this form will look like inside the Nexus?—What shall I do when time comes?

—It makes **no difference** how you imagine the **Creator**. It is **important** that you create **space** deep inside you that can accept the Creator. **That space is called pure heart.** Once the beam of **white-bluish** light strikes you, everything will come by itself.

—How will I know if Am I ready?

—**You are ready.** Otherwise the **Universe** wouldn't bring this information to you.

—Will someone survive on the physical plane?

—Around **2800** people.

—Who are they?

—The **Will** and **Choice** of the **Creator.**

—Will they be experiencing a better reality than the ones that will left the physical form?

—**Unknown.** Lets say that the both sides will be cleaned from the darkness, so the both sides will be affected by the **evolution.**

—So we don't have to worry about? At the end everything will be right?

—Yes, but I will recommend staying with full consciousness till the cosmic beams arrive.

—Are the beams coming this way, for sure?

—Yes, the beams of the **Nexus** are coming. The scientific community is just starting to understand the true nature of this phenomena. If you do not have the both **components** in **balance,** both intellectual and spiritual, you do not see the true reality of the Universe and how it functions. **With the beams also is coming the transformation of the Matrix.**

Put in other words, the beams will bring **extreme transformation** of the **configuration** of the **DNA molecule** of every **living thing** that is touched by them.

—Will there be some kind of adjustment or will it happen suddenly?

—Yes, there will be a **vibrational acclimatization**. That is why they will arrive one after another. The last one will arrive on **28 March 2013** and it will **mark** the completion of the evolution of the **DNA** of life in the region of the **Cosmos**.

—How long will the change last?

—Long enough, so we can forget that there was a time of sorrow, pain, anger, jealousy, hate. Long enough so we can start to love again and to start to respect the life in the Universe. At the end we are all **ONE**.

Respected friends, I have nothing else to add to this. You will probably not hear from me again. I'm hoping you will be able to see what is coming this way with open mind and open heart. At the end, as **David Icke** clearly put it : **Love is the Only Truth Everything Else is Illusion**.

My job is done. I have passed the information. Now it's up to you. **May you be blessed by the Supreme Love and Truth of the Cosmos** . . .

I love you all,
Astralwalker

The **left side** of the **rectangle** is the **breaking line**. The **dark green area** along the **east coast** is completely **flooded**. The **lighter greenish area** inside the borders of the rectangle in the previous picture is the **safe area** that will stay intact. The best area is around **Queensland fields**. It will sustain a **suitable climate** for life after all this is over.

Update >> Many people were asking if there is a possibility that I can deliver a more precise map, so here it is: Now, you don't have to do anything, it's your life and your choice. If NWO is not introduced and there still will be a freedom of movement, many people can find safety to those areas.

The Pole Shift will not happen instantly. It will probably happen in two or three days as a result of slower and slower rotational speed of the

Earth around its axis before it reaches the zero point time on 21 Dec 2012. The Earth will roar, and you will notice that despite it has passed midnight outside is still a day or vice versa.

If you are smart you will notice these changes long before this time frame. This is the logical culmination from the Climate Change that we all ready experiencing. Its so obvious and it stays before our eyes, but we just don't want to see.

When the Ice around the South Pole Entrance melts more and more, this water will have to go somewhere. http://www.nasa.gov/mov/133778main_FUV_640x480.mov

It's like a glass of water with ice cubes. When those ice cubes melt, the water will flood the top of the glass. The balance of the sea level it's already distorted on a planetary scale and even a small change in this equilibrium causes drastic climate changes. Anyway everyone knows what the huge melting of the Ice ****s leads to . . .

This is 1992-2002 . . . What about 2008?

The safe zone that is shown, when time comes, it's large enough for evacuation of the whole population of Australia and surrounding areas. Its lot better then the underground shelters that will be offered to some of you. Just know this, everyone that has to be there, will be there. The colour of the skin, nationality and all the rest of the social illusions, fill fall. For anyone who thinks that he/she is above the others and that he/she is more worthy to survive then some other one . . . its wrong approach that leads to doom!

Just remember this: When everything goes out of control, and our governments thinks only of themselves and you are wondering where it is smart to go and where to take shelter . . . **here is your answer**. If I'm wrong . . . it makes no difference. But think what if I'm right? In that case, this info will be essential to your **survival** . . .

Respected **Friends**, **Divine Love** will show all the **safe locations** everywhere and for every **continent**. Just fallow what appears in the **crops** around the world.

Don't worry and do not fear. If your heart is pure and you are good person you have nothing to fear. You are already safe and the Divine Light will show you the way and bring the right information that will lead to safety.

Take care
Astralwalker

A FORUM MEMBER SAYS . . . . WE ARE NOT!! PASSING
THROUGH THE EQUATOR OF THE MILKY WAY SO GOD
KNOWS WHAT THE REST OF THIS NEWS FLASH'S ACCURACY
IS GOING TO BE. WE ARE HOWEVER MOVING INTO THE
HYPER DIMENSIONAL POWER STREAM OF 19.5 DEGREES SO
YES WE MAY EVOLVE, TRANSCEND DEVOLVE??! SO PLEASE
EDIT YOUR ERRORS AND GIVE TRUE COORDINATES,
EVERYTHING HERE WE COULD ALL WRITE, MANY OF US HAVE
BEEN ON THE SAME JOURNEY, BUT ALL THIS IS PRESENTED
AS FACT WHEN IT IS VERY LOOSE. SORRY, THE CROSSING OF
THE DARK RIFT/ PHOTON STREAM WHATEVER WE WOULD
PREFER IS WRONG, IF YOU THINK IM WRONG ASK DAVID
WILCOCK (AS SOMEONE YOU MAY BELIEVE OVER ACADEMIA.

These are the strongest opposing "**mainstream science**" arguments
about **Dec 21, 2012**:—**Maya** never really made a prediction about
a **galactic alignment** for Dec 21, **2012**. This idea was **born** out of
attempting to figure out why the **Long Count ends** on that date.

—It is true that the winter solstice **Sun** will fall into the **Dark Rift** also
called Cygnus Rift on Dec, 21, 2012, but the closes to **zero degrees**
(longitude/latitude) that we will come, already happened on **December
21st, 1998**.

—Where will the **Sun** be relative to the **galactic equator** in 2012?

—**Precession** is moving the winter solstice Sun **further** form the galactic
equator. So it will not be in exact **alignment** on Dec 21, 2012. We are actually
at this point **78 light-years away** from the galactic equator and we are
moving **further away** from it. So actually, its only **3 million years** ago that
we passed the galactic equator and we wont be back to it for **60 million years**.

—The Dec 21, 2012 Winter Solstice Sun will not be in exact alignment
wit the Galactic Equator. Its actually moving further away! This alignment
already happened in 1998!

—No evidence that ancient Maya predicted the End of the World in 2012 (discussible)

—No exact alignment of the Winter Solstice December 21st 2012 Sun with the Galactic Equator and No Crossing over the Galactic Equator by our Solar System

Those are all Nice conclusions!

But they are all False.

Lets start with this:

—**According** to their arguments this alignment already happened in 1998.

—True . . . **but** that is not essential to this **2012** scenario.

This means that we already passed that point and we are moving away from the galactic center.

—True . . . but. There is **also** something else that **has to be** considered here. We are **influenced** by new energies from the **Galactic Center**. The proof is everywhere. http://www.crawford2000.co.uk/planetchange1.htm

The Sun entered the galactic equator during the Winter Solstice of 1980 and will have completely cleared the galactic equator on the Winter Solstice of 2016. The years between 1980 and 2016 is called the Galactic Alignment zone or **era-2012**.

>> Did this triggering of Hunab Ku energetic Nexus start in 1998 and not 26000 ago?

—Still unknown. Possibly that in the moment of the alignment in 1998, there was some kind of Aetheric portal opening, inside the center of the Sgr A*, from where the energy shot appeared, and is now traveling this way.

If this is true, this Aetheric energy is moving with speed that is way beyond of our current perception and understanding. As "**ETs**" clearly delivered, we will "**visually**" see this phenomenon from Earth on 13 Dec 2012.

On the other hand, not everyone in NASA is polluted. This is the update they put on about center: http://en.wikipedia.org/wiki/Image:2 . . .—annotated.jpg

Now, let's add some more **power** to this. I will **land** on the **scientific field** so that we can be completely "**equal and legitimate**" inside the ring. **Not to be misunderstood**, I'm not against science, but I'm against the **false science** or at least deliberately false presented science from **NASA** and **ESA** which as we all know are own by the **Elite**.

Hmm . . . here is the interesting link.
http://www.physicsforums.com/showthread.php?t=208935

You have to log in to see some of the things. Most of the physicists are ridiculing the 21 Dec 2012 and considered it as nonsense, but there is also a nice star chart of the night sky on 21 Dec 2012 seen from a location near Philadelphia in USA. And its about all that we need! Its a screenshot of Starry Night for 21 Dec 2012. The Sun, viewed from earth, is 6 degrees, 38 minutes away from being aligned with the galactic center. It is how the Milky Way will look like on the night sky.

Now let's get back to that **1998 alignment**. The real alignment is not going to happen by continuing on the same course away from the galactic center, but its going to be about the increasing of the mass of our star, increasing Sun's energy output and also extreme gravitational pull and than as a secondary result a change in the Earth's orbit.

Also there is evidence that in the all atmospheres of the planet there are drastic changes and this includes the increase of mass.

In any case, the change in the conditions as influence of the unknown form of galactic energy that radiates from coming Nexus which arrives in 2012 will generate the perfect alignment. I believe that NASA had already figured out but they are silent about this. In other words, in this scenario we have a straight line between our planet, the Sun and the Galactic center.

Once more lets see, the pictogram that appeared Avebury Manor on July 15, 2008. Now, let's compare the Update that arrived on 22 July 2008. And if we utileze the downloaded data . . . lets see: Hmm . . . we still have no perfect alignment.

True . . . but, if we add here that because the Sun will get so large scaling to a central point of planet Venus, we have unseen gravitational pull that certainly effect the geological and the magnetic poles of our planet. This new influence of increased gravity from the Sun will make Earth out of its nominal orbit, just as little so we have a straight line. 21 Dec 2012 Time: 11:11:11

Way . . . The **ETs** are clear about this. There is your perfect alignment.

The only **error** in all this that I presented can be that the **21 Dec 2012** is actually a 22,23 Dec 2012 as depicted in the Crop formation. That means that on the Gregorian Calendar the **Mayan Long Count Calendar** end on 23 Dec 2012 and not 21 Dec 2012.

However, we are talking about a difference of only two days.

Does two days difference makes any difference to you?

(No, Of Course Not!)

But, on the other hand, I have seen data in the crop formation that clearly depicts the Dec 21 2012.

Summary:

The **new input** is suggesting that very soon we will have a transmutation of genetic material of this solar system, including our own DNA, or if I can say some kind of "**special bath**" in galactic light emanating from our own Sun, acting as transducer of the Galactic Centre itself, passing directly to the Earth. In other words we are talking about **Evolution**!

Evolution of the **consciousness** that will take the life forms to a higher level of existence when the cosmic beams start to hit our Sun and when they did our Sun will project a white blue beam of beatufull light that will hit the Earth.

Respect to all,
Astralwalker

And little zoomed out **21 Dec 2012** night sky from another location: **Plus** add the **little change** of **Earth's nominal orbit** as a result of our **star** gravity, and there you have it—**the perfect alignment!**

Please take in consideration that **Mars** is not burned by the expansion of the **Sun**, but together as **Earth** stays dangerously close to the Sun. What you make of this info is your choice.

But this is what "**they**" insist that we understand!

And the dates are all there depicted in the crops. You can check them all on **NASA's Solar System Live**: http://www.fourmilab.ch/cgi-bin/Solar/action?sys=-Sf

I have **not much** to add to this information. **Please** if someone overthere has extended knowledge about this, he/she is absolutely welcome to extend this info.

Once more, we are only **transit phenomena**. The **Cosmos** is bigger then us, our reality and our lives. This isn't either good either bad. Its just is . . .

I believe that there is nothing that we have to fear . . . At the end of 2012 we are facing something remarkable. We are facing something beautiful and something extraordinary. We are facing our **Creator** and we are **going home**.

It was about time . . .

Respect,
Astralwalker

Quote:

I have a question regarding an apparent contradiction. If the energy emanating from the galactic centre is so high and as a consequence we are going to roasted, what's the point of the genetic reprogramming that is supposed to happen with the wave ?

Our **DNA** structure does not stand only for our **physical bodies**. It goes way beyond the physical. I will try to explain this. We are **consciousness**, an intelligent **life force** that exist in this **Universe**. Now . . . It doesn't matter if you believe that there are only **7 dimensions** or planes of existence or **22** or **infinitive** number of them it simple doesn't matter.

Every one of those **dimensions**, frequencies, worlds or planes, possess **different density** and different **speed of vibration**. We are on **lowest** which is most dense and with the lowest speed of vibration.

As we go higher, towards **etheric**, **astral**, **mental**, **causal plane** and higher, the **inviroment** gets **less dense** and more **subtle** and the **speed of the vibration** gets higher and higher. Every **next dimension** you are in, extends your abilities, understanding, it brings you more extended perception and wisdom if I may say. The more you climb up

2012 A FAMILY BRIEF

the more you understand yourself and the purpose of your existence in the **Universe**. The more you learn and grow the more you become aware of your responsibility and what is your task or how you can help in the **Divine Plan**.

Now . . . For our **consciousness** to move inside all of those planes of existence, it needs a **suitable vehicle** or a body that possess the same density and the same speed of vibration. For your consciousness to be capable to be consciously present on these higher realities, you need something which is called fully activated DNA structure. You can call it 12 strain DNA functionality, and you will not make a mistake. Because if you look at the crop circle data, they usually insist that we understand and pay attention to the germ, seed, flower, and the fruit of Life. That is . . . that we pay attention to the Sacred Geometry.

I'm sure that you have already noticed that this relation of forming the Germ of Life, Seed of Life, Flower of Life, and the Fruit of Life, it's exactly the same what is happening inside the DNA. The nature is manifesting its self through the mathematical language which has precise geometrical patterns starting from germ, seed, flower, and then developing to the fruit of Life and on and on to infinity.

It doest matter if you are a flower, fish, alien, tree, human, animal, bird or whatever you had been created from the same geometrical pattern starting from the germ of life to the fruit of life. The **Fibonacci sequence** is present everywhere and it is found in everything in nature. Anyway, I do not have time to go too far with this . . . what I mean is that when a human being physically dies, that does not mean that the DNA is destroyed. It means that the DNA stays with consciousness because it has many subtle bodies that are attached to it, and which the consciousness can use as vehicles for moving through the higher realms. The only part that fades when the event we call physical death happens is the **harmonic** that matches the physical vehicle. The higher DNA harmonics are still with the consciousness because as I said earlier is capable of multidimensional existence or capable of moving through different dimensions using suitable vehicles. The astral body matches the astral frequency, mental body the mental frequency or plane of existence and etc.

Note: Please consider that I'm using here terms that are understandable for most people. You can consider all this differently: The consciousness, possess a suitable subtle vehicle, to be in all this dimensions and all this is the DNA structure. I does not take long, that we can determine that the more DNA is activated the more we can reach towards other planes of existence and fully conscious.

Now, let's get to some heavy discussion. Scientific community had split the composition of our Universe to Normal Mater, Dark Mater and Dark Energy. They were comfortable with Normal Mater observations but when you ask about Dark Matter and Dark Energy they just stick their finger in the mouth and if they are honest they will tell you . . . we simple do not know what is going overthere. Also they call it a vacuum and they tell you that is empty. But, its not. How can a space vacuum can be empty when all around we see that all energy is radiating to. Vacuum is not empty but exactly opposite full of energy.

We now know that 99% of atom is empty space. This is where science usually stops. But there is much more beyond that. In fact we are living in the Universe which is contracting to infinity in a fractal matter and expanding to infinity by constantly getting larger and larger. I can elaborate on this much much more, but my time is limited.

Now how all this have to do with the coming **Nexus 2012**?

—Simple. All this Cosmos or Universe (if we use scientific terms) that consist of normal mater, dark mater and dark energy, is interwowen and all dimensions or planes are connected. The intelligent Force behind all this is what is called Aether. It creates galaxies, stars, planetary systems, comets, everything that we see etc. This Aether can be summarised as direct creation force that manage the balance between all the dimensions and express the will and the purpose of the Creator Himself.

Now, in normal conditions, when a human from Earth, dies, consciousness leaves the body but because of the low progress, still attached to consuming flesh, still used to bring pain to others, polluting and destroying the nature etc, the vibrational state is still low and can only use the DNA harmonic that matches the astral plane. In most cases the consciousness is unprepared for this new density and it takes a while before it get use to this new reality and starts to use the astral body.

Now . . . The Nexus represents the higher energy. The higher purity. The Most Divine. Because it's higher, for our consciousness to be able to sustain on that frequency in a nexus body, all the DNA harmonics including the higher must become functional. And the only way that someone can progress is through Love, Compassion, Self determination, self-respect and respect of all other life forms in Universe, respect to the Nature itself as living organism and understanding that we are all equal, we have the same potential and that we are all ONE.

Inside the NEXUS you can not exist as individuality. It's just Oneness! No one is better then the other . . . it's all illusion. We are all connected and we are all part of the same. If that is the case, who needs fighting and who needs contest? The only things that can free us is harmony with ourselves and with the nature. When you find a central point inside yourself, the point that is hollow and that contracts to infinitely small you know that in that small dot you are one with the Creator.

In that moment all the fears disappear and you are not afraid any more.

The Nexus is something extraordinary. I'm not sure about this, but there is a possibility that as soon we penetrate, we will teleported (the billions of consciousness will become ONE and that ONE will be transported back to the galactic center and through the point of the Singularity it will be taken to another Universe with even higher cosmic DNA harmonics from this one.

At the moment the atmospheres of all the planets in the solar system are changing. They are becoming bigger as we receive more and more out side energy. The bigger the atmosphere, the more prana or mana is present, and the more prana the more DNA strains can be awaken and utilized. But we have to stop destroying everything we touch, and turn to our cosmic nature.

In simple words, the reprogramming of the DNA structure is necessary for activation all DNA harmonics that can support our presence in all planes and most of all to support the fusing of the consciousness into one ocean of consciousness—into oneness. We will separate from the physical mater as a race and not as individuals. And that my friend is what makes all this so extraordinary.

Why do you think they give you E-codes, aspartame, fluoride, use frequency weapons as HAARP and similar, vaccines, mediums, creation of fear and low emotions? They are trying to stop what is happening my friend. We are in the middle of something extraordinary, so don't stay in the frequency that all that will happen is that you are going to be roasted. Its only physical harmonic, we are much more then that.

So do not fear . . . If we fear we vibrate low, and the quickening and evolution inside the DNA can not proceed. The beams will do the most part of the genetic engineering but small amounts of cosmic rays are already hitting the Earth and are changing everything in the Matrix.

Stop consuming drinks and food that pollutes your physical body and are stopping the dna awakening. Turn to meditation and to nature. Return to yourself.

As, I was saying I believe we are on the door of the Evolution of our race. However, this only my perspective. What you do with this info its your choice.

This is what I know and what I feel.

Respect,
Astralwalker

Koyaanisqatsi wrote:

Quote: as we approach and transcend 2012 and this cosmic energy imparts maximum effect during that time are you thinking that no humans will remain here? all present will be taken through singularity to another octave?

What is unclear is what will happen to the ones that will sustain to vibrate low.

Some beings have done terrible things and they are still not considering changing their behaviour no matter of the influence of the Great Awakening that is happening in this moment. They simply can not comprehend these new vibrations, and those are heavily attached to 3D reality and the old ways.

What will happen to them I do not know.

The rest, normal people will experience the Shift in Consciousness, a reprogramming of the DNA structure, will leave the physical matter and will join to one Consciousness. Than this ONE Consciousness will regain all its lost memory, it will assimilate all the experience that each individual previously had, it will regain its power and wisdom and in the end it will leave this Universe through the point of the Singularity that is located in the Center of the Sgr A*.

Regards,
Astralwalker

To figure out the **2012 event** we must figure out many things . . . Among them this also: The East Field 'DNA double helix' formation. 648' long and 89 circles. hmm . . .

A triple DNA helix. Interesting . . . http://www.youtube.com/watch?v=hekKx . . . eature=related http://www.youtube.com/watch?v=l_Lzb8fdJaA http://www.youtube.com/watch?v=-6Wim . . . eature=related http://www.youtube.com/watch?v=6M6vP8-SbU0 http://www.youtube.com/watch?v=KRvWR . . . eature=related

Respect,
Astralwalker

Despite the fact, that its only my perspective, if the human kind wants to understand the **2012 coming scenario**, which is what all those hidden messages in the crop circles are really about, the humans have to able to change their brain frequency, to trigger **DMT** creation in the **pineal gland** and to be able to tune with their consciousness into to this so called planetary **Psionic energy field** or should I say better to the **planetary Psionic internet**.

From here they can generate much more useful input, about what is going on here, and what to be done. I know that all this is difficult to comprehend for most of you, but I believe that in this state of only double helix of our DNA and the five sense limitation we are seating ducks . . . for races who are taking the advantage of this our current state.

Tuning into this frequency is only the beginning and humans can go deeper into this Universal Psionic Field, induce completely control out of the body experience, and bring useful data from there. This is very similar to what is commonly known as remote viewing, but what I'm talking about is incomparable much more then that.

Here how this can be done.

I will explain one of the fastest ways how you can train your consciousness to able to perform this. Let's start with basics: Remote viewing is a term in parapsychology that describes a process in which the psychically gifted person is watching some kind of astral television based on a program that he/she has chosen on his/her own.

This process enables a well-trained remote viewer to focus on his/her mental screen where psychic visions start to appear and the viewer uses mental efforts to catch those psychic pictures, which in fact are astral reflections of places that he/she wants to observe.

This ability is actually related to the human's third eye that makes it possible for the consciousness to see things that are not visible to physical eyes. Therefore, the simple definition of remote viewing would be—observing the physical, etheric and the astral area without leaving the physical body.

However, many anomalies that can distort much of the results of the remote viewing might appear in the early stages of the development of this extraordinary ability. That's because a beginner strongly focusing on scanning some area is unable to hold his/her consciousness unattached longer, and often becomes emotionally attached to his/her individual feelings and opinions concerning the things he or she is watching.

Those individual feelings and opinions will have major influence on the already very weak astral reflections of the chosen area, and it will distort them or even trigger observation of some other area, which suits mostly his/her suppressed desires. Therefore, a student on the start line must learn to shut down his/her own personal feelings and attitudes first if he/she wants to succeed mastering the remote viewing ability. One should become an empty receiver to be able to absorb major amount of data, classify them, filter or analyze them.

In other words, once one succeeds to catch the astral radiation of the target area, he/she has to remember them and analyze them later. Later the student will learn another approach to the analysis. All this etheric and astral radiation that streams through the mental screen is provided by the third eye and projected directly to the consciousness. Any invisible physical place becomes visible for the third eye if the consciousness succeeds to dive deep into the etheric or the astral matter and remains there long enough.

Nevertheless, as I mentioned before, the etheric and the astral plane are almost infinite areas full of infinite radiation. Thus, in the beginning, the mental screen of the remote viewer is flooded with hundreds of other reflections coming from those planes in the shape of psychic pictures or visions.

Anyway, practice has shown that as far as the radiation that appears on the remote viewer's mental screen is concerned, I am positive that mostly it has its origin in the astral plane and much rarely in the etheric plane. I know it sounds contradictory, and logically it should be the other way around, but the fact is, it is the truth. It is simple, because our third eye absorbs the sensitive images that come from the astral plane filled with thoughts better than the ones that come from the etheric plane mostly filled with vital energy.

However, every student must understand clearly that remote viewing is ability hard to achieve and that it will take some time before one is capable of selecting the target astral reflection from the infinite number of other astral reflections appearing on the mental screen. Thus, in the beginning, one's consciousness must become a clear mental surface open to all the astral radiation coming from the astral area. Then, by using the third eye one must learn how to focus the consciousness on chosen targets and block all other images visible to him/her.

This is a hard period for every student, but in time and with practice it can be achieved. Long ago, I personally had some difficult time mastering it and I will try to explain how the whole thing works: Before the student even tries to achieve the remote viewing, first of all he/she must learn to create mental shapes, forms and images of the places in the consciousness helped by the ability of visualization and concentration. Then, the student must learn how to keep them frozen for as long as he/she wants to, not letting them disappear from the mental screen.

Therefore, at the very beginning of the mental creation process, the observer must become capable of holding the selected target in his/her imagination. To achieve that, drawing the place on a sheet of paper for at least 30 times, and practicing the visualization of the place for at least 7 days for about 10-15 minutes would help a lot. That would help the observer to "imprint" the picture of the selected area deep in the etheric matter below the astral plane.

Anyway, after a few days of previous visualization exercises, the student should sit in some comfortable half lying or lying position, in some quiet place where he/she can work in peace and silence. The potential remote viewer should close the eyes and start to observe the mental screen. At this stage, the viewer does not have to imagine or force a mental creation of some kind, but to become a passive observer. The effect that will follow will be very similar to watching a movie in the cinema. The only difference is that this time the screen will be in his/her head.

The secret lies in becoming a silent witness, without getting emotionally attached to things that will start appearing on the mental screen. In time, the astral reflections will start to enter the consciousness and the remote viewer has to put some mental effort to remember them. The training with the zener cards can also help the student activate the mental screen. Besides the zener training, a constant focus on looking straight up with eyes closed can also trigger the awakening of the third eye (Ajna chakra), and because of that, the mental screen will become more visible.

In the beginning, in front of the student's eyes only total blackness will be visible, which in time will become foggy, and then as the third eye becomes more and more awaken, it will become an open mental screen where the astral pictures will be coming. It is interesting to mention that the third eye is the one doing all the work and the consciousness is only the boss in the whole process.

Anyway, it is not necessary for the student to visualize the third eye becoming shiny or to take some shape like circle, triangle, etc., only to remain a passive witness of the astral visions streaming into his/her consciousness.

Here how the blackness is altered by the open mental screen: From the depth of the blackness of the mental screen slowly but surely, some unclear figures and contours will start to appear. In time, the student will notice an unclear picture appearing and it will be gone in a split of a second.

Then, if the student's consciousness remains calm and only focused on the mental screen without any emotional attachment other pictures will come.

The color of the pictures will be unclear and mostly black and white in the beginning, and as the time passes by, images will become clearer and somewhat colorful. The duration of these astral pictures will vary and it will be determined by the student's will to hold them or reject them on the mental screen.

Later, when the remote viewer reaches an advanced level of psychic observation, the pictures will have clear colors. When the student reaches this level of development, he/she will start to hear sounds in the consciousness connected to the related astral vision almost instantly.

Eventually, the pictures and the sounds will become perfectly clear to the observer. When that starts to happen, the student will know he/she has reached the point where from he/she can try to find the selected target of the carefully chosen place. The whole process at this stage will appear as somebody presenting the student infinite series of pictures of people, animals, close and distant places, houses, buildings, space images, underwater images, pictures from another time, etc.

In time and with practice, those pictures will become motion pictures. Thus, the remote viewer faces another task and he/she has to learn how to freeze the chosen picture, and analyze it without vanishing from the mental screen. The analyzing process should be done strictly mentally and without using words in the thoughts and with no emotional response towards the astral picture; otherwise, the picture will vanish from the mental screen. It is not advisable to even think in pictures because that will surely influence the sensitive mental screen.

If the remote viewer is capable of reaching this point of observation, I suggest a further try for observing the astral radiation by using a higher level of consciousness. By a higher level of consciousness, I mean consciousness of the consciousness itself. One consciousness exists in the average consciousness, and observes filters and analyzes all the data coming and leaving the remote viewer's mental screen.

In other words, the remote viewer adapts to this new part of the consciousness and when this is achieved, the viewer can easily think and make choices in a higher and subtler way not touching the sensitive astral pictures. Then, all that needs to be done is to focus on the selected area and in a few seconds, it will become visible for the remote viewer. Once the viewer catches the astral radiation of the target area, he/she can easily move and observe everywhere and everything just by using a small mental effort.

Here is one example of how the remote viewing process is done: Let us assume that a picture of a large apartment appears on the mental screen of the student. To succeed with the remote observation of the apartment, the student should use some mental efforts to partly enter the mental screen by pulling the picture towards the third eye.

While this process is on, the remote viewer's consciousness must remain maximally calm and under no circumstances to become emotionally attached to the picture, because if it does, the picture will be altered by another one more suitable to his/her emotional charge.

The next step is to locate the astral reflection of the target and to "freeze" the picture on the mental screen. Once the student succeeds that, he/she has to pull the picture closer to the third eye using strong mental efforts to slip into the picture at the same time. In other words, the student has to magnetically pull the picture in and to use mental efforts to push him/herself out and walk through the picture at the same time.

The whole secret at this stage is in the balance of these two forces—the outer, which one pulls towards him/herself and the inner, which he/she forces out. In this particular case, one has to pull the picture of the entrance room of the selected apartment towards the dot between the physical eyes, and walk through the entrance of the apartment visible on the mental screen at the same time.

Once the student succeeds to pull that off, he/she is free to observe the other rooms in the apartment. In other words, the remote viewer adapts to this new part of the consciousness and when this is achieved, the viewer can easily think and make choices in a higher and subtler way not touching the sensitive astral pictures. Then, all that needs to be done is to focus on the selected area and in a few seconds, it will become visible for the remote viewer. Once the viewer catches the astral radiation of the target area, he/she can easily move and observe everywhere and everything just by using a small mental effort.

Here is one example of how the remote viewing process is done: Let us assume that a picture of a large apartment appears on the mental screen of the student. To succeed with the remote observation of the apartment, the student should use some mental efforts to partly enter the mental screen by pulling the picture towards the third eye.

While this process is on, the remote viewer's consciousness must remain maximally calm and under no circumstances to become emotionally attached to the picture, because if it does, the picture will be altered by another one more suitable to his/her emotional charge.

The next step is to locate the astral reflection of the target and to "freeze" the picture on the mental screen. Once the student succeeds that, he/she has to pull the picture closer to the third eye using strong mental efforts to slip into the picture at the same time. In other words, the student has to magnetically pull the picture in and to use mental efforts to push him/herself out and walk through the picture at the same time.

The whole secret at this stage is in the balance of these two forces—the outer, which one pulls towards him/herself and the inner, which he/she forces out. In this particular case, one has to pull the picture of the entrance room of the selected apartment towards the dot between the physical eyes, and walk through the entrance of the apartment visible on the mental screen at the same time.

Once the student succeeds to pull that off, he/she is free to observe the other rooms in the apartment. By using this method of achieving the extraordinary ability known as remote viewing, very soon you will become capable of observing distant places helped by your third eye with major precision without leaving your physical body.

Now, we were speaking about observation. Now let's speak about connection with intelligence that is making those crop formations and give us those instructions. If you become skilled enough with this, you will soon find out that from time to time some kind of download comes to take place to your consciousness. It will appear in a geometrical forms exactly as it appear on the crops.

The only difference will be that you will feel also a telepathic streaming and emotions that are describing those images. Those beings that are more awake, will start to get the download about 2012 sooner then the others. Then more and more individuals will start to receive them, because the DNA structure is changing as everything else in this solar system. We are already exposed to new outher energy and the most sensitive of us are starting to feel it. The more and more people are fusing and they intuitively are starting to be aware of this connecting and tuning into this Oneness, a responsibility becomes visible, that we all have to work together to figure the big picture before its too late.

On the other hand there are many that are still heavily attached to 3D reality and to the old ways, they are supposedly search for some kind of spirituality but the truth is that they do not want the change and they are still not ready. Anyway, besides many of you will probably disagree, but I feel that we have been given a choice to take a side.

One leads to evolution of Oneness that will be transformed to Light through the 2012 Nexus and the other leads to staying in the 3D reality and generating the experiences that are missing, although it possibly means a thousands years new manipulation from other being who have benefit from this option.

The option of Oneness means end of all fights, it means Evolution to Divine existence and to understanding of the deeper mysteries of the Cosmos.

The option of 3D means, staying to physical matter, experience of pain and suffering in infinitive forms, limited perception as five sense reality. Those beings will be probably be taken to other systems before the Nexus arrives because once it hits the Earth we have at least 5000 years time of divine vibration which is not compatible with the vibration of those beings.

That is why, we have split in the understanding of the human beings.

The ones that are trying to do everything they can to make some difference and who fight for freedom and harmony, for cosmic existence and balance, can absolutely not be understood by the beings who despite the DNA progression from the new coming outside cosmic energy, are still vibrating low and are still hunger for consuming flesh, cigarettes, alcohol, drugs, simple to the old ways.

Nevertheless, we as a race are growing. Both intellectually, both spiritually.

For all ones who disagree, and who are seeing blackness everywhere they look . . . my apology, please skip this post. For all ones, who sense that you can use something from this post, there is your way to tune to this psionic field of planet where you can come to direct contact with the **Gaya** and to find out who you are, what you doing here, what you have to do and where all this is going . . .

Anyway, the choice is yours . . .

Respect,
Astralwalker

Quote:

I was just wondering—when I go into remote viewing mode, and let the images form on the screen I am always transported right out into space somplace, really quickly.

>>

As soon as you **stop thinking** in words and you had made a strong command that your **consciousness** must remain completely empty and without thoughts, the **frequency** starts to **change**.

In front of the **blackness** of the **mental screen**, images start to appear.

You **do not go** to those places in this stage, you just tuning to this frequency as receiver with the antenna that tunes in different radio wavelengths. In fact your **pineal gland** starts extending its receiving frequency range and your consciousness can perceive higher realities.

If you maintain your concentration and exclude a logical and analytical thinking, more and more pictures will start to appear on your mental screen. It's like a **watching a movie**, but this time the screen is in front your closed eyes. The more you practice to maintain your consciousness in this frequency, the pictures become more and more clear, more colourful and alive, more and more vital with a sound, and after a while you get a sense that your vision is extending more then your mental screen.

Regarding your question, this time you are close to projecting to that place, and if your are skilled enough you can magnetically pull this picture to you and you can force yourself out through the picture, with intention to penetrate this place in the **etheric** or **astral dimension**.

Quote:

Is there a way I can slow down these images? Or even try and determine where I amviewing?

Yap, no point doing this if you can not control it. You can **slow** them with the power of your consciousness. You have to use will, to do this. In the beginning because the consciousness its not use to, it looses those pictures as soon as the concentration break down and you start to analyze them using word thinking.

The mental screen closes immediately; it becomes black again, because the pineal blocks the extended receiving frequency range. In fact the **pineal** stops producing **DMT compound** a necessary chemical compound responsible for **ESP,** as a result of your consciousness switching back to its nominal beta brain patterns (**analytical thinking in words**).

Now, to try to determine where you remote viewing, you have to pull this off:

While the basic observation process is on, and the pictures come and go in your mental screen, you have to concentrate to a place you want to visit. Not with words but with using simple visualisation or focus. **In fact, I'm talking about awareness inside awareness.** It may sounds like difficult to comprehend, but **its not**, when you get some practice with this, you will perfectly know what I'm talking about.

In basic, while you are observing those images on your mental screen, give command with your consciousness (not with words), or express a clear desire what do you want to see or visit.

For example, you are observing places of your city, you see a motion pictures of building, streets, people walking around, traffic jam, sounds of vehicles, etc, you want to skip this concrete reality, you just want (the desire has to be honest) to observe a lake or some similar landscape, and those images of concrete reality are immediately changed with another motion pictures of beautiful lake or the desired landscape.

That's also not so advanced stage. This mental control is still weak. The more practice and time you put into this tuning of the planetary psionic field, your consciousness is getting more and more use to this new frequency and the observation becomes more precise. Very soon on your mental screen the precise desired location will appear.

The more advance level provides tuning to the observation of the exact place of the planetary grid of Earth's Psionic field so precise, that is so attractive to the military, and its has to do a lot with marking the exact coordinates.

Anyway if you get this far, you will discover all this by default. The experience will be your teacher.

If you want to explore deeper, you can penetrate to the astral dimension through some desired image on your mental screen, which in most cases will trigger a kundalini rising accompanied by a strong zooming sound, but this time to move through this new density you will have to your another vehicle—known as astral body.

In other word, the Remote Viewing process stops, and you are there on that place completely, in one of your subtle bodies. What then it's a whole science.

But my point is that you can find out about your self, what is going on, and what has to be done, lot more that you will find out from any medium that exists on Earth. It's like an internet connection with the Universe. And I can assure you that the things look much different from there . . . I hope some of this data will be of some use to you.

Regards,
Astralwalker

Thank you AstralWalker, this helps greatly.

Quote: Not with words but with using simple visualisation or focus. In fact, I'm talking about awareness inside awareness. It may sounds like difficult to comprehend, but its not, when you get some practice with this, you will perfectly know what I'm talking about.

Awareness inside awareness. This immediately brought the thought of when you are asleep and dreaming, then you realise you are dreaming. Usually you wake up at that point or the dream ends but on a couple of occasions you get a chance to do some stuff that would not usually be possible.

Do you know where we are at that point? When we know we are dreaming?

Thanks
Best Regards
Iain

CONNECTION BETWEEN A.P. AND DREAMS

Great emotional outbalance when dreaming, can easily change the structure of a dream itself. In most cases, this automatically leads the consciousness of the dreamer to an instant conclusion that he/she is dreaming. When this happens, the person is only a few moments aware of being in the middle of a dream.

During these few moments, one has a chance to take an advantage of the situation and to awake on the astral plane because while dreaming one's soul is already in his or hers astral body. So, let us see what happens during these few moments while one becomes aware that he or she is dreaming:

—Instant physical awakening (If his or her consciousness is not trained and very weak)

—The dream is continuing as if nothing has happened and the awareness that he or she is dreaming is fading away (If his or her consciousness is not trained and very weak)

—Biolocation of his or her consciousness: A part of his or her consciousness is still in the dream and a part of it returns to the physical body. If one manages to stay awake somewhere in the middle, he/she will become conscious in the astral body that in most cases is very close to his/her physical body. (Fully conscious state that can usually be induced if he or she is trained, but it can happen to him or her naturally because of the momentary strong and vital consciousness)

—Disability of physical moving through which one's whole body is paralyzed but the consciousness is fully awake: The whole process is followed by a strong zooming sound as a result of the Kundalini energy rising. Although it is a creepy feeling, if the consciousness is not in panic it can easily slip out of the physical body. (Fully conscious state that can usually be induced if he or she is trained, but it can happen to him or her naturally because of the momentary strong and vital consciousness or as a result of some extraordinary circumstances on the astral plane near his or her physical body)

—Instant awakening on an astral plane: Let us suppose that someone is dreaming of falling down from a very high building. In the moment of the falling down, one becomes aware that he or she is dreaming and that in fact he/she is on the astral plane, and cannot be hurt. As soon as that happens, one fully awakes in the astral body, the falling stops and he or she is free to explore the astral plane. (Fully conscious state that can usually be induced if the individual is trained and very rarely happens spontaneously as a result of the instant astral awakening caused by strange circumstances in the dream)

All of the above mentioned states of consciousness can happen to us during our dreaming time. Unfortunately, it is a big question whether our brain is programmed or not to remember such experiences. In fact, the achievement of the astral projection is not so hard, compared to the difficulty of remembering the things we have seen and heard while being conscious on the astral plane.

From one point of view, the memory we bring from the astral world can be compared to the memory of our dreams. The moment we awake physically, we remember the whole or at least the biggest part of the dream, but as minutes go by, the dream starts fading away from our memory until it is completely gone. That is why keeping the diary is so necessary.

Our brain is usually not programmed to keep and archive the data that we bring from the astral plane. When we dream we are already on the astral plane and when we wake up physically, the brain recognizes the experiences we have collected from there, as not important, not real, etc. Classified as unimportant data, the brain erases them from our active memory.

Anyway, right exercises and keeping the diary can help us reprogram our brain to start registering these "out of the body" experiences or at least the events happening to us in our dreams on the same level it does with the experiences that happen to us in the physical reality.

This reprogramming of our brain is very important for a student who practices astral projection, because there are many methods for achieving the astral projection through the control of the dreams. Most of them are based on the programmed awareness about detecting the small anomalies that can be seen in a simple dream. Therefore, I will give you two simple dream control methods I have used so many times and I can freely say that theywork almost always.

The first method tracks for some anomalies in your dream. For example: you dream you are walking in your room, but something is not right. The television is upside down, and one of the windows is broken. These anomalies are the main switches that can awake you on the astral plane because as soon as your consciousness detects them it will become clear to you that something is not right and that you are probably dreaming. In that instant, everything will change and some of the states I have mentioned above will happen to you.

So, in those few moments that offer you a chance to awake in the astral world fully conscious, all you have to do is this—say to yourself: "I am (your name), I'm dreaming and I want to awake in the astral plane". If you do this, in those few moments of awareness you will awake in your astral body and from there, you will be free to take an astral adventure.

The American shamans have developed the second method, which can awake your dreaming consciousness. It consists of making some effort to see your own hand while dreaming. When you manage to do that, you will become aware that you are dreaming, and during those few moments in which the astral portal is still open, you will have to put just some small effort to stay in the astral world. Where do you go from there, is your own choice.

Take care,
Astralwalker

Quote:

Originally Posted by **Astralwalker**

Despite ridiculed by whole world this 1977 Vrillon message is authentic.

The message is authentic, and its your choice to believe is it genuine or not. There are many scientific and technical data that can be brought out to

*prove that hijacking five major UK transmitters in fashion that was done, is
extremely hard technological achievement in 1977 and that was not done by
us but from beings that have multidimensional flying crafts.*

The one place I have a concern, is that the "message" from Vrillon
was very non specific. I've always felt, if there ARE other entities wishing
to be in contact with us, they'd open a two-way continual dialog where
ANYONE could contact them, by phone, by radio, by email. If they're
advanced enough, that should be simple enough for them.

AstralWalker, I appreciate your post. I would be very interested in
knowing who you are by description . . . what occupation do you work
in, what education you've had and in what line of study, where do you
live (continent, general area by city). Could you also tell us how you
obtained all this information and where you got all the photos? Are they
your photos?—Kathy

Re: 2012 Nexus Event—Unknown Form of Energy comes our way!

Respect to you all,

Astralwalker very interesting post. enjoyable. found some of this in
the process of poring over the mountains of information.

250 acre 70 "string" cosmic antenna due to be complete by 2011 heart
of gold, snippet from discover magazine mentioning curious facts about
the makeup of earth's core and assumptions about planetary collisions.
use of silver and palladium nanoparticles in clothing to fight illness and
prevent ill effects of pollution. any thoughts?—Snodome

Respected Sno dome,

Thank you for the info. It's great.
I will share few thoughts with you about this issue, as soon as I can.

Kind regards,
Astralwalker

Respected Kathleen,

Thank you. I really appreciate.

It's not difficult to see that the increased flow of **Aetheric energy** into our solar system is changing the **DNA structure** of **life**.

We are all experiencing some form of awakening our DNA.

The more **pure emotions** a being can manifest the more it becomes the DNA starts to become **functional**.

The gate for allowing higher streams of this **Aetheric energy** to come to us, is to stop the emotional frequency we call fear, and to be able to sustain more and more into emotional frequency we call peace, love and friendship.

Put in other words, to achieve an equilibrium and harmony with the cosmic that is in every each of us. To be in the balance with ourselves and with the world around us, which leads to learning and understanding of the true nature of everything . . . \

In fact, it's huge Cosmos and the process of learning never stops. The more you learn the more you become aware how small you are and how big the Universe is.

It's a privilege to be here.

Let the Light shine upon you all,

Astralwalker

AstralWalker,

Here's what I think.

You've posted a long series of messages, with a large number of photographs. It is as if you were writing a book, or a magazine article. I was interested in knowing your background, your education and line of study. I also asked you privately where you got all the photos, and were they yours. Your answer to me was that the photos were NOT yours and that it didn't matter about all the rest.

I think it does matter. A good writer will give references to all things that don't originate with themselves, so that it doesn't have the appearance of plagiarism. You have a lot material, ideas, and photography, but you do not indicate on the specific items or photography as to where you got them.

If the ideas are yours alone, then we'll have to judge it as that.

This forum is accessible by the public. I think the moderator's need to advise you about copyright laws. Every photo on the internet, whether indicated or not, is copyrighted. Don't believe me? Check it out here. http://www.photosecrets.com/tips.copyright.html and http://www. photolaw.net/faq.html

Every one of the photographs that you are displaying should have a source reference listed.

I'd suggest you edit every one of your messages and 1) either remove the photography, or 2) obtain permission from the owner if necessary and cite the original source location. This will be the only way for any of us to verify the authenticity of what you write about.

Kathy T

Hmm . . . should I copy-right you for all that you have read here? Or for all that time and effort that take to create this to be understandable?

—Why do you come here?

—Do I ask some of you, who you are, what you do, where do you live, what's your number of shoes . . . do I study you under microscope?

—I hope you will have strength to look at all this with better eyes. Nothing that is posted here is for money or profit; it's for shearing the information. Information that is very important and useful to many.

So don't tune to that frequency . . . I risk a lot to put this on the net, because the data is lost on thousand places. If you do not solve the puzzle combining the peaces you do not see the big picture and what this reality is all about and what is coming. It's far from the true one, but it's something and its a start. We have to start figuring out this, because soon it will be too late.

Since we leave in the ocean of energy and thoughts, it's like 100 monkey syndrome. If one being figures out, through the DNA antenna, the rest in his/her species are getting the same download.

So please, let me do what I came to do . . .

Respect,
Astralwalker

AstralWalker, I agree with you that you are free and welcome to do what you came to do on the forum and please, let us enjoy you and what you offer as long as you want to share.

Thanks again for this wonderful thread,

Jenny. Project Avalon Moderator

Thanks Jenny. I really appreciate.

I have problem finding time, but as soon as I can I will share some other data that I come up with.

Regards,
Astralwalker

Greetings to All in Peace

Astralwalker, Your Presentation is most Remarkable. Very well researched. I also appreciate the additional Links you provided, which do give greater insight.

Allow me to share one more Link here for You, and for all other members and readers here . . .: http://futureofmankind.co.uk/Billy_M . . . act_Report_229

Within the events recorded in these PREDICTIONS of the Prophet Jeremia, is found the following passage . . . :

"But it will be too late for fear, for an enormous destruction will rage, and the earth will be covered far and wide with desert. And there will also be mighty waters that become deeper and deeper, and at certain times and days, the waters will flow violently to such an extent that, like a deluge, everything will be swept away, destroyed and annihilated, and the lives of innumerable human beings will be claimed. Through the human beings' destructive rage, the air and sun will become poisonous and dangerous, and for this reason the bodies of the weak will be burned." {Jeremia speaks here of OUR Time} J Rod 7

Thanks guys. I really appreciate.

As soon as I can, will post some other thoughts and conclusions. I believe they can be useful too.

I love you all
Astralwalker

This is the best thread I've read to date on this site . . . . Thank you AstralWalker . . . . You have helped me overcome my blocks in accepting that leaving our 3d boddies in the 2012 catastrophes is not a thing to be feared. I have read much about the "Cosmic wave" in the Barbara Marceniek "Bringers of the Dawn /Earth Changes series of books . . . She refers to the "Wave" as the "Cosmic party" and sugests it a party we definatly want to go to. I had forgotten it of late and become obsessed with trying to susrvive the floods or the solar flares or the winds whatever but now I'm reasured that there is work to be done on many dimentional levels and that maybe being in astral or fifth dimentional form is a better place and an easier place to get to that "Cosmic party" . . . obviously I'm not contemplating suicide here . . . and I will do all I can to survive each catistrophe as it arives but if I have to leave . . . then I can get up to all sorts of exciting things over there too. I'm looking forward it all . . . What an adventure eh? I'm feeling the love . . .

Antonia . . . .

CLAIRVOYANCE

If someone is interested in 2012 RV and tuning into the Planetary Psionic Field, here is a good method for developing clairvoyance, increasing the power of the consciousness and utilization of the higher harmonics in the human DNA structure.

It will make the **Remote Viewing** and **Astral travel** completely **under control** which is essential for the **2012 scenario**. As soon as the **mental screen** opens, you will be **connected** to the **Universe** and from there "**input**" that concerns the purpose of your **incarnation** here, will **come**.

It will come as a **contact** with **higher intelligences** or with the **Gaya** itself.

So here it is . . .

Clairvoyance is a mental ability closely tied with the remote viewing. The human's third eye has the leading role since it enables the consciousness to see through the mental screen. There are many methods to develop the third eye. I have chosen to present the method I have mastered a long time ago and am completely sure it works. The method is universal and most of the psychic abilities can be developed through it. This method, which in fact is a system of exercises, consists of seven stages starting with the zener cards and as each stage is completed, the training gets harder and harder.

However, before I go any deeper into the subject of the psychic training, it is better to clear up some things first concerning the zener cards. The zener cards, also known as ESP cards, are the basis of the modern parapsychology. Although most people consider them just a useful tool for ESP testing, they are much more than that.

The use of the zener cards is also a powerful tool through which individuals can awake and develop their third eye with a special training. One deck of the zener cards consists of five symbols multiplied five times—star, waves, plus, circle and square. All of the 25 cards are totally black on one side and on the other side there is white background with bolded black colored symbol. In the beginning, the task of the student helped by an inner sense is to feel or see the symbol on the card. Because during the training the student is blindfold, while the sense of the student's physical sight is off, the other senses increase and sharpen more. However, the purpose of this universal method is not to sharpen the other senses, but to awake and develop a (figuratively speaking) "new sense", which contains the other five senses in it and much more.

It is completely irrelevant how you call this new sense—third eye, sixth sense or Ajna chakra. You will not make a mistake, because it is the same thing. Once this new sense starts to awake, most of the psychic powers can be achieved, cultivated and made ready for everyday use. The system of exercises to follow will lead you to a point where you will become the master or your mental potentials.

http://www.youtube.com/watch?v=uoKGO . . . tija_3886.html

The universal method is so complete that each of you willing to sacrifice about two or three years of your life can achieve higher mental powers and develop your psychic abilities beyond words. Nevertheless, although some authors promise you that you will develop your psychic powers in weeks by some specific method, I will not give promises like that and I will give you only the time that took me to develop them. It must become perfectly clear to every true student that it takes major self-discipline, top determination and persistence to keep up with the everyday hard training.

STAGE ONE

As for almost everything else in psychic practice, the student must find a quiet and isolated place to be able to work undisturbed with an assistant. It must be done in pairs because the student must be blindfold all the time during the training, and will need somebody to put the zener cards behind him/her and inform the student about the accuracy of the answers.

The whole process in this stage works like this:

The blindfold student must sit on a chair and the assistant should sit two meters behind him and put the cards one by one on a smaller table. In the beginning, there must be some clear space between the student and the table where the zener cards with the symbol up will be put.

Since it is going to take about two to three hours of practicing, some instrumental music cannot do any harm to the training process and it will provide a smoother training. However, it is important that the music is with relatively slow rhythm and pleasant for hearing. Some sounds from the ambient or meditative music enjoyable to both of them will do just fine. However, the quiet and slow instrumental music can bring only benefit to the training and if somebody likes to practice in silence, it is perfectly OK, and can work in silence.

The assistant should reshuffle the deck well, and put the cards one by one. The table, except for the only one card, which has to be put in the middle point, must be completely clear of any other objects. For now, the side with the symbol of the card should be up because it will allow the student to feel or see it more easily. If during the process of practicing, the assistant forgets to remove the deck from the table, it will be a major mistake.

The rest of the cards in the deck, though turned on the side with the black background up, will surely influence the student's efforts to locate the symbol of the target card, which is in the middle of the table. For that reason, the assistant who can seat on a comfortable sofa should split the deck in three piles.

All those parts should be put on the sofa and by no excuse on the table where the target card is. The first pile should contain the wrong answers, the second the correct answers and the third one normally, the cards which remain to be put on the table.

When the mental scan of all cards of the deck is over, the assistant informs the student of the correct answers and notes the result in a notebook, which can be used for monitoring the student's progress.

Then, the assistant has to reshuffle the deck well and when that is done, the process goes on with the next deck of cards. The student is allowed only one answer, and the assistant (at least in this stage of the training) should confirm the result of the student's answer with a simple Yes/No, or True/False.

The black fold around the student's eyes should be neither too tight, nor too feeble. Student's clothes should be comfortable like tracksuit or so, to be able to sit in the chair for hours without feeling uncomfortably. Before the training starts, it is advisable to try to sit in one position on the chair that mostly suits the student's body and to "freeze" in that position. The student can put his/her arms on the knees or thighs, but he/she should try to hold the head, neck and spine in straight line as much as possible.

Until now, I have explained the basic rules of the training, and now I will explain the inner procedure to be performed in the consciousness. The inner process, which will be performed in the student's consciousness, requires great ability of visualization, concentration, and focus on the inner psychic force and by all means great ability of the consciousness to move out of the physical body.

Before the start with the first deck of zener cards, the student must calm the consciousness by entering the meditative state similar to the one I have mentioned in the "deep peace meditation". After having calmed his/her thoughts and reached inner peace, the student should concentrate on the inner energy, which flows through his/her body. When the student feels the streaming of pleasant bioenergy, he/she is ready for the training and can give the assistant a sign to put the first card on the table.

Then, the student must completely free his/her consciousness and to try to pull out the inner sense, which lies deep down in him/her to feel the objects that are around. The first thing the student must concentrate on is the streaming of the energy between him/her and the chair. The student has to feel the chair as a part of him/her the way a driver feels the car while driving it. Then, he/she should redirect the psychic feeling following the pattern:—between him/her and the carpet (if there is any); between him and the legs of the table; the whole table; the target card; and at the end between him/her and the card's symbol.

The secret lies in establishing some kind of emotional bridge between one's self and the symbol of the target card. The rest will come by itself. The establishing of this kind of an emotional bridge should be done very slowly and with no rush because it will produce contra effect if done otherwise. In other words, the student should not keep a feeling of impatience within since it will surely lead to a wrong answer. On the contrary, the correct answer comes as a result of patient psychic touch between the student and the symbol of the card.

In time and with practice, the student will surely reach the level of being capable to easily move his/her psychic touch onto the objects that are around. The inner feeling will develop greatly and it will become so sharp that sometimes the student will feel almost equal as a physical touch with the card. After the student reaches 70% of accuracy in the answers, he/she should repeat this score with the next five decks.

Note: Under no condition, no matter how well the training goes, it is not advisable for the student at this stage to start working with two cards at the same time. That would be a big mistake, which will cost the student much more than he/she can imagine. If the student tries to feel two or more cards in the same time, he/she will soon lose the psychic touch and will have to work very hard again to reach the same point of development.

In other words, the student will lose two months of training just in a few minutes and will have to practice even longer to reach the same level of accuracy.

Please, believe my words, because I have passed the whole training myself and encountered the same problem. The practice with more cards will follow in the next stages when the student is ready for it. Thus, the discipline of practicing is one of the primary objectives that will lead the student to progress.

When the student succeeds to reach the result of 70% accuracy in the previous five decks only by using the psychic touch, which usually takes two or three months to complete, he/she is ready for the next stage. Still, this is only the weaker manifestation of the third eye and now the student must work on the visual part of it, which is much harder to achieve.

STAGE TWO

In the second stage, training should be extended from three to four hours a day. After the completion of the first stage, average time to complete the second stage varies from four to six months.

The training starts exactly the same way like in the first stage, with the student's achievement of 70% accuracy only by using the inner sense of psychic touch. A new element, which the student has to add to the training, is focusing on the mental screen in a similar way previously described in the remote viewing process.

The student should become a passive observer and wait for the mental screen to become visible. When the level of concentration is reached and the mental screen is visible, the student has to change its location. That can be done by using mental efforts to pull the mental screen from the location in front of the student's closed eyes to the back of the head. Perhaps, in the beginning, the student will encounter some difficulty to achieve that, but with practice, he/she will succeed to hold the mental screen on the back of the head.

The best way is to visualize that the mental screen has become transparent, to look for the counters of the table and to try to catch the rectangle form of the zener card, which is in the middle of it. The most important thing for a student is not to forget to lose the psychic touch with the table behind even for a second.

Both processes should be done in parallel mode, and the mental screen should be frozen in the right direction by the guidance of the psychic touch. Next thing that the student should do is to try to observe through the mental screen towards the table where the card is. At the same time, the student has to free him/herself from all the other thoughts or emotions and to concentrate on the observation as best as possible.

With the intensive concentration, the contours of the table and the card, which is on it, will start to emerge from the depths of the student's mental screen. The student's consciousness must be completely stable and frozen on those very weak counters appearing on the mental screen on the back of his head.

The primary process in this stage will be similar to making a hole in the back of one's head. The stronger the mental pressure on the back of the head becomes the stronger and more visible the mental screen becomes. By mental pressure, I do not mean physical grip of the neck or the head. I mean psychic pressure which is psychical 100 %, and which is a natural result of the strength of the consciousness reached by previous training with the zener cards.

In other words, the student guided by the psychical bridge between him/her and the table behind must force the consciousness with enormous amount of mental strength to make a psychic fissure in the back of the head. The appearance of this psychic fissure is the first sign that the third eye of the student has started awakening.

This period of training is very important because the actual thing that is happening is creation of the fissure in the etheric matter and the third eye will be learning to observe through it. Anyway, the student will know when this fissure is present if he/she experiences these effects in the consciousness: Because of the major mental effort during focusing on observation through the mental screen in the direction of the table where the zener card is with the symbol up, one will soon feel that something has changed in his/her inner sight.

The darkness of the mental screen will be altered by some illumination. Then, as the mental pressure on the zener card starts to hit stronger, it will become gray. Next, the gray color of the mental screen will start to express some foggy characteristics. Stronger concentration with the mental force through this foggy background will make the dot with less density visible on the mental screen.

Finally, in this small dot something similar to a small fissure will start to appear. Everything around this small fissure will remain foggy, but some light will start to appear in it. If the student continues pushing the mental force through this small fissure, he/she will realize that he/she is seeing something. At first, one will see unclear contours but in time and with practice he/she will develop his/her psychic observation.

Once the psychic fissure appears, the consciousness of the student must not become weaker not even for a second, but to keep the fissure open all the time. The fissure can also appear in some other direction like table legs or so, and the student should put some mental efforts to relocate the fissure directly above the surface of the zener card. It is very important not to become over-exited from the newborn situation but to remain perfectly calm, stable and focused on the training.

In case the student becomes overwhelmed with excitement, starts to fool or jump around, the psychic fissure will close instantly, and the student will have to work twice harder to achieve the same level of development. Very soon after the student freezes the fissure above the zener card, the symbol on it will start to become visible.

However, in the beginning the images will be only black and white. With further concentration on observing through the fissure on the mental screen, the color will start to appear. It will be coming slowly, but it will come. The moment the color starts accompanying the psychic sight of the student is a good point from which it will not take long before the student manages to scan all of the 25 cards correctly.

Perhaps, in the beginning through the fissure, some anomalies will be visible like seeing a square, which in reality is a circle, but they will vanish in time. Under no circumstance, the student should try to extend the psychic fissure until he/she has reached the score of 25 correct answers with five decks in a row. If the result is 24, it is not good enough and one has to work until he/she reaches the top score of 25.

Note: Jumping into the next stage of the third's eye development, without previously achieved perfection is a major mistake. Because of that, the student must be patient and wait until the result shows full readiness for the next level.

STAGE THREE

This stage covers the gradual extension of the psychic fissure and observation improvement. The training time extends from four to five hours a day. It is very important for the student not to miss a single day of training because the painstakingly achieved level of development is still very sensitive and it can easily retreat. Average time for completing the third stage varies but usually it takes two weeks to a month.

Stage 3a

After the student has achieved the result of 100% of accuracy in minimum five decks watching through the tiny fissure, the assistant is allowed to put two cards at the same time in the middle of the table. As before, perfection is top priority and the student has a chance to no more than just a single answer.

Now, the assistant puts two cards on the table and the student has to put strong mental efforts to extend the psychic fissure for at least as much as to have both cards in his/her psychic sight. Soon, though maybe not at once, after the student succeeds to extend and freeze the psychic fissure, he/she will reach the result of 100% accuracy with two cards at the same time.

Because his/her inner sight will be constantly frozen on the surface of the table, the student will soon notice that from time to time he/she is able to catch the movement of the hands of the assistant. In contrast to the previous experience, now the student can watch strictly in color.

At this level, the fissure will allow the student to feel and watch in more subtle way. Although the third eye is developing very fast and the student is capable of feeling and seeing the next cards in the deck, it is advisable (at least in this stage) to concentrate only on the two target cards.

Sometimes, the psychic fissure will get foggy but with the right mental pressure, it will get clear again. When the student has managed to perfectly see the five decks with two cards, the assistant can add a card more and now the student has to observe three cards at the same time. As before, one has to use mental efforts to extend the psychic fissure even more, at least as much as to be able to have all three cards in the psychic sight.

Usually, this is the part of the stage the student achieves very easy, because now his/her third eye has become very active and the consciousness has only to use just some small mental effort to focus on the task and to solve it. The rule of five-deck accuracy with 100% applies here as well, and before achieving this result, one cannot go any further.

The moment the student has reached the needed result he/she is ready for nine cards at the same time. Now, the assistant has to put nine cards on the table in a formation of three rows with three cards. Both the student and the assistant have to agree and clearly determine the rules of the next task. Whether it is going to be from left to right of the row or the opposite, or the first row is going to be the row closest to the student or to the assistant is completely up to them.

Again, using some mental effort, the student has to extend the psychic fissure, for at least as much as to have all nine cards in his/her psychic sight. At this level of the stage, the student has made a larger psychic hole and is no longer experiencing an observation through the mental screen, which is on the back of his/her head, but is experiencing an experience as if watching directly above the table.

In a way, the student's consciousness has departed very close to the border side of the psychic fissure and the student is watching through it. Once the part with the nine cards is over, the student is ready for the fourth stage.

Stage 3b

STAGE FOUR

When the student is done with the nine cards at the same time, put in a formation of three rows with three cards, the assistant has to put all cards on the table. It has to be in a formation of five rows with five cards. Logically, bigger extension of the psychic fissure is required for the student to be able to see all of the 25 cards.

It is different now, since the student now gains the ability to move the psychic fissure on his/ her own will. In this stage, the psychic fissure does not retreat and because of that, the student is capable to freeze or unfreeze it wherever he/she likes in the room.

Soon, other effects will appear in the student's consciousness like hearing the assistant's thoughts, feeling of floating, all sorts of astral visions coming from the astral plane, etc. All these effects in this stage of development have to be put aside and the student has to focus only on the observation through the expanded psychic fissure.

When the student has finished with all of the 25 cards in a formation of five rows with five cards with at least five decks, the next task is three rows, which contain 10 cards in the outer rows and 5 in the inner one. To test the student's ability to perfectly see with his/her inner vision, the assistant should turn a few cards in different rows upside down.

The student has to locate the cards turned onto the other side and to tell which cards in which rows are with the black background up, with their symbols, of course. It will be a good checkpoint for both of them to get the real picture of the accomplished so far.

Watching through the black background of the zener card will not be difficult for the student at this stage. However, if it is, the student can solve the task only by using stronger mental pressure on the black surface of the card and the symbol will immediately become visible for him/her.

As the training advances, the assistant should test the student's clairvoyance more often by putting harder and harder combinations in the rows like five circles in one group in the same row, or lots of cards turned upside down, etc., and to ask the student: "What do you see now?"

In other words, the assistant has to put the hardest combination of all of the 25 cards, but now, the student can do amazing stuff and probably, he/she will solve every task that the assistant assigns very easy and very fast. Once they are done with the testing and the student does not miss any more, the assistant should go in another room and put all the zener cards on the floor.

To achieve the observation in the other room, the student has to move the psychic fissure through the wall, which separates the rooms and to freeze it above the floor where the zener cards are. When he/she successfully solves all 25 cards, the assistant should take a pen and point to different cards. Now, the student has to see in which row the card is, and normally, what is the symbol touched by the peak of the assistant's pen.

Stage 4

When all this is achieved, the student is ready for the next stage of the training. The time necessary for completion of the fifth stage varies from one to three weeks.

STAGE FIVE

In this stage of the training, the student continues to further develop the inner sight, and if everything goes, as it should, eventually, he/she will be capable to read with the physical eyes closed. To complete this stage it usually takes 7 to 9 days. Before one begins with this stage, he/she has to repeat all of the previous stages.

Since the student is now in great shape, he/she will finish them incredibly fast. I assume this so, because having reached this level of clairvoyance to repeat all of the four previous stages it will take the student less than 20 minutes. When that is done, the assistant should bring other deck of cards, which will contain all the letters of the alphabet. By size and shape, they have to be the same with the zener cards with the only difference that where the symbols of the zener cards have been, the letters stand on the letter cards now.

Just the same, as with the zener cards, the assistant has to reshuffle the letter cards well and to put one letter card in the middle of the table. The task of the student is to try to see the letter helped by the psychic fissure.

The procedure is just the same, as with the zener cards. When the student finishes with 100% accuracy with one letter card in minimum five decks, the assistant can put two cards and when that also is completed, they can start with 3 letter cards at the same time. When they reach this level of the fifth stage the assistant can combine some shorter words like day, sun, one, car, etc.

The student should not have any problems with the reading now, because his/her third eye is very active and the inner sight is very close to the clearness of the physical eyes. Then, the assistant has to combine some four or five lettered words and when that is done as well, to start with more complex and longer words. It is amazing for both of them to participate in live in this kind of event, and it is an unbreakable proof for presence of the pure mental power in the room.

The training continues with long and complex sentences, which will require other decks with letter cards because sometimes (for example) the letter E will repeat itself in the sentence more than 7 times. Very important to be mentioned is that the student's consciousness will surely be exposed to all sorts of phenomenal effects like direct hearing of every assistant's thought, many astral visions that come and go incredibly fast, the loss of the physical feeling, etc.

The student will have to restrain from exploring those effects, which come because of the high level of the third's eye function. Also, the student will experience a strong spinning feeling around the chair to the level that it will look like he/she is going to lose consciousness and fall into a complete blackness. The above normal mental power awaken by the third eye produces that effect and the student has to get used to it and to learn to control it.

Now the student's consciousness is very strong and the student will soon discover awareness of the processes that are running through his/her physical body never dreamed of before as possible to be sensed.

Stage 5

After the student manages to perfectly solve all the tasks assigned in the form of sentences to be read, the assistant should mix the two decks, the one with the zener cards and the one with the letter cards and put them on the table in many rows. This time all the cards have to be closed or with the black background up, and the next task for the student will be to see them all.

The assistant now has to take the notebook and record all the data the student tells. The successfully solved task of this kind points clearly that the student is ready for the next stage.

STAGE SIX

In the sixth stage, the student will experience something as never before. He/she will be able to see exactly the same as with his/her physical eyes and much more. The student will achieve what most people consider impossible in this very advanced stage.

Example:

Shaun Tan—a Mentalist was totally blindfolded when he demonstrated his skills. To ensure that he really cannot see, a thick black hood was placed over his entire head. He was blindfolded at the wheels of a Pajero and drove the 5km distance—from Dataran Merdeka to KLCC—in 15 minutes.

The mentalist, who undertook the fund-raising daredevil attempt to prove his power of "positive thinking" and raise funds for the National Council for the Blind Malaysia, was assisted by 40 traffic policemen. They guided him at every turn of the route, and half of them escorted him along the way.

Shaun arrived at KLCC around 9.30am and walked, still blindfolded, to the fourth floor, to thunderous applause. It was his first attempt and he cautioned others not to mimic him as "it is very dangerous". However, he said his achievement proved everyone has the power to overcome and achieve.

This event is just one of so many. This one happened at the end of January 1999. To be more precise, he drove through the heart of Kuala Lumpur on 24th January 1999 wearing a blindfold and with a hood over my head! Over 40 policemen were involved, with beautiful model and actress Angeline Tan in the passenger seat. "It was just like a normal drive!", she said.

This is not special. You can learn to do it too. They only thing that stops you is your belief system and a solid mental practice. There are at lest 200 well documented events about similar ESP performances in all parts of the world but all this is blocked and continue to be blocked by the Elite because they simply do not allow this to come to worlds leading media, or they simply ridicules those persons . . .

Anyway, lets go on . . .

However, before the student becomes capable of doing far greater extraordinary things, he/she will have to develop the inner sight even more by learning to observe deeper into the physical matter.

The stage begins with the assistant reshuffling the whole zener deck and putting it in the middle of the table. Now, the student is assigned another task and will have to see all of the 25 of the zener cards at the same time penetrating through the whole deck. To achieve that, the student will have to change the frequency of the inner sight and to visualize all 25 cards placed in the astral space one above another.

When the vision of the whole deck levitating in the dark astral space in front of him/her appears in the consciousness, the student will have to use some strong mental effort to deeply penetrate through the whole deck. He/she will have to tell the assistant the answers for all of the 25 cards starting from the first to the last one and the assistant should not touch the deck, not even once. The assistant will have to record the student's answers in the notebook and when all the 25 answers are given check the results.

It is relevant to be mentioned that this is a difficult task and in the beginning, the student will surely encounter some errors. That is because the student does not use psychic fissure but uses another frequency of perception instead. A little patience will enable the student to get used to this new way of seeing and he/she will surely learn how to do it. He/she can achieve the deeper scanning visualizing his/her astral hands removing and seeing one by one all the cards on the astral reflection of the physical deck, which is in the middle of the table behind.

In other words, the student will have to visualize him/herself astrally picking one card from the deck visible on the dark astral space in his/her consciousness.

Then, turning it upside down the student has to try to see the symbol if it is not already illuminated and to throw the card away. Then, he/she has to repeat the same procedure with the remaining 24 cards.

When the student reaches the level with 100% accuracy with the deep scanning in minimum five decks one after another, he/she is ready for the next thing to be mastered. Now, the assistant will have to reshuffle the deck again and to ask the student to tell which card is (for example) 19th, counting from the top or from the bottom card. To solve this task the student will have to reach again within his/her consciousness and to find the whole deck, which will be levitating in the astral space.

Then, the only remaining thing to do is to count and when he/she reaches the 19th card, the student will have to remove it from the deck and to see the symbol on it. It is as simple as that and there is no other secret for this task. The assistant will have to ask the student then about the symbols of the rest of the cards in the deck and when that is nicely done, the training continues with 2-5 cards at the same time. For example, the assistant can assign the student to tell the symbols on the 5th, 11th, 18th, 21st and the 23rd card.

Usually, when the student reaches this level of the sixth stage, the astral light will illuminate the symbols of the target cards and he/she will not have to count them one by one to solve the tasks. I have mentioned the way the procedure can be done anyway. The next step in this stage is deep scanning through the deck of letter cards.

The procedure is the same as with the deep scanning of the deck with the zener cards. Once this is also accomplished, the assistant mixes those two decks into one and the student has to see them all starting from the top to the bottom card. This is a very good exercise or should I better say test, and when it is completed the student is in a real good psychic shape.

The next level of the sixth stage is mastering the zoom ability of the student's inner sight. Including a book into the training should do the trick, so the assistant should put it in the middle of the table and open it on some page. Then, the assistant will have to choose one sentence from the page and to encircle it with a pen. The task of the student is to move the inner sight through the psychic fissure above the surface of the book and to start to zoom it until he/she reaches clearness to read from the page. To pull this off, the student will have to work and train hard because maximum concentration and full control of the psychic fissure is required. The student will have to find out the number of the page of the book and to read the sentence circled with the pen.

When the student achieves this, he/she is capable of seeing so much clearer and better even to pull this off when the task of reading is in other rooms. In this stage, the student has to tear apart the psychic fissure by using the maximum strength of the consciousness. In other words, to do that, the student will have to use ultimate mental efforts to extend the psychic fissure to infinity until he/she experiences that the psychic fissure is starting to tear apart. Furthermore, he/she has to put final pressure until the fissure is torn apart completely and until the student gains an ability of seeing exactly the same as with his/her physical eyes.

Once the psychic fissure is torn apart, the student will see completely normal no matter to where the student points his/her head. It is an amazing experience and only by using certain mental efforts, this new inner sight will be accompanied with other supernatural abilities.

This is a very high achievement of the student and his/her third eye is awake. From now on, the real effects of the astral force will start to appear and the student will have to learn to master them all.

STAGE SEVEN

The seventh stage is the most interesting. Further, the training continues with the student's practice to project him/herself into his/her astral body and to perform an observation in other locations. In this stage, the student needs two assistants to give the assignments, one to stay in the same room with him/her and one situated in another apartment.

The first thing that the assistant in the other apartment should do is to take one photo and put it on a table with the image up. Now that the psychic fissure is torn apart and the student watches with the consciousness just the same he/she would observe with physical eyes the student faces another task. He/she has to leave the physical body and to take a short astral trip to the other apartment with the purpose to see the photo that is on the table.

Then, he/she has to return to the physical body and to tell the first assistant all the data collected from the out of the body experience. Both assistants should be in touch by cellphone all the time during the seventh stage to exchange the data and clarify the accuracy of the student's observations.

The best part is that the time has come for the student to explore all of the psychic effects that have been coming into his/her consciousness a long time. Because of the great strength of his/her consciousness, the student is capable of leaving the physical body in less than five seconds. It is enough for him or her to use some mental effort to astrally project at least two meters up front. Then if one turns around he/she will find him/herself sitting on the chair blindfold.

As soon as the student does that, the perspective of the inner observation changes, he/she will lose the physical sense and will find him/herself in a new body standing two meters in front of him/herself. If he/she looks upon him/herself, the student will see that he/she is in a transparent body, which is a complete copy of the physical carrier.

The student will also notice the tiny white line, which is representing the human contour and some smaller or larger blue illumination around it. The whole room will be perfectly visible to his/her astral vision and he/she will soon discover that he/she can see much better and deeper than being ever able with his/her physical eyes, of course if that is what the individual wants.

Perhaps in the beginning, the student will experience biolocation because of not being experienced enough to hold his/her consciousness on the astral plane but in time one will learn to control the new, immaterial body. Now is the right time for the student to learn to use the astral body, starting with a strong command to leave the apartment. As soon as he/she does that, the student will realize that his/her astral body reacts on even tiniest wish or command.

To reach the apartment where the second assistant is, the student can concentrate to find him/herself directly in front of the table where the task is and his/her astral body will teleport there immediately. It is also possible to do the same by the standard way of penetrating through walls, entering into other apartments, going out and flying until reaching the target apartment, etc.

However, after the student has reached the target apartment, he/she has to look around and walk to the table where the photo has been placed. It is appropriate to try to remember as many details as he/she can about the apartment and all the things seen on the way there and back, before he/she returns to the physical body.

After the student regains the physical feeling of sitting on the chair again, he/she has to describe the photo to the first assistant with all the data noticed during the astral excursion. Then, the assistants have to compare all the data from the student's statement by their mobile phones. When the second assistant confirms the accuracy of the data in the student's statement, the first assistant is writing down the result in the notebook or a laptop.

The student has to repeat the same procedure by reading a sentence from some book, which will be put on the table of another apartment. The reading should not be difficult for the astral vision of the student because of the previous experience with it. However, it can take some time to master this to perfection. Then, the tasks will become increasingly difficult and chosen on the second assistant's free choice. For example, the assistant can put some thing in a closed box somewhere in the apartment and the student using the astral body has to find where the box is and what is in it. All the results will have to be constantly monitored.

A good exercise for the astral development of the student would be to work with both assistants at the same time. Reading 10 digit numbers, one that will be in the same room in front of the first assistant and one that will be written on a sheet of paper in front of the second assistant is the content of this exercise. The best test for the limits of the student's power will be to solve tasks that are very distant from his/her physical body by taking astral trips to other countries if he/she has some friend to assist him/her with the training.

Note: When the student has reached the level to read with his/her physical eyes closed and by using only the inner eye, the student will have to develop his/her mental skills more and more. In time and with practice these abilities will increase and the effects of the mental power will be beyond average human understanding.

http://www.youtube.com/watch?v=iUD2W . . . eature=related
http://www.youtube.com/watch?v=9pdpy . . . eature=related
http://www.metacafe.com/watch/1061954//

As far as clairvoyance is concerned, I will have to note that it is not complete until the third eye is fully open and active. However, now the very advanced student who has mastered all these seven stages has also made tremendous step forward, but a long and unknown path still has to be walked ahead.

I have presented all seven stages designed for the awakening and developing of the third eye. In this particular case it is the clairvoyance ability through the step by step opening of the psychic eye from which the sixth and the seventh are the most appropriate for developing many other psychic abilities like telepathy, psychometry, psychokinesis, etc.
http://english.pravda.ru/topic/Boriska-30
http://farshores.org/pmag2.htm
http://www.harryprice.co.uk/Famous%2 . . . harryprice.htm

Presently, our solar system is gaining powerful energy from outside, and that outer energy is changing entirely the Matrix structure of this region. And I mean on every level and on every harmonic. This change as was previously presented will finish on 28 March 2013. So probably this method that took me long time to complete, under this new conditions that we are exposed to, and the obvious DNA awakening on a global scale . . . it will probably be achieved much, much faster. I hope that all this that was posted, will be useful to someone.

Best regards,
Astralwalker

Hollow Planets

Respected Sno dome,

Still having difficulties finding time. I'm almost done here.

Here is some material that can trigger some thoughts. The data is not filtered and analyzed. But I will try to do that as soon as I can.

. . . . Under Construction . . . .

Saturn's Polar Aurora

This image of the northern polar region of Saturn shows both the aurora and underlying atmosphere, seen at two different wavelengths of infrared light as captured by NASA's Cassini spacecraft. This image is a composite captured with Cassini's visual and infrared mapping spectrometer. The aurora image was taken in the near-infrared on Nov.

10, 2006, from a distance of 1,061,000 kilometers (659,000 miles), with a phase angle of 157 degrees and a sub-spacecraft planetocentric latitude of 52 degrees north. The image of the clouds was obtained by Cassini on June 15, 2008, from a distance of 602,000 kilometers (374,000 miles) and a sub-spacecraft planetocentric latitude of 73 degrees north.

Jupiter Bright, star-like beacon that sends rays streaming from the exact center of Jupiter's north pole in the midst of the circling polar winds.

Cloud formations on earth's north pole analyzed by NASA, leaving an unexplained dark hole . . .

This image shows newly discovered "hot spot" on Saturn's north pole and the mysterious hexagon that encircles the pole. Image credit: NASA/ JPL/GSFC/Oxford University While scientists already knew about the hot spot at Saturn's south pole from previous observations by the W. M. Keck Observatory in Hawaii, the north pole vortex was a surprise.

These are conclusions from respected Sepp Hasslberger:

The concept of a "hollow earth" has been around for centuries. Today, it is debunked as a mere scientific curiosity, as in this article by John H. Lienhard, from which also comes the following illustration by William Reed published in 1906.
http://blog.hasslberger.com/2007/12/ . . . e_whirlpo.html

Is there any evidence for this and why haven't we seen it?

If this line of thought is to turn out correct, we should be finding evidence that planets and stars are hollow shells that possess polar openings and a luminous point at the center.

There are some strong indications that this may indeed be so, but it seems that data is not freely released. Where some evidence does become available, it is explained away or simply not discussed—it does not fit the prevailing view.

We should see polar openings leading to the hollow interior of our own and other planets, only we don't—at least normally. Very few published images actually show the poles directly.

Many of those present signs of having been manipulated. Try and find images that show the earth's north pole. You will see a blob of "white-out" or the electronic equivalent of it—a white spot that has been airbrushed into the picture.

Here is an example of an image of the north pole, where a large portion of the polar area is covered by . . . ice you might say, but it really isn't ice as you can see by comparing the completely white area of the polar ice with the much more structured and darker color ice on adjacent Greenland.

I realize that there will be strong feelings both for and against, but I am bringing this forth as a matter for consideration, and in the firm expectation that time will prove that I am not out on a limb here.

1) both planets and stars show variously sized openings at the poles
2) a central luminous feature will be visible when looking straight at the polar opening of a planet and reflections of this "internal light source" will cause luminous phenomena visible under certain conditions to observers who are not aligned with the planet's axis.
3) Planets are habitable (hospitable for life) not only on the outside but also on the inside of their "shell".
4) More heat is radiated by planets than is received from the outside.
5) As the planetary vortex continually collects and brings in more particles of matter, planetary shells will be found to be slowly expanding.

. . . .

Mars pictured with South pole either cut off (first four images) or covered by a large circular black patch in the last image

Another image of Mars showing a polar feature that is said to be ice, but may well be a polar opening . . .

Earth

The Eskimos say the Vikings had migrated further and further north, then one day their men found a paradise in the north—a place the Eskimo had always known about but stayed away from because they believed it to be inhabited by evil spirits.

The Viking explorer parties had came back and had told the rest of their Greenland colony of their wonderful discovery. All promptly packed their bags, and singing songs, departed suddenly northward and never returned. The eskimo tradition is that over the ice towards the northwest, in the direction

Admiral Peary sighted Crocker land and Cook sighted Bradley land, is a . . . "land that is warm; is clothed in summer verdure the year around;

is populated by fat caribou and muskox. It lies," they say even to this day, "in the direction of the coastal trail-route north."

Lt. Green shows that trail on his map. It is located on west side of Greenland, and goes up around Ellesmere Island, and out over the pack ice in a northwest direction towards the land he claimed exists in the "Unexplored Area." That is the same area I have estimate the North Polar Opening is located.

Jan Lamprecht has a map on page 193 of his book, HOLLOW PLANETS, showing the location and directions of sightings of mirages of anomalous lands in the arctic, which are not found on any current day maps. Taking these sightings and triangulating them towards the points towards which they were observed and from the locations in which they were sighted, gives us another indication where the center of the North Polar Opening most likely is located.

North from the New Siberian Islands the Russians sighted the mirage of Sannikov land. Northwest of Ellesmere Island was the sighting of Crocker land by Admiral Peary, Lt. Green and MacMillan. Bradley land was also sighted in that same general direction by Polar explorer Dr. Cook on his way to discover the pole. From Alaska, from Harrison Bay, was sighted land also towards the north west by Captain Keenan. If we draw a line from each of these locations towards their respective directions in which anomalous land was sighted, we arrive once again at the location I have determined must be the location of the North Polar

Opening centered on 141 E Longitude, 84.4 deg N Latitude.

This is a comment from a new energy list I am copying here as very relevant to the subject:

The statement is I suppose referenced to the Sun. The basic assumption for the past about 80 years has been that the sun was doing thermonuclear fusion. Unfortunately there is absolutely no data to back this up. Here goes for the proofs.

(1)   If the sun is a thermonuclear fusion furnace as postulated it will produce the neutrino output associated with the reactions. (The have not been found. I know the excuses but fudging doesn't fix a theory!)

(2)   If the sun core is the source of solar heat, it will be hotter near the core and cooler as one extends outward. (In fact the sun gets hotter going out towards and past the photosphere. The temperature of the surface that is seen is about 5,000 C to 10,000 C. The corona

temperature reaches 2,000,000 C. The heat is going the wrong way folks!)

(3)   If the sun core is the source of solar heat, the "solar wind" will be fastest as it leaves the sun and slow down as the ionized material exits the sun and goes far into space. This is defined by the laws of motion and by the laws vapor pressures, ionization etc. (In fact the solar wind accelerates going outward to a great extent. The solar wind varies but has been observed to leave at about 10,000 km/hr and to pass earth going about 83,000 km/sec. The same event saw the material pass Saturn going nearly 150,000 km/sec. And yes I do know that the speed of light is 300,000 km/sec.)—That ain't possible by the fusion theory of the sun!

So we have 3 big undeniable violations of the process. But it gets deeper. The concept of the construction of a star by nebular condensation which follows the rules in (3) says that as soon as a star went critical on its fusion reactions, the result would blow the star to bits. We would never see anything but a trivial emission from a star because it would stop collecting matter if it cranked up.

Now the spectrograph has actually told us quite well what the sun was for 250 years. It is in fact a body largely of Iron and Nickel with a thin Xenon atmosphere with hydrogen and helium in it. Thin here is relative to size.

The issues then come up. How do we power the sun and what evidence do we have of this power and then what are the results of it.

If you divert to the FARADAY basic design for a generator you find that you have the basic prototype for any body in space that is spinning. It generates current centered on a mid latitude band near the equator and at both poles. These currents will enter the disc as vortex structures of plasma. This is exactly what we have seen with the Stereo B mission to the Sun. It is exactly what we see with the aurora on earth and it has been observed on the sun as the concentration of the opposite charge appears on the polar zone also somewhat separate of the pole. This is typical plasma physics. The only problem with the Faraday design is that it provides a massive torque load on the spinning body. However; De Palma solved this in the 1950's. If the magnetic field bias is allowed to float with the spinning disk, you get all the current and none of the torque loading. So now you can get the energy of a spinning earth, sun or whatever without a torque stopping the spin. This is the observed Storms and Aurora on the earth and is the observed sun spots and polar aurora on the sun.

It has been observed on Saturn, Jupiter and Uranus.

This predicts the flow patterns observed and temperatures observed. The surface of the sun would be much cooler as it would actually have little current passing it. The exterior would be hotter and the solar wind would sustain increasing velocity so long as it is connected to the disc. Similarly the earth would have such high altitude heating and ionization. It does!

The sun and all bodies in the universe operate the same way and this is standard electrical physics and plasma physics industrial rules. This explains a lot and doesn't need the funky physics and dark matter etc that says we cannot see what is going on.

Now there is only one problem with this prediction set. According to the Gravity theories of the Universe, the high mass of the sun predicted by the intensity of the discharges on it and the mass predicted by the spectrograph demands a mass of the sun be about 100,000 times that observed. This would seem to be a theory killer until you run into the work which has been experimentally demonstrated and is published on CERN. This is the work of Fran De Aquino of Brazil. It turns out that these massive plasma flows in space gravity shield the mass of the sun from us. As a result the sun can and probably does have 100,000 times the observed mass by Gravity effects.

Thought you might like to know. There is absolutely no evidence to support the nuclear theory of stars. I know that this blows up nearly all the pet theories of cosmology and such.

But it opens the door to understanding the structure and operations of the universe. This is what Tesla called the Wheel Work of Nature.

One of the curious facts of Gravity shielding is that the inertial mass of an object is proportional to the G. That is if you shield to 1/100th G you have 1/100th inertial mass. The oddity here is well described on Fran De Aquino's website. If you read it, it tells how to build an interstellar space ship motor and power it. This is pretty simple stuff so don't get lost in his heavy math.

. . . .

Vortex as an 'accretion' force

I am proposing a different, more or less diametrically opposed view of star and galaxy formation to the theory held today. Stars are formed by vortex action. It is the spin that initiates a vortex that is the seed for star formation, not gravity.

What is today called an "accretion disk" is actually a sign of dissipation. The work of accretion in star formation is done by the concentrating forces of a double vortex. That vortex accretes interstellar gases like a giant whirlpool and it is vortex action that is causing the accumulation of matter—not gravity as generally assumed.

Spin is the causative force in forming any agglomeration of matter. It is the seed for the formation of galaxies, stars and planets. Spin forms a double vortex, which is responsible for concentrating matter that is finely distributed in the universe. Spin is also a telltale signature of the energy of life.

Spin stresses and distorts the fabric of space forming a helixactually a pair of vortices.

Those vortices, like two huge whirlpools, induce a flow of space and the matter contained in it, towards the center point of torsion, the star seed. At the point of impact of the two opposing in-flowing vortices, incoming matter collides and, in an explosive fireworks display that forms a plasma ball, is thrown outwards in roughly spherical symmetry.

At a certain distance out from this central point of impact—how far depends on the rotational strength of the star seed—the outward motion of the particles accreted by the vortices is now stopped by growing gravitational influence. The matter so collected starts to form a hollow shell of first gaseous and later solid matter. That shell is roughly spherical—with openings at the poles where the vortices are free to bring in further material.

What is generally called an "accretion disk" today is formed by matter that escapes along the shell's equator due to overwhelming centrifugal forces. In the case of a star, this becomes a protoplanetary disk, a flat equatorial disk of rotating matter that provides the material needed for the formation of planets. Planets coagulate around their own planetary seeds of spin, in a similar manner as the original star. A planetary vortex pair attracts matter from the protoplanetary disk to form a planetary shell. This hollow planetary structure coagulates around the point of equilibrium between centrifugal forces and gravity.

Instead of only gravitation, we have several forces at work in star and planet formation: the centripetal (concentrating and accelerating) action of vortex, the explosive, expanding action resulting from two opposing streams of matter violently impacting at the central point, gravitation which tends to stop and compact matter that tries to escape from the exploding center and finally centrifugal force which provides a counter to gravitation, and which is responsible for the formation of any "accretion disk" which is really a disk of matter that escaped gravity through overwhelming centrifugal force.

The firework at the point of impact of the two opposing vortices is a permanent feature, a source of light and other radiation located inside both planets and stars. In planets, it remains a hidden feature only occasionally giving rise to a halo of light we see as an aurora around the polar openings. In stars, that firework is what determines a star's luminosity.

The postulated fusion reaction of hydrogen into helium is not what drives heat and luminosity of stars. It may be a secondary reaction to the real source of the star's luminosity, which is the violent impact of two vortices at the center of the star.

. . . . Under Construction . . . .

An image of the earth's north pole where a large area appears "airbrushed" in pristine white—to cover up . . . what?

The Cassini spacecraft has sent back to earth an interesting image of Saturn's north polar aurora. According to the Hollow Planet theory, auroras on all planets are caused by the inner suns within each planet emanating their solar winds out through their polar openings. Until scientists realize that all planets are hollow with interior suns, they will be mystified by these auroral lights on other planets. Read their report here.

There are four reasons that the auroras cannot be caused by our outer sun. First, the solar wind from our outer sun is not powerful enough to light up the auroras on any planet including earth. Second, the magnetic fields of each planet prevents the outer sun solar wind from entering the atmosphere and lighting up the auroras.

Third, the auroral radiation has been observed to emanate FROM the planet following the electromagnetic field lines of the planet AWAY from the planet. Fourth, variations in the auroral displays occur at both poles at the same time strongly indicating that the solar radiation causing the auroral displays originates from the planet's core.

(Ed Note: The Cassini image included by Astralwalker on Project Avalon would not copy to this document).

. . . .

Mars

Map showing the thickness of Mars' south polar layered ice deposits(purple represents the thinnest areas; red the thickest). The dark circle is an area where data could not be collected.

Regards,
Astralwalker

Thanks Oneness. I really appreciate.
http://www.youtube.com/watch?v=0wE5J . . . t=44268&page=5

Respect,
Astralwalker

Nexus Part Two

More Crop Circles

Hi Marian-Librarian,

Thanks for the link. I really appreciate.

We had been monitoring this site from the very beginning. It's a nice website.

You can also check the crop formation archive: http://www.x-cosmos.it/cropcircles/

Regards,
Astralwalker

Hi Tez,

Thank you for the link. I really appreciate.

I had analyzed the data and my opinion is this:

I have been monitoring the scientific results that comes Russian Academy of Science, especially the ones from Russian physicists Alexej Dimitrijev who was in charge of a project concerning changes in the all planets of our solar system and 2012 scenario.

Also I know about the bio-chemical research of Dr. W. C. Levengood concerning the crop samples inside the crop formation.

I totally agree with the magnetic anomaly present inside the formation.

I agree with laser beaming the DNA sample, mapping the Junk DNA and founding language patterns and the results that they come up with.

But I disagree with these conclusions:

According to Fosar and Bludorf therefore this natural energy is brought about by hypercommunication via magnetic wormholes in the DNA. The crop circle is a group consciousness phenomenon.

The geometrical and philosophically significant shapes of the crop circles can be explained, as pointed out by Fosar and Bludorf, in terms of Jungian archetypes in the collective unconscious.

This also of course explains a very essential element of crop circle phenomena, namely that they seem to interact with the psyche of the human observers.

———————————————

—The crop circles are mainly made by extraterrestrial intelligence capable of multidimensional travel. Despite that the messages have hypercommunication nature they are not made by human collective unconscious but from intelligent beings.

—If they are materializing via thought manifestation, then how come they send helicopters and jet to chaise them away. And they are traceable on the radar.

If you notice, they are visible on infrared before they enter the atmosphere and because of the Star Wars weapons targeting them, they dematerialize and vanish from the infrared which means they disappear from the physical-etherical plane to astral or higher and when they passed this line of "defence" they materialize above the fields when they again become traceable for radar.

They usually use etheric portals which are marked with large stone horses on the ground (I'm speaking mostly UK).

Other thing is . . . from time to time I receive this download too . . . so I can assure you that are solid and not a product of archetypes in the collective unconscious.

http://www.youtube.com/watch?v=cM2_dcH_NNA
http://www.youtube.com/watch?v=X5bG0UVVEEk
http://www.youtube.com/watch?v=e0jpU . . . eature=related

http://www.youtube.com/watch?v=kdeJr . . . eature=related
http://www.youtube.com/watch?v=ATd3-2sslbs&NR=1
http://www.youtube.com/watch?v=d1h_0ZWLECs&NR=1

Namaste!!!

As you are so knowledgeable in astronomy, my question is have you heard of Coral Castle in Homestead, Florida?? The builder of this was from Russia and built this using levitation.

Many are wondering what multiple reasons it was built. Many feel that he hid Tessla's coil in shape of walls.(To be used as a free energy capacitor) Pls see this site: www.code144.com by Jeremy Stride

Then I came across another clip of Jack schmidt from apollo 17 flight to moon and shows shapes of a Hedgestone and a similar structure to Coral castle!!! (On moon surface!!!)
http://www.youtube.com/watch?v=q7enBCfQ1uc

If there is more to this than a free energy capacitor, as he has alignments in stone, pls let me know!!! I feel he knew (ED Leedskolnin) that there was another structure on the moon. It was not randomn sculptures!!!

I am awaiting what you may think!!

Blessings!!

Aloha Astral Walker,

I've spent quite a few hours going through your thread as I was so impressed by your research data that I wanted to download it and give it to my husband as a Christmas gift.

Subsequently, I've also spent many hours editing it down into a readable format. All 157 pages so far.

I've also been going through the pictures collecting the original links to post under the picture for reference and can easily do that here if you wish.

After reading all of this data my question is this . . . how do you know it is a blue white light eminating out from the galactic center on 13Dec2012 and that one would be safe in the open with hopes of the light coming in contact with them. Given the angle of the planet during that time of year I would think only portions of the planet may be exposed to the galactice rays while others would not be exposed.

In addition, I'm also curious as to what happens upon contact. Does one become transfigured where the physical body gets to come along on the journey (only transformed), or is it dropped and the soul is free to soar into the astral plane while waiting for the glacitic dimensional shift to occur?

In fact, my curiosity is aroused as I wondered just where did this particular information come from?

I love science but also studied as a mystic years ago so am personally familiar with interdimensional travel. And during a few near death experiences, as my soul was beginning to flow out and away . . . it did seem to know where to go.

However, your information tends to conflict with other information I've read about portals. As I also have friends who are contactees, the information is very varied depending upon the type of contact experience they've each had.

As I've had personal conversations with a number of these people, my curiousity is stoked as to who is really giving correct info.

How do we know that this information with respect to the harvesting of souls is accuracte?

Again, I want to thank you and let you know how much I've appreciated what you have shared here as it is clearly evident to me this has taken many, many hours of research and is a labor of love.

Mahalo nui loa,
Carol

Aloha, thank you, do jeh, toda, arigato, merci, grazie, salamat po, gracias, tack, sukria, danke schoen, dank u, mahalo nui loa

Aloha Carol,

Thank you that you have use from the material. I really appreciate.

Quote:

After reading all of this data my question is this . . . how do you know it is a blue white light eminating out from the galactic center on

13 Dec 2012 and that one would be safe in the open with hopes of the light coming in contact with them.

Scientific research have shown that the core of the **active galaxy** can shine brighter then the **galaxy** itself. We are talking about the **cosmic rays electrons braking away** from the **galactic core**, close to the speed of light which produces a **bluish bright light**.

When this happens the bluish bright light is so bright that it starts to produce white light also.

The **Divine Creation** expresses its self in this holographic Universe mostly through light that can sustain the most high vibration. The most pure energy can stream in bluishwhite light. It's a colour of the **Creator** as **777** is his mathematical divine code.

Despite the fact that Divine Creation exist in all colours, its everywhere and in everything, other colours are only less or more, conductors for energy of Love.

For example **Cosmic Love** can not easily move through black and red colour because the vibration is not suitable, but can move more easily through yellow, light green, and the most though light blue and white.

You see, whoever created this animation can not feel from the hearth. The Sry Yantra is the depiction of everything we have to know about our Universe, but the person who created this, has put light flashes that do not resonate with peace and harmony but it disturbs you.

On the other hand, this simple image emanates with much calmer and balanced energies and resonates with your inner divine.

The observations of so many other galaxies through the **Hubble Telescope** and **Chandra Xray Observatory**, we can see that bright bluish light from the active galactic cores.

On the spiritual side, the **Hopi Indians** spiked about blue star in sky that shines so bright that illuminates everything and that was visible on the sky for a long time.

As the results of long years existing research of my friend Dr. Paul, what he calls a **Super Wave Event** and I'm calling it a **Nexus**, triggers extreme climate changes.

Dr. Paul is aware that Super Wave Event is coming but he has no evidence what date in the future they will arrive.

I on the other hand, went on different areas of research and I get different input from most of the people.

Quote:

After reading all of this data my question is this . . . how do you know it is a blue white light eminating out from the galactic center on 13 Dec 2012 and that one would be safe in the open with hopes of the light coming in contact with them.

—The pictograms had clearly shown that **on this date** from Earth we will see those beams in the sky coming from the direction of the center of Milky Way. Also on this day we will se a bright comet in the sky. How long it will take before they reach our Sun and this expands to the scale of the centar inner sun of Venus and projects a beams toward our planet I can not tell precisely. My guess is **21 Dec 2012** or **23-24 Dec 2012** like depicted in the crop circle formation.

—I do believe that we will have better input next crop circle about this event and lot more. Now people starts to look more openly to this messages and we are going somewhere. I believe that we will be ready on time and do what we have to do.

Quote:

After reading all of this data my question is this . . . how do you know it is a blue white light eminating out from the galactic center on 13Dec2012 **and that one would be safe in the open with hopes of the light coming in contact with them**.

—The **Ancient records** clearly stated that those who are one with the Creator, fears nothing, no man, no evil, no phenomena, no event etc.

—You are possibly right, it will be safer inside because climate change and other things will start to happen long before December 2012. Its posible (its only my opinion) many of us will experience unpleasant moments.

The Elite will try to reduce the population and they already started to. They are shooting at us with everything they got. From vaccines, aspartame, fluoride, electronic pollution, subliminal technologies, and everything you can and can not think off. So we have to build a balance to that force a grid of consciousness a new awareness that will connect humanity. Time is speeding up and everything starts to unroll much much faster and soon it will be obvious what I'm talking about. More and more people will start to receive the same download and what have we do. I will elaborate on this when some time off comes.

However, its smart that people start to preparing and exchange experience and ideas about eventual survival groups. I believe it will be a one or two save zones on every continent and the ETs will show the correct ones. Will people choose to believe them or not, its their choice and everyone has to reach deep down where the heart shines the most brighter and to feel what choice it points. In the times that are coming, our only compass will be our heart and our saviour will be combining into one stream of consciousness which aligns with the stream of the Universe.

Summarised, do what you feel that is right choice for you . . . in any case the Matrix will show you the way . . . that's how this info come to you . . .

Quote:

Given the angle of the planet during that time of year I would think only portions of the planet may be exposed to the galactic rays while others would not be exposed.

—The whole planet will be exposed on **Hunab Ku** bath through our **Sun**, so it will not make difference about which side you are on the planet. Besides if you look around you in the solar system you will notice that all the planets are getting brighter and brither and this will continue to a point where our solar system will become so bright that will be incredible. I know I can not back this up in the moment, but it will come.

Quote:

In addition, I'm also curious as to what happens upon contact. Does one become transfigured where the physical body gets to come along on the journey (only transformed), or is it dropped and the soul is free to soar into the astral plane while waiting for the galactic dimensional shift to occur?

—There are opinions that because of change in the Matrix, the physical vehicles will become half etheric which is probably true for ones that will stay on this 3D plane of existence. All the rest will fuse into Oneness and that's what makes all this so unique. You can feel it in your heart and you know it is true.

Quote:

In fact, my curiousity is aroused as I wondered just where did this particular information come from?

—Its coming from everywhere. The nature speak to us, the Gaya speaks to us, the changes of the solar system speak to us, higher intelligent speak to us, the Cosmos is speaking to us. But we have to tune into higher wavelength to be able to hear see and emotionally touch all this that I'm talking about. Some people are awaken and already connected with the Universe, and everyday more and more are awakening.

The Dark Side is powerful but it can not stop what is happening. And what is coming is remarkable. As I stated before, if you see with the analytical eye or yourself you see disasters as product of natural cycles and energy movement but if you see with you heart the things look different. We are much more then this. We belong to the Cosmos and its our true home. What will happen after the Nexus brings as back through the center of the Galaxy . . . I just do not know.

Quote:

However, your information tends to conflict with other information I've read about portals. As I also have friends who are contactees, the information is very varied depending upon the type of contact experience they've each had. As I've had personal conversations with a number of these people, my curiosity is stoked as to who is really giving correct info.

—I only share what I know and what I feel in my heart that is truth and nothing but the truth. What other people get as information I can not comment. Its their existence and everything they do its their path and I respect that. I'm telling what I concluded, what my star friends are telling me and that is what I'm passing.

Quote:

How do we know that this information with regard to the harvesting of souls is accurate?

—You don't. I'm just shearing my conclusions based on my experience and my knowledge and the download that I receive from my friends. But your heart knows what is the truth and what is not. You can not lie to your heart. It can feel what is truth and what is false. From my perspective, the most important thing in the moment is to return to yourself, and to connect to the **Planetary Psionic Field**. When this happens you will receive all the download you need.

But your heart has to be clean and you must be ready for it from inside. That means that you will start separating with this physical reality already. The more you dive into this Psionic Field the more a new higher world opens before you. You have to become a Fire of Love, Oneness with everything, a Power of Everything and ability to read what is known as hidden codes inside the Matrix.

And most of all you have to be moral and understand that everything that exist has the equal potential. Everyone's experience and knowledge is precious and contains a missing peace of the huge mosaic that we all are trying to assemble. We do not see this purpose but we feel it deep inside ourselves and we feel it the Divine Plan that is behind.

Everything in nature is in balance. The two primary forces in this Universe are Shadow and Light. We had enough from the Darkness and its time to move towards Light.

It is interesting, respected **Arthur C. Clarke** figured out long ago and presented in his novel "**Childhood's End**" this **oneness . . . a separation from the mater not as an individual but as a race.**

And that is what I believe awaits us in the time frame between 13 Dec 2012-28 March 2013.

Mahalo nui loa,
Astralwalker

Ed Leedskalnin—Rock Castle

Namaste

Respected **Samarkis**,

Thank you for your words my friend. I really appreciate.

About the case of **ED Leedskalnin** . . . I had read about his connection with levitation and psychokinetic influence on the rocks and all this is quite extraordinary.

Honestly it was a long day for me, and this subject deserves more analyzing so I will analyze the links you gave me about him and the moon connection tomorrow and I will be able to discus more about this issue.

Thanks for shearing this with me. Its very interesting. I will back to you as soon as possible.

Blessings!!!
Astralwalker

"AussieG says) I pirated this from David Icke's forum Your thoughs please A Giant Breach in Earth's Magnetic Field.

Dec. 16, 2008:

NASA's five THEMIS spacecraft have discovered a breach in Earth's magnetic field ten times larger than anything previously thought to exist. Solar wind can flow in through the opening to "load up" the magnetosphere for powerful geomagnetic storms. But the breach itself is not the biggest surprise. Researchers are even more amazed at the strange and unexpected way it forms, overturning long-held ideas of space physics.

"At first I didn't believe it," says THEMIS project scientist David Sibeck of the Goddard Space Flight Center. "This finding fundamentally alters our understanding of the solar windmagnetosphere interaction."

The magnetosphere is a bubble of magnetism that surrounds Earth and protects us from solar wind. Exploring the bubble is a key goal of the THEMIS mission, launched in February 2007.

The big discovery came on June 3, 2007, when the five probes serendipitously flew through the breach just as it was opening. Onboard sensors recorded a torrent of solar wind particles streaming into the magnetosphere, signaling an event of unexpected size and importance.

"The opening was huge—four times wider than Earth itself," says Wenhui Li, a space physicist at the University of New Hampshire who has been analyzing the data. Li's colleague Jimmy Raeder, also of New Hampshire, says "1027 particles per second were flowing into the magnetosphere—that's a 1 followed by 27 zeros. This kind of influx is an order of magnitude greater than what we thought was possible."

The event began with little warning when a gentle gust of solar wind delivered a bundle of magnetic fields from the Sun to Earth. Like an octopus wrapping its tentacles around a big clam, solar magnetic fields draped themselves around the magnetosphere and cracked it open. The cracking was accomplished by means of a process called "magnetic reconnection."

High above Earth's poles, solar and terrestrial magnetic fields linked up (reconnected) to form conduits for solar wind. Conduits over the Arctic and Antarctic quickly expanded; within minutes they overlapped over Earth's equator to create the biggest magnetic breach ever recorded by Earth-orbiting spacecraft.

The size of the breach took researchers by surprise. "We've seen things like this before," says Raeder, "but never on such a large scale. The entire day-side of the magnetosphere was open to the solar wind."

The circumstances were even more surprising. Space physicists have long believed that holes in Earth's magnetosphere open only in response to solar magnetic fields that point south. The great breach of June 2007, however, opened in response to a solar magnetic field that pointed north.

"To the lay person, this may sound like a quibble, but to a space physicist, it is almost seismic," says Sibeck. "When I tell my colleagues, most react with skepticism, as if I'm trying to convince them that the sun rises in the west."

Here is why they can't believe their ears: The solar wind presses against Earth's magnetosphere almost directly above the equator where our planet's magnetic field points north. Suppose a bundle of solar magnetism comes along, and it points north, too. The two fields should reinforce one another, strengthening Earth's magnetic defenses and slamming the door shut on the solar wind. In the language of space physics, a north-pointing solar magnetic field is called a "northern IMF" and it is synonymous with shields up!

"So, you can imagine our surprise when a northern IMF came along and shields went down instead," says Sibeck. "This completely overturns our understanding of things."

Northern IMF events don't actually trigger geomagnetic storms, notes Raeder, but they do set the stage for storms by loading the magnetosphere with plasma. A loaded magnetosphere is primed for auroras, power outages, and other disturbances that can result when, say, a CME (coronal mass ejection) hits.

The years ahead could be especially lively. Raeder explains: "We're entering Solar Cycle 24.

For reasons not fully understood, CMEs in even-numbered solar cycles (like 24) tend to hit Earth with a leading edge that is magnetized north. Such a CME should open a breach and load the magnetosphere with plasma just before the storm gets underway. It's the perfect sequence for a really big event."

Sibeck agrees. "This could result in stronger geomagnetic storms than we have seen in many years."

Impressive input. Thanks for shearing. I really appreciate.

The **presentation** clearly shows the presence of **sacred geometric science** all over the **Coral Castle**. The same knowledge can be seen in **Giza** and originates from **Atlantis**, and is also depicted on the walls of the **Freemason temples**. www.code144.com by Jeremy Stride

We are facing the sacred geometry again, architecture that is defined by precise numerical patterns. In the **Coral Castle** there is a heavy presence of **144 code, Phi ratio, Golden Ratio** etc.

Meanwhile you can also check this links. It may interest you>>
http://www.themeasuringsystemofthego . . . es/Page380.htm
http://ourworld.compuserve.com/homep . . . 5/gravity2.htm

I had read about **Ed** previously, but the presentation you suggested triggered my interest and new perspective to his work again.

Just for the record his **Moon block** weights 23 tons and his **Jupiter block** weights around 30 tons.

All the stones in the structure have a combined weight of over 1,100 tons.

My opinion is that he was some kind of **alchemist**. No matter that he obviously used magnetics, he understood very clearly that this works only if the correct components are balanced: sacred sound, sacred geometry and sacred sequence.

http://www.youtube.com/watch?v=64NR8gIcaD0

Still, the option of using **mental power** is not totally excluded. The use of technology based on ancient knowledge is obvious, but even so, it points to a person who also was initiated in hidden knowledge of psionic abilities. The witness testimonies are speaking of transporting huge stones through air like they are balloons without touching them.

This is just an observation of mine, nothing more:

He was using his **technological devices** to generate **electromagnetic field** that altered the **electro-atomical properties** of the stones he was moving. Then they started to **levitate**, he was pushing them with the hands and placing them on the desired locations and when the place was much higher than he could reach he was using sticks or ropes to put them where he wanted.

This option has many **holes** and one of them is how he managed the stones to regain their normal electro-atomical properties and to become heavy again on his command.

Was he using some kind of controlled electromagnetic field to pull this off, I can not say for sure, but it's possible. I will read his book and I will try to look through his work and the evidence from witnesses again to see what can I find. or . . .

He managed to generate what is called **prana** or **mana energy** from the surrounding space by using sacred ancient technological knowledge, stepped in the center of this force field, which provided Edward a **mental power** that allowed him to use it to move, shape and rotate stones that weight many tons according to his desire.

It is interesting to be mentioned that only this gate of the Coral Castle weights 9 tons.

This 8-foot-tall gate is perfectly mounted and balanced so that a child can open it with the touch of a finger.

Impressive indeed . . .

Regards,
Astralwalker

Humbly submitted: the enclosed link . . . scroll down the page until you see Montalk's post on 2005-11-10 19:05:00
http://forum.noblerealms.org/viewtopic.php?id=2425

He writes: "the link is HERE" . . . click the word 'here' and print out the pdf formatted article . . . it's entirely a worthy read while reclining and at your leisure w/hi-lighter pen in hand.

The dots are well connected.

It's 47 printed pages.

My humble best.

S

_____

The Truth Will Set You Free But First It Might Tick You Off

Respected AussieG,

Thank you for shearing. I really appreciate.

The original link is from here:

http://science.nasa.gov/headlines/y2 . . . tm?list1049613

http://www.nasa.gov/mpg/297403main_THEMIS_svsLG.mpg

They are changing their theories in NASA so often, lately, that its starting to look that in fact many previous and present theories are false, and that we are just on the begging stage of understanding the true dynamics of the Universe.

All the solid theories are starting to fall one by one, as new input comes along.

http://blog.wired.com/wiredscience/2 . . . breathing.html

The other reason why their theories are starting to malfunction is because the conditions in our solar system including our Sun are changing and we are getting more outside energy.

This outside energy influences the dynamics of our star, not to mention the obvious and detected changes in the magnetic structure of all planets in our solar system including Earth, producing increased brightness—the planets are starting to shine brighter in the night sky, then visible climate changes, this outside energy also is altering the structure of all DNA harmonics etc.

A part from everything, I believe that we have a small time to figure out what we have to accomplish.

They **burn** us a CD in the 2002 crop formation and using ASCII computer code people succeeded to decode the message.

Deciphered, it reads:

"**Beware the bearers of FALSE gifts & their BROKEN PROMISES.
Much PAIN but still time.
There is GOOD out there.
We OPpose DECEPTION.
Conduit CLOSING [bell sound]**"
Much PAIN but still time.
The question is: **time for what?**
What do we have to do before it's too late?
I believe the first steps for all humanity is to awake, to stop all conflicts, to unite, to start rejecting the ruling system, to become aware that our world is changing and so are we, to come together and to discus what we have as a solid evidence of coming events, to become aware that the ETs are worning us about the expansion of our star and that cosmic beams are coming towards us from the galactic center in 2012.

We have to do this people . . . we have to get together and solve this out before its too late . . .

Regards,
Astralwalker

Respected,
Humble Janitor, Mudra, Carmen, EpiphaMe, Efields, Mulder, Thanks my friends. I really appreciate.

It was not yesterday, it is not in the future . . . it is now that we must step forward, unite and do what we as humans can do, to protect Gaya from the Evil that is upon her . . .

I love you all
Astralwalker

http://www.youtube.com/watch?v=LjrKM . . . ICR-080803.php
http://www.youtube.com/watch?v=tKp7Dm1Y2KE
http://www.youtube.com/watch?v=Vyyn6 . . . eature=related
http://www.youtube.com/watch?v=0mesS . . . eature=related
http://www.youtube.com/watch?v=y9T5b . . . eature=related
http://www.youtube.com/watch?v=ZhA3W . . . eature=related

"Unlimited Mind" Wrote:

thank you astralwalker. i have spent several days studying this and want to thank you for the care at which you have shared your wisdom and experiences.

it is always nice to hear the 'how-to' from someone who has the wisdom of their actions, as the subconscious mind immediately registers it as truth and the information has little resistance. here is a link that folks might find useful . . .

http://cassiopaea.org/

here is a link to read the first of 6 books about The Wave

http://cassiopaea.org/cass/wave_i.htm

thanks for the spin!

much love

---

http://www.thesynchronicitygrid.com/ . . . othetribe.html

Much PAIN but still time.

I believe this is what they are referring to . . .

Present Longitude is WRONG

First . . .

We have to **map** all the **correct places** of the ancient energy grid.

The present **0° longitude** is wrong.

The **Elite** moved the **starting meridian** from the center of the **Great Pyramid** to **Greenwich** (current planetary 0° longitude).

If we **can not see** the connection, we do not bother to see the **geometry** and the **geomathematics** which is far advanced then the one we all use in the public arena today. Not to mention the connection to the monuments on other planets in our solar system and beyond.

As clearly showed in **The Code**, the ancients used completely different method then we today in the public arena. As we today can not duplicate those **achievements** with all the precise angles you soon discover which **mathematical approach** is more advanced.

Anyway, the **truth** is this: The real **zero Meridian** is in the center of the base of the **Great Pyramid**.

The **ancient Egyptians** were calling the Great Pyramid the **center of the World**.

Seen from space, the **Great Pyramid** is located on the exact center our planet, or if you prefer better, in the center of all the landmasses.

In any case, if you start from Great Pyramid as correct 's center (true 0° longitude), and you start to draw lines, all the ancient sites like:

> Stonehenge,
> Pueblo Bonito,
> Teotihuacan (Mexico)
> Cuicuilco Pyramid,
> Angkor Watt,
> Pyramids in China,
> Eastern Island structures,

Machu Picchu, and many other **ancient sites**, **temples** and **pyramids** around the world are in the exact place forming some kind of **energetic acupuncture** of the **planet**.

Those places are very important for all of us. In fact much more than we realize. They all speak to us . . . and we are just about ready to hear what they have to say . . . **And to learn once more**.

The evidence from decades of analyzing the sat images, geolocations, geo-mathematics, correlation between the temples, pyramids and ancient sites, had shown that the ancients were aware of those magnetic streams bellow the planet's surface and they build the monuments on the spots where this energy is the strongest.

The ruling Elite knows all about this ancient knowledge and had put obelisks on the exact locations around the world to gain their evil acupuncture results.

Now let's proceed . . .

Those places are very important for human beings who are on the side of Light because they can be utilized to stop the evil of our planet and to unleash healing process of our planet. The statistics is showing that our planet is dying, the leading media are suppressing the catastrophic percentage of dying species, and every month more and more are gone, and more and more death ocean zones are appearing. The climate is almost out of control, so we have to take steps, now.

Fallowing the pattern of their chess moves, they are forcing Shadow Government to proceed their plan of destroying the planetary economy possibly around spring 2009, so that means that it want be long before they start showing us their real face.

http://www.youtube.com/watch?v=4Q5I7AqBwWw
http://www.youtube.com/watch?v=Mjee-...eature=related
http://www.youtube.com/watch?v=79qeC...ext=1&index=31
http://www.youtube.com/watch?v=ee3bl...next=1&index=1

—We have a little time to unite, organize, figure out what can we do and act.

Waiting for the final minute I believe is a **wrong option**.

Now let's forget everything else and concentrate what we as humans can do.

As you probably conclude, I believe that one of peaces of the puzzle lies in this knowledge:

Info of geo-mathematics "The Code", by Carl Munck.

We have to learn **math** from the first grade again, and this time the correct math not the one that brings confusion and not logic in all universal laws and expressions.

http://vids.myspace.com/index.cfm?fu . . . deoid=21513766
http://www.pyramidmatrix.com/

The only way out this is through unity and team work, so I believe it is smart that people create a **map** of all **sacred places** that are part of the true **ancient planetary grid**.

We don't even have to go very deep into this **mathematical knowledge**, it's enough if we look for the location of every ancient sacred site on every continent and start marking the spots, draw lines toward others and find new ones that was in the shadow before.

Once the new planetary acupuncture map is finished we can proceed further.

Second:

Another peace of the puzzle I believe can be extracted from this info: **James Paul Furia**—Geomusic expert.

I will explain about him shortly, but let's continue further.

The **next step**, which I believe is smart to do, is gathering of the people who feel and resonate with this idea and purpose, in these **ancient sites**, which are part of the planetary energy grid.

It is better that most of the people have already experience in the **meditative arts** and altered states of consciousness.

There are many who can lead the group meditation and they should organize the group and explain them the basics and the purpose that we try to accomplish. When thousands of people come to one of the locations marked using the Code, they will all have to seat and have to enter a state of inertia and stillness. In fact this will work only if millions gather on the same day and on the same time—adjusted to time zones, and all tune to this psionic field of our planet with the focus to make a change.

And it will work!

On the sites also we have to assemble some audio equipment.

Then the audio engineers can press the play button so the music according to the Latitude and Longitude of their exact geographical location can be heard by the meditators loud and clear. See the work of James Paul Furia—fusing geomusic with geomathematics of the ancient sacred sites.

Why do you think the Elite is desperately trying to suppress what happened in the History?

This is why, and if we discover this and utilize this ancient knowledge the game will be over soon.

The exact **audio frequency** will generate the vibration that all meditators will soon tune to, and a powerful field of clean mental energy will manifest as result of that. In this point this force can be focused.

In other words, with previous experience in meditation and helped by the precise audio frequency that will activate the pineal gland, the group will be able to tune into the same wavelength, to materialize a clean psychic force in the surrounding space. We have to the best of science and the best of Esoteric knowledge to achieve this. The balance between the intellectual power and spiritual power.

The key lies in the middle of those two. Ones you can see it in your consciousness you start to understand its dynamics and you tune into this new frequency.

The nature recognize this and its start to help you and you become a channel of Divine Plan.

>>

This has to be done on every spot of the planet. The **audio engineers** will create the correct musical file that corresponds for every place in the continent. I think, at least 7 places in one continent are minimum that we can accomplish this.

The **best ratio** is **7000 people** on **7 biggest ancient sacred sites** on every one of the **7 continents**.

In other words: 777

The balance to . . . you know which mathematical number and which vibration.

But, that's not important. We have to **gather** as much as people we can. The more come, the better. This **777 ratio** will surface in the end because simple in this holographic universe everything is defined by numbers and codes.

We just have to remember that everything is energy.

We are energy and we are flowing in the ocean of energy.

If done how it should be, we can change the state of the four elements and lot more. Only if we understand this on time. Every thought contains energy. Imagine if we can do this on planetary scale.

We do not need guns to defeat evil, we just have to stop excepting evil to be a part of our reality.

http://www.youtube.com/watch?v=GpkfC . . . AD050&index=14

Now let's get back to **Mr. James Paul Furia**—Geomusic expert!

Geomusic is a **formula** connecting **geometry** with **sacred sites, earth measures** and music knowledge.

Inspired by the research of Carl Munck, researcher and composer James Paul Furia demonstrates a relationship between the notes of the musical scale, the locations and shapes of hundreds of ancient monuments, and the exact distance and location of stars, planets and constellations.

Furia combines the art of music, mathematics, astronomy, the study of the pyramids and ancient cultures revealing stunning correlations between colour, frequency, physics and consciousness. His program reveals a glimpse of the interconnectedness of everything in our reality reflected in our awareness, our ideas, how we express ourselves and shape the world around us.

In the link bellow you can see video file of his work. He is musician and composer but he has some extreme powerful leads that humanity can use in the constructive purpose if can understand and utilize the true meaning behind between sacred geometry (the correct location on the planetary grid) and the sacred sound.

In the **Part 1** if interested you can see his presentation:
http://maya12-21-2012.com/2012forum/ . . . _next=next#new

I also have a friend in **Canada** who is **audio engineer** and who is building something remarkable on the same **Geomusic formula** principles as respected **James Paul Furia**.

He is working on something called a **harmonic lattice**, the first prototype of which will cross the tactile, auditory, and visual sensory bands. The periods of the **waveforms** will be phaselocked and the intervals will all be tuned to whole number ratios, none of this Equal Temperament nonsense.

It's not exactly the same as Pythagorean tuning or Just Intonation, and he had to work his own system out from something called a **Lambdoma Matrix**, all the intervals can be found in a 16x16 matrix, except for the illustrious tritone (if you know the history), it tunes interestingly enough best to the square root of 2.

The **frequencies** in the tactile and auditory vibrations correlate to colours in the visual band, whereby violet wraps around back to red so that all frequencies have a unique hue.

Some **auditory parameters** correspond to **texture** (ie a sawtooth waveform would have a rougher texture, square less, sine would be very smooth), while other parameters like numbers of notes played, spacing, correspond to visual geometries in **2D** and **3D**.

He had a **near death experience**, and he calls it total synesthesia. He was **astraly travel** to the center of the Milky Way and before he reached Sgr* he was stop by higher intelegence and explained that he had to go back to Earth to do some important things concerning **2012 Nexus scenario**.

We discussed many aspects of the coming events and we agree in more then 98% of them.

I have seen his sound laboratory, the presentation that he gave me and the ideas what his work is trying to accomplish. I can conclude that its impressive and that he will finish this technology on time.

His work also included using high-tech to induce **controlled ESP** and **out of the body experience**. In other words to use latest technology like retinal display imaging or video glasses for visual, in-ear reference monitors for audio, and say a chair with subwoofers built directly into the frame, so someone would not have just a virtual reality experience, but an full conscious out-of-body experience. **Astral excursion** that can be used to bring important input from the higher planes of existence. And to be functional every time when this **technology** is used.

But this is just a small range of his ideas.

He is targeting something on a planetary scale something that has to do with the correct planetary locations found by the **Carl Munck** in his work known as **The Code**.

So we have few top audio engineers, who understand the purpose and can create the necessary audio frequency which is needed.

Let's go on . . .

Using the best of what pineal gland can provide, a single group will have to convert this energy into a clean Light. A strong beam of blue-white light filled with creative emotions of love and harmony and directed as protection of GAYA, will have to penetrate the Earth's surface and reach the highest point in the Earth's atmosphere.

Remember the research of Japanese scientist Dr. Masaru Emoto who discovered that crystals formed in frozen water revealed changes when specific thoughts were direct towards it. We carry extreme natural mental strength but we are just programmed from the social system not to use it.

People just have to break the old belief systems to peaces, remove all chess figures and start from the empty board.

To start from the beginning.

Unfortunately it has come to that. But we can still do it.

Much pain, but still time . . .

Let's proceed . . .

This energy is similar to a one that you experience on major concerts when thousands people are singing as one.

You can feel the power that is generating in the surrounding space.

In places that are part of the planetary grid, where there will be no people who are introduced to those arts it will be enough that they all sing the same song loudly and as one.

The song has to be emotional and noble and to generate the most divine emotions of the person. The only difference from the previously mentioned concert experience, is that you are not singing for fun, but for the survival of Earth and the protection of the all life that exist here.

But this would be a weak point in the energy field so experienced instructors who knows what are doing, have to reorganize and went to those places and it has to be done by the right audio sound which I mentioned before.

It's like music, no matter you have 100 professional musicians, if one or two of them does not know to play the instrument like it should, the listeners experience this musical off balance immediately.

But if all play their instruments very good, the public greatly enjoys the musical experience and vibrate with the same.

We have no time for mistake. If people agree with this idea I can assist with all this.

Luckily, today there are rare ones who have not heard what meditation is and how to tune into at least as little as deep relaxation.

What ever I do, from time to time these lines from the 1977 message are coming back to me:

We come to warn you of the destiny of your race and your world so that you may communicate to your fellow beings the course you must take to avoid the disaster which threatens your world, and the beings on our worlds around you.

This is in order that you may share in the great awakening, as the planet passes into the New Age of Aquarius.

The New Age can be a time of great peace and evolution for your race, but only if your rulers are made aware of the evil forces that can overshadow their judgments.

Be still now and listen, for your chance may not come again.

All your weapons of evil must be removed.

The time for conflict is now past and the race of which you are a part may proceed to the higher stages of its evolution if you show yourselves worthy to do this.

You have but a short time to learn to live together in peace and goodwill.

Anyway, the Nexus is coming for sure, but we have lot of work to do, before it arrives.

And no one else can do it for us.

It just doesn't work that way.

Therefore, to achieve this I believe we have to do this:

This beam of light, visible in the consciousness of the few thousands human beings that are meditating on the same place, will have to connect to all the rest locations in the planet where other groups of human beings will be in deep meditation.

If done properly, every group will feel emotional and telepathic connection with every other group and in the end, all groups will tune into one giant rhythm of conscious energy.

As previously explained, this has to be preformed in the exact same time no matter the time zone you are in.

First as described before, the audio engineers have to play the music sound according to the Latitude and Longitude of the exact geographical location.

Then after an hour or so, they all have to tune to Shuman resonance and to start influencing the reality of the four elements. Please remember we can use technology to generate sound or light but it is the most divine inside our beings that will save us and our planet.

Many of you are still asleep and you are on the opinion that we don't have to do anything . . . that everything will work out by default . . . but that's wrong. There is much that we have to do and in such a small period of time.

The awakening is quickening, and the Dark Side is aware that very soon the critical mass of people will start to see through the clouds of their deception.

That is why they are in hurry to brake the economy around spring 2009, to cause a problems on a global scale, and normally send the military to solve the problems they caused themselves and that's how it will started with the primary goal—>NOW< one world ruling government which will wipe out 5/6 of the current population, small group on the top, and all the rest as slaves and turning this planet to a real dark place.

That is why we have to **start preparing** and grouping now.

Tomorrow will be too late. Today is tomorrow. If we do not act today, tomorrow is difficult to be seen. But the Divine Intelligences are watching us, waiting for us to grow, to remember, to understand and to solve our puzzle.

So, there is hope . . .

We have to step forward, not with guns, not with their ways, but with higher consciousness, with transcendence of the Law of Attraction, with visualising that we no longer want any evil in any form or shape to be part of our reality.

If two argue and fight, and one realizes that the fighting is useless, no matter what the opponent is saying, the words does not reach him/her because he stepped out of it, and has erased this person from his/her life permanently. It's a plastic example but that's how simple is, if we awake and do this.

The higher Divine does not expect from us to get guns, that leads to the Dark Side, it expects from us that we awake, that we remember, understand, transcendent and to evolve and to stop the Dark Side.

Make no mistake; the Dark Side is aware what is coming, so if we DO NOT DO THIS it will probably be over for most of us and our planet, before the Nexus arrives.

That is why they are trying showing us the way through the crop messages.

I hope you people will have wisdom and clear sight to see what I'm seeing.

I believe that, this is what we as humans, have to do, with all my heart and all my soul.

Once more,

—People have to map the exact sacred places. It's easy they are all over the place as ancient giant temples and pyramids.

—To discuss this idea and purpose in every continent where there is Light workers and who understand the importance of this.

—Once the beginning time is set accordingly to all time zones, the planetary group meditation can begin. It can not be done at first. People have to practise this many times before reaching the satisfactory level. We have to ask the audio engineers to construct the sounds for every place so that we all can tune into the higher harmonics of Gaya. So we have to do this now. Before the audio files are ready, it can be practice with one melody that has pure emotions but it has to be played the same in all world and in the same exact time, no mater if it is the day or night in the time zone you are in.

Groups has to consist of thousands of people who are able to meditate or at least able to seat in stillness with their closed eyes. The melody that brings emotions to the surface will do its magic. That is why the melody has to be the correct one and which brings inner fulfilment, emotions of love and friendship to the surface.

—Then the people have to visualize the blue-white beam materializing above them towards the highest point of Earth's atmosphere and penetrating down through the surface to the inner Sun that is in the center of Shamballa.

—The meditation has to proceed with connection with all other groups around the world.

That is why most of the people have to know all the spots where energy will be coming and connecting. In practice, you will tune into this field by default because our planet is hallow and the space between the surface of the Earth and the conductive ionosphere acts as a waveguide which is nothing more but a structure which guides waves, such as electromagnetic waves or sound waves. So we will build a wave of thought-sound-light energy filled with positive emotions that will also penetrate into the space attracting similar vibrations.

—Once one rhythm of consciousness is established, people have to visualize Earth flowing in this energy, and all the evil fading away from here. To just visualize everything that you believe its causing harm in every form and shape, and to use mental effort to make it transparent, and slowly fading to total disappearance and the planet is left into Light.

—The GAYA is alive and if we do this it will response back. Also the crop circle makers and others will respond to the signal. We have to understand the importance of this people.

—I believe that we can stop the Evil that is upon our planet with the power of our feelings and that is the ultimate message that Nature and our higher brothers and sisters are trying to tell us. I believe that only then we will be truly ready for the Evolution as a race that awaits us at the end of Dec 2012.

Respected ones:

This is only the basic form and it's still shaping and rearranging in my consciousness.

Perhaps many of you will think of even better solutions. In fact I'm hoping that you do.

In any case, it can be altered as long the purpose and the result is the same.

As more and more people will start to change, they will start to separate more and more from this 3D reality and become more receivers for this new vibrations, gathering powerful amount of input, from many sources.

When that happens, many will understand what was known to the ancients . . . that everything is connected, everything vibrates, even when you read this you are tuning into my wavelength and we are making connection.

We have to stop seeing blackness everywhere we look and we have to start seeing our selves as living fire of Love, Intelligence and Light.

Because, that's who we are . . . Oneness of Love, Intelligence and Light . . . so common please wake up and do this . . .

I have started this thread . . . now you finish it.
Spread the word out . . .
God bless you all
Astralwalker

Last edited by Astralwalker; 12-22-2008 at 07:35 PM

—piers2210>>

Quote:

At last I understand there are multiple dimensions beyond our low-vibrational earth plane and as we raise our vibrational level we can reach these higher dimensions, as we move towards 2012.

I think the following 10 minute video ties into everything you are saying . . . please can you confirm/comment? It was posted on the wonderful website www.fourwinds10.com

Its called "The Matrix Matter Creation" . . . .
http://www.youtube.com/watch?v=i3cHPqIq6Rk love, light and thanks to you as always, Piers

*In a **Hologram**, the whole **pattern** is **whole** and **complete** on its self. If you take any **little portion** of this whole **out**, and examine it closely, you will*

*see entire pattern **repeating** its self again and again to **infinity**. **Anywhere** in this pattern if we **change** one little aspect of the **Hologram**, that change will be reflected throughout the **entire system**.*

Amen to that . . .

All the **physical mater**, everything we have around us, is the result of a **frequency . . .**

If we gather **together** all around the world and in the same time, if we tune into **one stream of consciousness** we can manage to **amplify** this frequency and **change** the **structure of the matter**.

In other words we can change this reality . . .

Thanks **Piers2210**. Thanks for shearing. I really appreciate.

We can do this people!

Regards,
Astralwalker

Quote:

WHEN DO WE GET STARTED??????
We are working on it.

As respected Mudra said, it takes about a month or so for people to gather in all parts of the word. We will have to provide a map of the exact ancient sites on every continent where this will be performed. We have contact one of the audio engineers and he agreed to create the correct musical files for the grid of sacred sites we will determine. I hope more will come to assist us.

The best is, if someone on the Forum has will and time, to arrange the list of people who will like to take a part in the worldwide meditation.

So everyone who is interested to send pm to this person and to give his name and location.

Anyway, it has to be done as soon as possible.

Respect to you all,
Astralwalker

Meditation—How To

Respected **StClair**,

Thanks for shearing your thoughts my friend. I really appreciate.

I'm most grateful that we have you with us. Your experience is highly appreciated and we can all learn much from you.

I have some experience with this also, so please allow me to share few thoughts with you also >

Perhaps many of you will disagree, but for me, the best posture for meditation is Lotus position. The truth is that only a small percent of practitioners are able to seat in this posture comfortably.

It takes long time and effort to master this position.

It is the most respected position by yogis, because it helps them to reach the deepest level of meditation and to achieve the topmost harmony of the consciousness and its subtle bodies in a way that no other meditative position does.

To keep it short, because we have to concentrate on the task what has to be done, when you master the padmasana, you are in a position to maintain your physical body stable for a long time.

Because of the fact that the consciousness and the body are connected and influence each other, the stability of the physical body will in time bring the stability of the consciousness.

If the stability of the consciousness has been achieved, the first step towards the deep meditation has been completed.

The Lotus position can also provide the best control of the breathing rhythm, known as pranayama in Yoga, and practicing it, a yogi can gain a full control over consciousness and its bodies with intention to understand and become one with the Divine Creation and Purpose.

Practicing special pranayama techniques, a yogi can absorb and use the prana, which is a free vital energy that floats everywhere.

The circuit of prana circulates best when one sits in padmasana and its strength is the most powerful.

But, currently it can not be achieved by the most, so we have to use another posture.

Crossed legs had to go like this:

First you bend the left leg and then the right leg above it. The left hand should be upwards like in the image and the right on the right leg as close as the point of crossing with the biggest finger of the left leg.

The backbone has to be erected as possible and in that position you mast maintain stillness.

Many of you will probably think that the consciousness is only important and if can transcendent its experience it doesn't matter in which position the physical body is. I have concluded that this is wrong. Please take this only as my opinion and nothing more.

Anyway, the Ancient Rama Kingdom was far advanced in this knowledge.

That is why they all seat mostly in the same position.

They are all tuning into one energy field and vibrate on the same frequency with the nature.

Not to mention that they were generating high amounts of free energy, chi, prana, mana, life force, whatever you want to call it and were focusing it to constructive purposes.

We have to learn the same dynamics again.

Together with my wife, we have analyzed over thousands pictures and video files that are connected with the ancient temples of Rama civilization and what we found is truly amazing.

They left us so many useful leads on the walls of their temples that we only have to look what they are showing and utilize that knowledge. I had study their ways and techniques for more then two decades and I can only say that we have so much to learn from them.

Anyway, I will construct few meditative methods that I believe are suitable for what we are trying to accomplish but the correct sound will do its effect and if the consciousness is ready and opens itself to true inner emotions, we will do it.

Not much knowledge is needed.

Just the correct geological place that has the correct proportion, geometrical design and location, all this combined with the correct sound, people seating relaxed in stillness and just open themselves to the most divine in them and reflecting all that to one beam that goes up and down from the meditative platform, and connecting with other groups and creating one energy field of consciousness that will want the change the Matrix of this reality.

Its useless we all talk about this on and on, and we are not doing it.

It is now. We are ready and we all wanted from the bottom of our hearts. We do not need someone to save us, we become aware that we can do it ourselves.

In fact people do not have to have a lot of experience with this. Its certainly better option if they do, but if not, it's ok. We will do it anyway.

It's like you go somewhere and all of the sudden you hear some music that is so emotional that whole inner being vibrates with it. You didn't prepare for it. You have it all along. You just need a proper tool to bring it out. The music in this case.

Anyway, not to distant myself, too much from what we are truing to do here.

For anyone who is interested in meditation here is one simple way how can be done:

After you have seat in the position I mentioned, become frozen and totally relaxed in the same time. It's like you had stopped the time and everything in the matrix, you have entered total stillness and inertia but in the same time your inner being is calm, relaxed and vital.

After few minutes in that state, become aware of what is going on in your consciousness.

Shortly after you have achieved that kind of awareness, you will notice that your consciousness is full of inner storms, tensions and all kinds of mental conflicts.

The stronger you concentrate on them intending to calm them down or even neutralize them, the stronger and more intensive they will become.

To calm all this manifestations in your consciousness you must build correct relationships with them.

That can be done in a very simple and effective way.

Start observing the thoughts that appear in your consciousness with the eye of a witness.

Do not play with them or even try to analyze them, just observe them.

Watch how they form and pass in their silent way, and maintain the perspective that does not have at least one crossing point with them. After a while, you will feel the source they come from.

At that moment you would be able to clearly see how shortly after they took form, they started to regroup themselves into chains of thoughts. One idea will pull at least few other ideas to the surface of your consciousness as a logical response to the first one, until some other idea with completely different context does not appear in your consciousness and again induces manifestation of other ones related to that one.

These thought chains will constantly be coming from the depths of your consciousness to the surface always taking different shapes on their way up.

Therefore, do sink even deeper into meditation, it is useful if you start to monitor the changes that your thought chains are going through from the moment they appear in your consciousness to the moment they are gone and altered by other ones.

Sooner or later, if you continue to observe these thought chains manifesting one after another still from the perspective of the silent witness somewhere in your consciousness, you will notice that they will reduce and after a while they will slowly start disappearing.

This is a completely normal effect because you are not playing their game anymore and you are no longer attached to them emotionally. If you continue to maintain this concentration your brain patterns will soon change and that will lead you to an altered state of your consciousness.

In the end, the last thought chain will disappear from your inner sight and you will find yourself in a completely new situation. Your consciousness will be moved directly to the present moment. Once your consciousness finds itself in the present moment, it will instantly discover that it is empty. Still, the emptiness that your consciousness will see in that present moment will not be a complete emptiness, but it will contain potential of all known shapes in the Universe. It might sound strange, but if you manage to get this far, your consciousness will constantly be aware of this fact.

You will see that in the emptiness in which your consciousness is present, a hidden reality is becoming visible for you. It might seem as if you are being pushed into the pure reality and you will experience your consciousness sliding from moment to moment as never before. This experience will make it clear that reality has a characteristic of sliding from moment to moment.

When this moment comes, all you have to do next is to attach your consciousness directly to that present moment and you will find that you are remaining in the pure reality all the time, by sliding from that moment to the next one. The feeling will be so vivid and so amazing that your whole being will float in the Inner Light.

You will be here and now, and that's the starting point for every advanced meditation. From this point you can proceed with the suggestions that I have gave for tuning into the Gaya's Planetary Psionic Field. This is the only true and solid way that you are connected and tuning into higher realities.

But, we will deliver what and how we can achieve what we plan to, so every comment and opinion on the meditative methods is most welcome.

Best regards,
Astralwalker

thuras>>

Quote:

What's the purpose of all this if this Nexus-Event happens anyway?

Is it just about warning people to not enter a tunnel if they die in the next 4 years? Or getting rid of the evil "soul collectors" so people can enter tunnels? Or is it just for "karma reasons"? Sry. if this sounds stupid or sarcastic.

Because, the **Entity** known as "**All Seeing Eye**", will not allow us to live. It will do everything in its power to finish this before the **Nexus** arrives.

Trust me, I perfectly know what I'm talking about.

If we fail with this . . . everything that we done before its for nothing.

Quote:

What's the purpose of all this if this Nexus-Event happens anyway?

No, you got it all wrong. The **Galactic Nexus** is the **ultimate purpose**.

Dear friend, your questions are logical and everyone has them.

Anyway, everyone has a **choice**. I certainly will respect yours . . .

But if you remember, I have point out clearly, that this is for everyone who vibrates with this idea. For ones that are seeing the purpose of this.

And it's **not useless**.

You see, if every **water molecule** responds to our **frequency** and **emotional charge**, you can imagine what this generated **energy field** of **fused consciousness** can do to this **Matrix** and **Mother Earth**.

Take care
Astralwalker

Piers 2210 wrote:

Scientific proof of realms beyond our physical realm:

Physicist Tom Campbell is interviewed in the link below by George Noory on Coast to Coast am. Campbell worked with Robert Monroe at the Robert Monroe Institute and laboratories from the 70's, helping to establish Monroe's laboratory for the study of consciousness and to develop the Hemi-Sync technology used to attain specific altered states. He tells us about the scientific proof of non physical realms.
http://uk.youtube.com/watch?v=BcLWTu . . . eature=related more information is available at www.mybigtoe.com where books and other interviews can be accessed.

Firstly listen to the 50 second clip of the late Robert Monroe on surviving physical death
http://uk.youtube.com/watch?v=VoZWOL . . . eature=related so there's really nothing to fear . . . so lets get on and follow Astralwalker's thread!

And hope you all enjoyed a day off today . . . .

Thuras wrote:

Quote:

Originally Posted by **Astralwalker t*huras*>>**

*Because, the **Entity** known as "**All Seeing Eye**", will not allow us to live. It will do everything in its power to finish this before the **Nexus** arrives.*
Trust me, I perfectly know what I'm talking about.
But as you said, we're going to die anyway. What's the point on this?
Is it important to be alive the moment the Nexus arrives?
Else it doesn't matter if TPTB stay in business for the next few years.

Quote:

If we fail with this . . . everything that we done before its for nothing. Why? It doesnt make any sense.
The game will start again for those, who were collected by the "all seeing eye". All the others are going to be safed as you mentioned.
And if it's not a fight to safe the other souls, what else is it for?

Quote:

No, you got it all wrong. The **Galactic Nexus** is the **ultimate purpose**.

Yeah, ok. I got this. But it arrives anyway, no matter if we're smashing our heads, dancing naked around a fireplace or meditate.

So whats the point when the Nexus arrives? Is it important the earth is vibrating in a specific frequency?

Quote:

But if you remember, I have point out clearly, that this is for everyone who vibrates with this idea. For ones that are seeing the purpose of this.

Just because I'm asking questions to understand the meanig of all this, it doesn't say that it doesn't vibrate with me?

Else I wouldn't be here asking "dumb" questions.

Quote:

And it's **not useless**.

I've never said it seems useless. I was just asking questions to understand why we should do this.

I am willing to help, but I'm not Rambo, running around shooting everybody an asking questions later. (so to say)

Quote:

You see, if every **water molecule** responds to our **frequency** and **emotional charge**, you can imagine what this generated **energy field** of **fused consciousness** can do to this **Matrix** and **Mother Earth**.

So it DOES make a difference in which frequency the earth is vibrating the moment the Nexus arrives?

But why? If the Nexus is wiping away all evil, than it doesn't make any difference if TPTB are in charge or chased away before.

The game has to start again, because we haven't finished our business in this dimension? But that wouldn't make any sense either, because this happens to the people who are collected by TPTB and not to them who are safed by the Nexus.

Thx. for your answers.

Astralwalker:

Quote:

But as you said, we're going to die anyway. What's the point on this planetary meditation?

My friend, look deep inside your heart and you will see the point . . . Otherwise you can not.

Quote:

Is it important to be alive the moment the Nexus arrives?

We are always alive.

If you think of physical existence—It is. But if that isn't possible, it's important to sustain in a fully conscious state inside one of immaterial bodies, of who you are and what is going on.

As I previously pointed out, the lowest layers of Earth's astral dimension are not safe at the moment, which is not the case everywhere. This solar system seems to be quarantined.

Anyway, I will try to share few more thoughts and conclusions that I come up throughout the years.

Please consider this only as my perspective and treat it only as such . . .

The lowest layers of the Earth's astral plane (we are talking about huge area) are not safe because something had happened long ago here on Earth.

The evidence that I have trucking for so long, points (once again it's only my observation, nothing more) that this anomaly happened as a result of using extreme advance technology in dark purposes.

Atlantis Lemuria

The closest that I come up with, was some kind of destructive technology that was used by the evil beings from Atlantis. Does those beings responsible for this originated from the Atlantis or did they come from the stars and took over, I just do not know at this moment.

Anyway, the result was a major break-up, fissure, opening, some kind of sharp cut or what ever you want to call it, in the lower region of what we today in our terminology we refer to as Astral Plane.

It became known as Great Void.

It also wiped out almost everything, cities, civilizations and most important of all it almost destroyed all life on Earth and imprisoned Gaya.

They used large obelisks on the exact planetary spots with a purpose to block the Divine Energy of what we are referring as life force of the Planet or Gaya.

Just as the same as acupuncture needles on a physical body—the same knowledge just on the planetary level.

However, let's continue . . .

As far as I can tell it happened somewhere between 10500 BC-35000 BC.

The damage was so great that the planet is experiencing its effect even today.

Compare today's HAARP technology development and multiply it by 20 centuries further research and development, and you will get a good picture what kind of weapon I'm talking about.

In fact this weapon made a large cut in the astral plane so the Divine Creation can not shine on this plane of existence and that is why now we have so much Evil present on the planet.

But the main damage was that the weapon opened a vortex which allowed powerful Entity of Darkness to come to this planet.

It allowed a whole legion of mighty dark beings to come over, but this powerful Entity of Darkness was the worst. And still is.

I believe you are guessing—The All Seeing Eye.

That is why the ones who are "CONDUIT CLOSING" (which gives you the answer for: Is it important the earth is vibrating in a specific frequency when Nexus arrives?) are insisting that we remember or at least we understand this.

That is why in the most ancient secret esoteric symbolism is always depicted as being inside the deep black empty space shining with beams.

After that, what was left on the physical plane of existence, split as Rama from the east side, and Olmec and Egyptian Empire from the west side.

Rama originated directly from Lemuria and Egypt and Olmec in Central America directly from Atlantis.

The beings from Rama lived in harmony and balance with the nature, the other ones were attracted to technology, hierarchy, total control, slavery, blood sacrifices and similar that is clearly depicted in the Inca and Maya temples. Beings from Rama were living according the Law of One, and the others were living according that everything is divided, separate from the source and that everything that exist around serves to fulfil their needs and pleasures, or simple—the strongest rules.

All this Illuminati vibration, if you start to trace it back in time, in fact originates from Ancient Atlantis. What we dealing at this moment is building of a New Atlantis based on a idea of the first one.

I have no time to go further into this huge subject but my point is this:

The Nexus will fix this astral fissure and the Light from higher harmonics of the Universe will be able to penetrate and shine again.

That is why the ancient records are referring as the Return of the Golden Age. The ones who will be left in 3D reality after the Nexus is gone, will face a New Era of Light.

It's like a wire cut off in electrical circuit.

But if you connect the wire again, you have voltage back and electrical current as result of that. In other words I'm talking about fixing the damage that was done, by rearranging the whole structure of the Matrix (Normal Mater, Dark Mater, Dark Energy) and allowing electrical spiritual current to flow top to the bottom of the Universe and vice versa—again.

How the Creator will deal with this All Seeing Eye, it's above my reach of understanding. But it will surely reprogram the whole structure of the Matrix where it reaches. And what is more important we have to play our role in this entire scenario.

For some reason, the Creator usually reprograms the Matrix from the centers of the Galaxies, and I have combined some of the peaces gathered from everywhere so that we can see some picture and sound. It's far from the exact one but we are getting somewhere.

I can assure you that not in any moment I do not consider myself special. In fact I have more questions that probably you do.

As I stated before, there are no masters in this, no prophets, no messiahs.

We are all equal and we are all ONE.

Its interesting, in this moment, the more you learn about this, the more you realize how small we are and how big the Universe is.

Anyway, the problem with this Great Void was, and still is, that in most cases when someone physically dies, the life force had to face this

Great Void which obviously has some unusual properties of cutting the memory of the previous experience, so when you reincarnate again you do not remember anything before and all your previous assimilated experience and knowledge is lost.

You still keep it in your DNA structure, in you subconscious, but its still a blackout and in most cases unreachable.

This is probably very complex for most of you, but is not. Once you learn the basic moves, you are starting to see a pattern and you learn more and more. But it's a long way home, and that is why those cosmic beams coming from the Galactic centre depicted in the crop formations, are so important for the Life here on Earth.

Quote:

Is it important the earth is vibrating in a specific frequency?
Yes it is. It is essential.

Please consider that I have explained only the basics of the nature of the Nexus. There is plenty more that I didn't mention at all, or I only had touched huge subjects on the surface.

Plus there are things about Nexus that I still don't understand and I'm waiting for further input.

Quote:

The game will start again for those, who were collected by the "all seeing eye". All the others are going to be safed as you mentioned.

I hope I'm wrong about this, but the logic points that the dark entities under the control of the All Seeing Eye, have to realize a Secret Plan behind the scene.

Btw, all that we see in the public domain is nothing but the theatre. Some are tricked by it completely, some see through the deception more or less, but the fact is that more are clearing their sights. When you remove all the layers you start to face what I'm talking about.

As I was saying earlier, we have visitations from distant places but Orion constellation surfaces back and back. The Giza Pyramids were exact match of the star map of Orion 10500 BC.

There is one good video file about the investigation of KGB in 1962 in the Giza Plato.

It's called "The Secret KGB Abduction Files". Here is a part of the documentary.
http://www.youtube.com/watch?v=v7IpokMjEF8

Lets proceed . . . The analysis from the M12 papers, thousands of other top secret papers etc, points that the malevolent race from Orion will try to trap the consciousness of most of the humanity into giant force fields, souls will be put into blackness or in a dream like hologram appearance, put in large containers and dislocated on other systems far away from the Nexus.

The performance and the evidence had shown that those multidimensional beings have the technology and knowledge to do pull this off easily.

That is, departing the souls of 5/6 of the human kind to the astral dimension and picking them up from there into huge electro-magnetic force fields, then shipped to large ships and transported to other systems far away from this galaxy with the purpose to put them in other cloned bio-computers and continue the needed exploitation.

Most beings stop at this investigation. But there are some who went for it . . . just to see how really deep the rabbit hole is. And what they discovered . . . it doesn't look nice.

That is why the **Nexus** is so important. And yes this has much to do with saving as much souls we can.

And much more . . .

That is why they started to map the higher planes of existence. That's how Bardo Thodol—The Tibetan Book of the Dead", and The Egyptian Book of the Dead were created.

They both talking about the Great Void but they are pointing to different directions as they were referring different realities of immaterial word.

In first sight, it seems that one of these books is false.

—No, they were both right.

In this physical reality if you go to Bangkok, Moscow, Kathmandu, Las Vegas, etc., you will face a different reality, different culture, different life stile, different achievements and intentions etc.

The Life on Astral Plane above East and West was different.

I know that this differs from everything you have read in the esoteric books but it's true. But it's another major story, that we have no time to get into at the moment.

Note: Not to be misunderstood. This Great Void is not visible in the average out of the body experience. A person cad perform astral travel all his/her life in the closest astral layer to the physical plane and never to detect it, because is simple on other wavelength. But the Consciousness that has permanently left the physical body is usually by default in a vehicle that matches the Great Void vibrations and that is why in most cases is pulled here after the physical ride is over. It is usually experienced as an astral flight through a dark, deep and long tunnel with the Light on the End.

You are probably guessing what that Bright Light is.
The Eye on the Top of the Pyramid—The All-Seeing Eye!
It's starting to make sense? Does it?
This is not happening to every case, but . . . That is why the Nexus is so important.
Anyway, I do not want to focus my energy in this direction too much. Lets live it for some other time.
Let's go back . . .
The both described this Great Void as not a safe place for souls and give instruction how to pass it safely.
The ancient knowledge that was left in the Bardo Thodol describes the nature of the beams that are shining in the Great Void and where they lead.
Anyway, to solve the problem of loosing the memory and the assimilated knowledge of the being that was at the end of his physical path, they used ancient knowledge and technology to transfer the consciousness from one body directly to another one with bigger proportions.
O yes, there were giants those days.
The atmosphere was bigger; the mana was present in abundance, there were beings with four, six arms, some half fish-half human, animals grow bigger and the plants were gigantic.
But the evidence points that there was truly five different levels of human consciousness possible here on Earth.
I came across this in few books but it was mentioned the best in the book The Ancient Secret Of The Flower Of Life by Melchizedek Drunvalo.

He was saying that he got this input from entity by the name Thoth, which at the time I took with great reserve, but when we started our own research trying to find the evidence of the same in the East, and when we started to find this everywhere on the territory of Ancient Rama it started to spin in our heads.

In short, this is excerpt of his book:

The Five Levels of Human Consciousness and Their Chromosomal Differences According to Thoth, there are five different levels of human consciousness possible here on Earth.

These are people who have different DNA, completely different bodies and different ways of perceiving the Reality.

Each level of consciousness grows from the last one, until finally on the fifth level human being learns how to translate into a whole new manner of expressing life, reaching Nirvana and leaving Earth forever!

The primary visual difference between these types is their height.

—The first-level people are about 4 to 6 feet tall.

—The second-level people are about 5 to 7 feet tall, where we are at now.

—Third-level people are about 10 to 16 feet tall.

—The fourth-level being is about 30 to 35 feet tall.

—The fifth and the last is about 50 to 60 feet.

This may seem strange at first, but do we not begin as a microscopic egg and get larger and larger until we are born?

Then we continue to grow taller and taller until we are adults. According to this theory, the human adult is not the end of our growth pattern.

We continue through DNA steps until we are 50 to 60 feet tall.

Metatron, the Hebrew archangel who is the perfection of what humanity is supposed to become, is 55 feet tall!

Remember the giants who lived here on Earth referred to in chapter 6 of Genesis?

According to the Sumerian records, they were about 10 to 16 feet tall. When we look at a three-year-old and a ten-year-old, we know that they have different levels of consciousness, and it is primarily by their height that we make this judgment.

According to Thoth, each level of consciousness has different DNA; however, the primary difference is the number of chromosomes.

According to this, we are now on the second level and have 44+2 chromosomes. An example of the first level is certain aboriginal tribes in Australia where they have 42+2 chromosomes.

On the third level, which we are about to move to, people have 46+2 chromosomes. The next two levels have 48+2 and 50 + 2, respectively.

We're now going to focus on Egypt because Egypt happens to be where the one of two main mystery schools was located and where evidence of the different-sized humans, and levels of consciousness, still remain, though generally unrecognized.

Egypt was the area they chose where they would ultimately restore our consciousness, and the primary area where survivors from Atlantis and the ascended masters were in one place. Abu Simbel, would be about 35 feet tall, representing the fourth level of consciousness.

They built rooms for these different heights.

This doorway is made for the Venusians—the Hathor race—who are on the third level of consciousness.

These third-level beings are about 16 feet tall, indicating they are male, as the females of this race are about 10 to 12 feet tall.

In their section of the building the rooms are around 20 feet high, with ceilings and beams in proportion to 10- to 16-foot-tall beings.

Next to that room, through a little doorway that looks like it's made for us, is a little room with a much lower ceiling.

The Egyptians didn't make these statues arbitrarily—they never did anything arbitrarily.

There isn't a single scratch on a single stone; there is not even one, I believe, that was done unconsciously.

There was a reason and a purpose for everything. And usually it was created on many, many different levels. The Emerald Tablets, for example, are written on one hundred levels of consciousness.

Depending on who you are, you'll understand something utterly and completely different from other people. If you should go through a consciousness change, go back and reread The Emerald Tablets again.

ROGUE SABER

You won't believe it's the same book, because it'll talk to you in a different way, depending on your physical body.

In their section of the building the rooms are around 20 feet high, with ceilings and beams in proportion to 10- to 16-foot-tall beings.

Next to that room, through a little doorway that looks like it's made for us, is a little room with a much lower ceiling.

The Egyptians didn't make these statues arbitrarily—they never did anything arbitrarily.

There isn't a single scratch on a single stone; there is not even one, I believe, that was done unconsciously.

There was a reason and a purpose for everything. And usually it was created on many, many different levels. The Emerald Tablets, for example, are written on one hundred levels of consciousness.

Depending on who you are, you'll understand something utterly and completely different from other people. If you should go through a consciousness change, go back and reread The Emerald Tablets again. You won't believe it's the same book, because it'll talk to you in a different way, depending on your physical body.

These are Earth beings passing through the various levels of consciousness. In this photo you see a huge 55-foot-tall being with a statue our size standing by his leg.

This is the king and queen.

Archaeologists don't know how to interpret this, so they just say that the kings were more important than the queens, and that's why they made her little.

But it didn't have anything to do with that.

The statues are showing the five levels of consciousness.

Every king and Pharaoh who ever lived in Egypt had five names, representing the five levels of consciousness.

On the wall of ancient Egyptians wall they are depicting the technology they were using to transfer one consciousness from its body to the next one without loosing any memory and knowledge of the previous assimilated experience . . .

Anyway, when we start searching in the East, we found the same exact body sizes and similar technology for the transfer.

In fact the all Buddha myth is twisted up side down. There was whole race of Buddhas with long ears and different body sizes.

They build huge structures with a places for meditation of every meditant and through their sharp antennas they were cutting the natural scalar waves or natural Chi energy and tuned into Oneness. As I said earlier we have still much to learn from them.

It's all there; you just have to start searching for answers on ancient sacred sites and temples. Especially jungles in Thailand are very impressive.

All this knowledge is suppressed by the Elite, but its coming back. Its all math and ritual for them, so in the critical point of mass awakening they will strike with force. And that is why we have remember, understand this very soon and to start to utilize this ancient knowledge. So we do not have to fear, but we have to put some solid input and to see with what exactly we are dealing with and what can do about it.

Quote:

Yeah, ok. I got this. But it arrives anyway, no matter if we're smashing our heads, dancing naked around a fireplace or meditate.

So what's the point when the Nexus arrives?

We have to buy some more time for the planet and for the life that lives here. They are crushing the economy already which means its already starting.

I believe it's our responsibility to step forward, to unite and to act now. Not with guns or violence but with Higher Consciousness.

The pictograms had shown us the way.

They are showing the fractals, the geometry, the **STRUCTURE OF THE UNIVERSE!**

AND MOST IMPORTANT—HOW WE CAN CHANGE IT!

So, yes it's essential that we start perfoming planetary meditations.

This is the **Checkmate move** they are referring to. I hope you will understand this before its too late.

That's why I believe we have to start doing those planetary meditations.

And if done how it should be—**IT WILL WORK!**

Respect to you all,
Astralwalker

Responding to a question about the maps of Australia posted earlier Astralwalker said:

Hey Pep,

The previous images were the closest that I could find on the net to depict the safe area I received as download from the Crop Circle makers.

The last one I have created after you and other people requested if I can provide more precise area, since the previous images are just Australia images for showing map of Australia, Queensland Mining Fields, Tasman Orogen etc.

http://www.international-travel-tour . . . australia.html

http://www.earthscrust.org/earthscru . . . /east_aus.html

http://www.codyopal.com/cody0405/aboutopals.htm

As I said earlier, please consider this only as download of mine. I hope I'm wrong about this, but it is the same that physically appears in the crop fields around the world. That is why we have to pay attention to the new input that will come in the crop formations.

And most important, to get together, unite in one rhythm of consciousness, and influence the basic fractal structure of our Planetary Matrix with the intention to prevent any further damage to our planet and the life on it.

Take care,
Astralwalker

"Samarkis" post: Just wanted to remind all of you-we are fractals of divine creator therefore, we can never be lost EVER. (Pls see Law of one, pls see www.michaelsharp.org, pls see www.ramtha.com, www.divinecosmos.com)

Pls do not set limitations-claim it. Claim Ascension.(Pls see Ramtha)

Pls do not fear-Lean towards light & Change!! Pls know that Paladians & Andromedans have done this Ascension Many, many times before.(Pls see Bringers of the Dawn-Youtube & Google)

Yes—I feel that the Meditations that Astralwalker wants to set up will "ease" our birthing pains, and yes it will be a good thing to meditate as a combined Unity, even those whom cannot make the trip may meditate in sync.

Blessings to all!!

Quote:

Just wanted to remind all of you-we are fractals of divine creator-therefore, we can never be lost EVER. (Pls see Law of one, pls see www.michaelsharp.org, pls see www.ramtha.com, www.divinecosmos.com)

Pls do not set limitations-claim it. Claim Ascension.(Pls see Ramtha)

Pls do not fear-Lean towards light & Change!! Pls know that Paladians & Andromedans have done this Ascension Many, many times before.(Pls see Bringers of the Dawn-Youtube & Google)

You are totally right **Samarkis**. We are moving towards the **Light** and there is **Great Awakening** everywhere. We all start to feel the connection between everything. So I totally agree that we are all **Light**, that we are all fractals of **Divine Creator** and that we can never be **Lost**. That is why we have to maintain the higher vibrational state, to get together, to tune into the exact wavelength of the Divine Creation and to assist in the Cosmic Plan and Purpose as much as we can.

Quote:

Yes—I feel that the Meditations that Astralwalker wants to set up will "ease" our birthing pains, and yes it will be a good thing to meditate as a combined Unity, even those whom cannot make the trip may meditate in sync.

Yes. Everyone is welcomed to join the Meditations. If someone can not come to the sacred planetary sites physically we will greatly appreciate if he/she will join the field of planetary consciousness from his/her home.

But I still hold the opinion, that those acupuncture points of the planet are very very important because they are chakras of the Gaya.

Btw, as soon as we will determine the dates, times and finish the technical aspects of the planetary meditation we will make announcement so everyone who feels that we have to do this or vibrates with the idea, can join us.

Blessings to all!!

Astralwalker

Firstfruit writes:

Brothers and sisters all the ancient monuments on this world point to something in the Pleiades system,

The monumental constructions on mars are a star map of the Pleiades system why?

Have a look at the giant stones at Baalbek from the air and you will see a man holding a key why?

The same key can be found as the ground plan for Vatican City Stonehenge seen from the air is a da vinci manstar and it represents a star in the Pleiades star system. As does all ancient monuments here and on mars

The big picture is astonishing

Please look here:

http://www.keyofsolomon.net/
http://oneism.org/
http://www.thehiddenrecords.com/

    Peace love and light to all
    Lak-ech my friends

Quote:

    Blessings to all!!!!

Astralwalker and I have been discussing this. All that would like to go to a specific Point on the grid, once points are picked may contact me by private msging in Avalon or by my Email at Samarkis@yahoo.com. Pls put Astralwalker in the subject line so I know right away this is for the Chakra point meditations.

In Joy and Hope and great Love,
Sara

Thank you Sara. We all appreciate.

This is where this part of the copy and paste finished. But there is much more that follows and it can be picked up at this point on the Project Avalon forum

Additional Resources

This documentary on DVD—"Contact Has Begun"—is James Gilliland's true story about his life, his near death experiences, and his contact with higher civilizations.

It includes additional information on Earth Changes including **graphic animations of what will happen to our planet when the sun unleashes her next round of Solar Storms.**

Formatted to play in all regions, shot in High Definition—this is a DVD you will treasure because it helps alert others to the concept of abrupt changes that are happening around the globe.

Earth Changes Are Real—And Serious.

"Earth Changes—Mind Matters" is a downloadable ebook that will save you countless hours of research while explaining exactly what you can expect to experience in the next few short years . . . and how to survive what's coming. **The tools you can use to create a totally different personal future are revealed in this work.**

"I've read the first page and it looks like the book will be a "doozie", as they used to say when I was a kid." Dodie—Australia

"God—at an initial scan—this is an incredible work!"

Yvonne—South Africa.

"Enjoying the book thoroughly; it's like an 'E-pedia' on what's happening in front of our eyes and behind closed doors." Jace—USA.

"How To Survive . . ." is an instantly downloadable ebook that answers questions from "Earth Change Report" subscribers around the world

* about earth changes, economic depression, and martial law.

* Provides factual information that is NOT being presented by the mainstream outlets.

* includes video links to far more information than you'd possibly get in a simple text document.

"Basic Preparedness "—**The Six-hour 3 DVD Home Study Course**" takes you inside the Whys and the How-To of Being Prepared . . . whether it's because of job loss, a shortterm natural disaster, or an extended period of time.

This One of a Kind Home Study Course was filmed before a live class with presenters who are top specialists in their fields.

The 12 DVD Chapters include:—How to look after children;

What you need to know about Bird Flu & Pandemics;

The essentials of General Preparedness; **Food Storage—what's available, what you need, where to get it;**
Grain Mills (you want the best, and at the best price);
Water, Water Storage, Water Filters—(think about volcanic ash and acid rain pollution—be aware and prepare . . . and much much more.
Earth Change Report is a regular free international newsletter that keeps you up to speed on what is happening around the world . . . from melting ice caps to increasing volcanic activity; from strange activity on the sun to "where is Planet X?"
If you have not already done so, you can subscribe here.
Most Importantly, YOU are your greatest resource. With Knowledge and Free Will we all have the power to create and change our personal destiny.
This information has been made available free of charge by Astralwalker and Project Camelot. Those of us who can sense truth as opposed to "spin" are more than happy to show our appreciation by becoming paid subscribers and active members.

Help Yourself and Help The Planet—Subscribe to The Forum Footnote—Press Release Re Black Hole Under Study

Astralwalker writes:

I come across this link:
http://eu.spaceref.com/news/viewpr.html?pid=26952

PRESS RELEASE

Date Released: Tuesday, November 18, 2008

Source: European Southern Observatory

VLT and APEX Team Up to Study Flares from the Black Hole at the Milky Way's Core

Astronomers have used two different telescopes simultaneously to study the violent flares from the supermassive black hole in the centre of the Milky Way. They have detected outbursts from this region, known as Sagittarius A*, which reveal material being stretched out as it orbits in the intense gravity close to the central black hole.

The team of European and US astronomers used ESO's Very Large Telescope (VLT) and the Atacama Pathfinder Experiment (APEX) telescope, both in Chile, to study light from Sagittarius A* at near-infrared wavelengths and the longer submillimetre wavelengths respectively. This is the first time that astronomers have caught a flare with these telescopes simultaneously. The telescopes' location in the southern hemisphere provides the best vantage point for studying the Galactic Centre.

"Observations like this, over a range of wavelengths, are really the only way to understand what's going on close to the black hole," says Andreas Eckart of the University of Cologne, who led the team.

Sagittarius A* is located at the centre of our own Milky Way Galaxy at a distance from Earth of about 26 000 light-years. It is a supermassive black hole with a mass of about four million times that of the Sun.

Most, if not all, galaxies are thought to have a supermassive black hole in their centre.

"Sagittarius A* is unique, because it is the nearest of these monster black holes, lying within our own galaxy," explains team member Frederick K. Baganoff of the Massachusetts Institute of Technology (MIT) in Cambridge, USA. "Only for this one object can our current telescopes detect these relatively faint flares from material orbiting just outside the event horizon."

The emission from Sagittarius A* is thought to come from gas thrown off by stars, which then orbits and falls into the black hole. Making the simultaneous observations required careful planning between teams at the two telescopes. After several nights waiting at the two observatory sites, they struck lucky.

"At the VLT, as soon as we pointed the telescope at **Sagittarius A*** we saw it was **active**, and getting **brighter** by the **minute**.

We immediately picked up the phone and alerted our colleagues at the APEX telescope," says Gunther Witzel, a PhD student from the University of Cologne.

Macarena Garcia-Marin, also from Cologne, was waiting at APEX, where the observatory team had made a special effort to keep the instrument on standby. "As soon as we got the call we were very excited and had to work really fast so as not to lose crucial data from Sagittarius A*. We took over from the regular observations, and were in time to catch the flares," she explains.

Over the next six hours, the team detected violently variable infrared emission, with four major flares from Sagittarius A*. The submillimetre-wavelength results also showed flares, but, crucially, this occurred about one and a half hours after the infrared flares.

The researchers explain that this time delay is probably caused by the rapid expansion, at speeds of about 5 million km/h, of the clouds of gas that are emitting the flares. This expansion causes changes in the character of the emission over time, and hence the time delay between the infrared and submillimetre flares.

Although speeds of 5 million km/h may seem fast, this is only 0.5% of the speed of light. To escape from the very strong gravity so close to the black hole, the gas would have to be travelling at half the speed of light—100 times faster than detected—and so the researchers believe that the gas cannot be streaming out in a jet. Instead, they suspect that a blob of gas orbiting close to the black hole is being stretched out, like dough in a mixing bowl, and this is causing the expansion.

The simultaneous combination of the VLT and APEX telescopes has proved to be a powerful way to study the flares at multiple wavelengths. The team hope that future observations will let them prove their proposed model, and discover more about this mysterious region at the centre of our Galaxy.

Notes for Editors

Sagittarius A* is a compact object located at the centre of our own Milky Way Galaxy, at a distance of about 26 000 light-years from Earth. In recent years, observations of stars orbiting in its strong gravitational grip have convincingly proven that Sagittarius A* must be a supermassive black hole with a mass of about four million times that of the Sun.

The 12 m Atacama Pathfinder Experiment (APEX) telescope is located on the 5000 m high plateau of Chajnantor in the Chilean Atacama desert. APEX is a collaboration between the Max-Planck-Institute for Radio Astronomy (MPIfR), the Onsala Space Observatory (OSO) and ESO. The telescope is based on a prototype antenna constructed for the ALMA project.

Operation of APEX at Chajnantor is entrusted to ESO. For this project, the researchers used the LABOCA bolometer camera on APEX.

The Very Large Telescope (VLT) at the 2600 m high Cerro Paranal is ESO's premier site for observations in visible and infrared light. The VLT has four "Unit Telescopes", 8.2 m in diameter, operating with a large collection of instruments. For this project, the researchers used the NACO adaptive optics instrument on the fourth Unit Telescope, "Yepun".

This research is presented in the paper by Eckart et al., "Simultaneous NIR/sub-mm observation of flare emission from Sgr A*", to appear in Astronomy and Astrophysics. It is available online at http://arxiv.org/abs/0811.2753

More and more evidence will be coming that something extraordinary is happening in the center of Milky Way. The cosmic beams are coming; they will be visible on 13 Dec 2012—just as depicted in the crop formations.

The situation is becoming ripe, so We have to start preparing!

We are on the door of something wonderful and something extraordinary! So do not fear.

Awake, remember, step forward, get together, unite and let's do this! Let the **Light** from the **Creator** shine upon all of you!

Astralwalker

-END NOTES-

The following statement is courtesy of Project Camelot

Although we support these meditations—from what we understand via testimony from our witnesses such as Jake Simpson and Dan Burisch, our solar system has *already* entered this area of the galaxy encountering waves of energy i.e. a galactic superwave based on the research of *Paul LaViolette*: *"(1981-82): LaViolette was the first to measure the extraterrestrial material content of prehistoric polar ice. Using the neutron activation analysis technique, he found high levels of iridium and nickel in 6 out of the 8 polar ice dust samples (35k to 73k yrs BP), an indication that they contain high levels of cosmic dust. This showed that Galactic superwaves may have affected our solar system in the recent past."* . . . and are now affecting our planet.

Some of the proposed effects are thought to act upon our consciousness as well as our DNA. As we move into alignment with the galactic center these effects are expected to increase and to reach a sort of peak around the years 2012-2013.

- **Nexus 2012 Series of Meditations, aka *The Gathering:*** As many of you are aware, we are assisting Astralwalker and *www.HealingExperiment.com* by announcing a series of meditations for peace and oneness, for healing of the Earth and creating a new paradigm . . . featuring what they have termed *Rainbow Warriors* around the world; this includes Ground Crew members, indigos and all other gifted children (all children are gifted), indigenous people and anyone else who wants to participate.

These meditations are happening starting on February 8th and will continue every Saturday, **culminating in a serioes of global sacred site meditations that will begin on May 9, 2009.**

This next section provides information concerning a couple of the types of crystals located at this site along with an obscure location of these crystals. If you know anything about 19.5 degrees then these colossal deposits might give you some insight as to what is behind the physics that have empowered some of the anomalies associated with mystic sites at these latitudes, like the Bermuda Triangle.

It has been stated that 6000 years into the future the earth is barren of life. In response to that I would hypothesize that perhaps by sucking and moving our planets natural resources on a never ending basis we are rendering it unable to call forth, harbor, or sustain the ability to receive the immortal spirit that is the living force within all sentient life forms. This planet is alive, and like anything else, if you continue to remove its internal workings it is going to die.—(Sabre)

The following information is provided courtesy of Project Camelot

- A few notes from a recent discussion with our witness Jake Simpson. We talked about the so-called "hollow Earth" and a technology in the public domain called *Metalstorm*.

Re the hollow Earth (or—more precisely—an air-pocketed Earth's crust that may be more cavernous than previously suspected or known in the public domain):

- One clue is the relatively recent find in *Mexico of large caverns filled with giant crystals.*

- He mentioned that two German ships disappeared in the Arctic . . . and were gone longer than they could account for.
- Within the Earth's crust are large pockets of empty gaseous space that were created as the planet cooled after formation.
- He also talked about plugs that were built and mounted on the sea floor by the Navy, around which were built undersea cities currently in use that will provide safe sanctuaries for select groups in the event of Earth changes to ensure the survival of the human race. (Note that *Henry Deacon* also told us that large undersea bases exist.)

-END NOTE-

Crystal Cave of Giants in Mexico

The Naica Mine of Chihuahua, Mexico, is a working mine that is known for its extraordinary crystals. Naica is a lead, zinc and silver mine in which large voids have been found, containing crystals of selenite (gypsum) as large as 4 feet in diameter and 50 feet long. The chamber holding these crystals is known as the Crystal Cave of Giants, and is approximately 1000 feet down in the limestone host rock of the mine.

The crystals were formed by hydrothermal fluids emanating from the magma chambers below. The cavern was discovered while the miners were drilling through the Naica fault, which they were worried would flood the mine. The Cave of Swords is another chamber in the Naica Mine, containing similar large crystals.

The Naica mine was first discovered by early prospectors in 1794 south of Chihuahua City. They struck a vein of silver at the base of a range of hills called Naica by the Tarahumara Indians. The origin in the Tarahumara language seems to mean "a shady place". Perhaps here in the small canyon there was a grove of trees tucked away by a small canyon spring.

From that discovery, until around 1900, the primary interest was silver and gold. Around 1900 large-scale mining began as zinc and lead became more valuable.

During the Mexican Revolution the mine was producing a great deal of wealth. Revolutionary troops entered the town and demanded money from the owners. One of them was assassinated when he refused to pay, causing the mine to shut down from 1911 to 1922.

Just before the mine was closed, the famous Cave of Swords was discovered at a depth of 400 feet. Due to the incredible crystals, it was decided to try to preserve this cave. While many of the crystals have been collected, this is still a fascinating cave to visit. In one part there are so many crystals on one of the walls, they appear to be like an underwater reef moving in a gentle undulating motion in an ocean current.

In April 2000, brothers Juan and Pedro Sanchez were drilling a new tunnel when they made a truly spectacular discovery. While Naica miners are accustomed to finding crystals, Juan and Pedro were absolutely amazed by the cavern that they found. The brothers immediately informed the engineer in charge, Roberto Gonzalez. Ing. Gonzalez realized that they had discovered a natural treasure and quickly rerouted the tunnel. During this phase some damage was done as several miners tried to remove pieces of the mega-crystals, so the mining company soon installed an iron door to protect the find. Later, one of the workers, with the intention of stealing crystals, managed to get in through a narrow hole. He tried to take some plastic bags filled with fresh air inside, but the strategy didn't work. He lost consciousness and later was found thoroughly baked.

When entering the cave our group is issued helmets, lanterns, rubber boots, and gloves. One must then be driven by truck into the main mining tunnel called Rampa Sn. Francisco. While the vertical drop is approximately 1000 feet, the drive is almost a half mile long. The heat steadily increases and women have been observed to begin "glowing". The truck stops in front of a concrete wall with a steel door. The intense heat can prevent brain functioning.

At the end of the tunnel there are three or four steps into the aperture of the cavern itself. It is in this short tunnel. In this short distance the temperature and humidity goes from being uncomfortably warm to literally a blast furnace.

Momentarily, the penetrating heat is forgotten as the crystals pop into view on the other side of the "Eye of the Queen". The entire panorama is now lighted and the cavern has a depth and impressive cathedral-like appearance that was not visible on earlier trips with just our headlamps.

When inside the great cathedral of crystals, the pressure of intense heat create a gamut of emotions and perhaps hallucinations. One can only remain for a short period of time.

Geologists report that these natural crystal formations are incredibly complex, yet so simple. They have a magical or metaphysical personality independent of their chemical structures. There is a magma chamber two to three miles below the mountain and that heat from this compressed lava travels through the faults up into the area of the mine. Super heated fluids carry the minerals the miners are seeking as well as form the crystals. The mine is ventilated; otherwise, it could not be worked. Some parts, however, are not air-conditioned, such as the Cave of the Crystals, and there you feel the heat from the magma deep below. The fluids travel along the Naica fault, enter voids in the bedrock, and then form entirely natural structures that are not easily explained scientifically.

In April 2000, the mining company became confident that the water table on the other side of the fault had been lowered sufficiently to drill.

When they did this, it is almost as if a magical veil of reality was breached and an entirely new world was discovered. Two caverns filled with the Earth's largest crystals were immediately revealed. More discoveries are expected to be made in this magical kingdom of intense natural beauty.

Selenite, the gypsum crystal, named after the Greek goddess of the moon, *Selene*, due to its soft white light, is said to have many metaphysical and healing benefits. Selenite powder has been used cosmetically for thousands of years to enhance one's natural beauty. It is believed that this crystal assists with mental focus, growth, luck, immunity, and soothes the emotions.

In the News . . .

Giant Crystal Cave's Mystery Solved National Geographic—April 7, 2007

It's "the Sistine Chapel of crystals," says Juan Manuel García—Ruiz. The geologist announced this week that he and a team of researchers have unlocked the mystery of just how the minerals in Mexico's Cueva de los Cristales (Cave of Crystals) achieved their monumental forms. Buried a thousand feet (300 meters) below Naica mountain in the Chihuahuan Desert, the cave was discovered by two miners excavating a new tunnel for the Industrias Peñoles company in 2000. To learn how the crystals grew to such gigantic sizes, García-Ruiz studied tiny pockets of fluid trapped inside.

The cave contains some of the largest natural crystals ever found: translucent gypsum beams measuring up to 36 feet (11 meters) long and weighing up to 55 tons. The crystals, he said, thrived because they were submerged in mineral-rich water with a very narrow, stable temperature range—around 136 degrees Fahrenheit (58 degrees Celsius). At this temperature the mineral anhydrite, which was abundant in the water, dissolved into gypsum, a soft mineral that can take the form of the crystals in the Naica cave.

Giant crystals enjoyed perfection BBC—April 6, 2007

With lengths over 11m, the giant gypsum crystals found in Mexico's Cueva de los Cristales are a great natural wonder

February 8, 2001—Discovery News

The largest natural crystals on Earth have been discovered in two caves within a silver and zinc mine near Naica, in Chihuahua, Mexico, according to mine officials. Reaching lengths of over 20 feet, the clear, faceted crystals are composed of selenite, a crystalline form of the mineral gypsum.

"Walking into either of these caves is like stepping into a gigantic geode," said Richard D. Fisher, an American consultant with the mining company to develop the discoveries as tourist attractions. Fisher said that most people can endure only a few minutes in the caves due to their high temperatures. The smaller of the two, which is about the size of two-bedroom apartment, is 100 Fahrenheit. The large chamber, which Fisher describes as the size of a Cathedral, is 150 F. Both are located approximately 1200 feet below the surface. The mining company plans to air-condition the caves before opening them to the public next year, Fisher said. He adds that reducing the heat gradually will not harm the crystals.

The largest previously known crystals were found in the nearby "Cave of the Swords", part of the same mine system. Some of these are now on display at the Smithsonian Institution. The local government and mine owners hope to avoid removing any of the new discoveries for museum displays or private collections, Fisher said.

Images: (c) Javier Tru

eba / Madrid Scientific Films /

The following information is provided courtesy of Project Camelot

- A few notes from a recent discussion with our witness Jake Simpson. We talked about the so-called "hollow Earth" and a technology in the public domain called *Metalstorm*.

Re *Metalstorm*:

- He said it was developed in Australia and is technology that will be used in the future by the military.
- It allows for high speed chamberless ammunition: metal tubes full of projectiles designed to be fired electronically.
- This technology allows the military to target *only* the enemy . . . and avoid collateral damage. If two men are sitting in a cafe talking it can take one out and leave the other unhurt.
- It can be used by drones or UAVs to intelligently target populations and then only specific individuals.
- It can acquire targets at light speed.
- It can be used in distributing viruses or in other lethal or non lethal applications.
- All of the above information he said is in the public domain but that there is less information out there now than in the past.
- • Mars anecdote: I heard from a source today who asked me to relate the following anecdote about his experience as a former NASA intern:

. . . He said that during a meeting he commented on a poster which showed people on Mars. When he asked if it was a depiction of a future planned mission, the person he was meeting with (who had worked in government for over 30 years) said it was a mission which had already been done. He then asked, "Do you mean the rover?"

And she replied, "No, we've sent people", as if she thought it was common knowledge.

So then he asked if she was sure, then she paused and looked up at the poster again and said, "You never know around here".

-END NOTES-

-D.U.M.B.S REPORT-

The intelligence data that follows was compounded and written as you see it here in approximately 2003. Although classified documents have been written much earlier in reference to projects that were being primarily conducted at the DULCE lab, this information was sourced from "The DULCE Book", written by an individual under the pseudonym "Branton". I have read this book and I recommend that everyone should read it. If there is any truth to what is contained within it, then it contains detailed information about the different types of aliens currently visiting this planet, and also describes a great war that is taking place all around us within our solar system and abroad.

I can confirm that the DULCE LAB does exist when read through the pages of the Air-Force Report that I read in 1994/95. Currently DULCE is the central hub for the US and all of these bases have been finalized to serve their purpose as the time is at hand for them to receive those who have been chosen to inhabit them during the difficult times and Earth changes ahead.

However, I must point out that I cannot personally confirm 100% that these bases exist at all, as I have not physically been to one and seen it with my own eyes. That does not mean I deny their existence, however what I can say is, as the information contained herein was sourced from other information that is now known to be 25 years old, the chances of it still being 100% accurate concerning the current usage of these bases is not probable.

-Sabre

The following information was souced from the Dulce Book by Branton

**My sources of information include people who worked in the labs, abductees taken to the bases, people who assisted in the construction of them, intelligence personnel (NSA, CIA, etc.) and UFO-Inner Earth Researchers . . . The Dulce Book by Branton

Joint Human and Alien Underground Bases

Pine Gap—Alice Springs, Australia. This base is a massive multi-leveled facility run by the Club of Rome which, like the Bilderberger organization, is reputedly a cover for the Bavarian Illuminati. Pine Gap is to be a major control center for the New World Order Dictatorship and is equipped with levels of computer terminals tied-in to the major computer mainframes of the world.

Dulce Base—under Mt. Archuleta, Dulce, New Mexico. Located close to the Colorado border and situated on the Jicarella Apache Indian Reservation. The town of Dulce is located off U.S. Route 64 population 900-1,700. A small town with one motel and a gas station. The base is located 2.5 miles northwest of Dulce and almost overlooks the town. Joint CIA-Alien base. 95 miles northwest of Los Alamos. Biogenetics Laboratory including but not limited to: Atomic Manipulation, cloning, studies of the human aura, advanced mind control applications, animal/human crossbreeding, visual and audio human chip implantation, abduction and feeding off of humans including children.

The Second Largest Reptilian and Grey Base in North America. The Central Hub.

- 1st Level—contains the garage for Street Maintenance.
- 2nd Level—contains the garage for trains, shuttles, tunnel-boring machines and UFO maintenance.
- 3rd Level—the first 3 levels contain government offices.
- 4th Level—Human Aura Research as well as aspects of Dream Manipulation, Hypnosis, and Telepathy. They can lower your heartbeat with Delta Waves and introduce data and programmed reactions into your mind (for those implanted with brain chips). Most people already are, they just don't know it.
- 5th Level—witnesses have described huge vats with amber liquid with parts of human bodies being stirred inside. Rows and rows of

cages holding men, women and children to be used as food. Perhaps thousands.

- 6th Level= privately called "Nightmare Hall." It contains the genetic labs. Here are where the crossbreeding experiments of human/animal are done on fish, seals, birds, and mice that are vastly altered from their original forms. There are multi-armed and multi-legged humans and several cages and vats of humanoid bat-like creatures up to 7 feet tall.
- 7th level—Row after row of 1,000s of humans in cold storage including children.

Wright Air Force Base—Dayton, Ohio. Includes a Warehouse with multi-levels underground packed with alien craft, hardware, and even bodies on ice. Headquarters of the infamous Project Blue Book.

Groom Lake, Area 51—Nevada. Includes the Nellis AFB test range but has nothing to do with underground nuclear testing. Huge underground facility where the exchange of technology takes place.

Located approx 125 miles north-northwest of Las Vegas and consists of the **Groom Lake and the Papoose Lake** Complexes. The expanded eastern portion of the property is known as the **S-4 site**, others allege its the southwest corner of Area 51. Either way the S-4 area is where the UFOs are stored. Also **Dreamland** (Data Repository Establishment and Maintenance) under the Groom Mountains is controlled by Greys.

Groom Facility—Large storage area in the tunnels that holds thousands of alien craft parts.

Dougway, Utah—underground shuttle system link between Dreamland and Dulce.

Page, Arizona—underground shuttle system link between Dreamland and Dulce

Mercury, Nevada—underground shuttle system link between Dreamland and Dulce

Burley, Idaho—underground shuttle system link between Dreamland and Dulce

Denver International Airport—Huge multilevel "city" under the airport.

Oklahoma City—multi-level base and underground shuttle system link between Dreamland and Dulce

Madigan, Fort Lewis, WA—Grey nest

Lakeport-Hopland, Montana—nest of Greys

Lassen & Deep Springs CA—nest of Greys

** It is the bases and connection link bases of Mercury, Nevada; Burley, Idaho; Dougway, Utah; Page, Arizona; the underground systems below the Denver International Airport and also Oklahoma City that are strategic sites that the NWO MUST maintain control over if they are to force America to submit to a one-world government. So it's these that need to come down!

** Other connecting facilities for the Tube shuttle system that link all these bases together are: Taos, N.M., Datil, N.M., Colorado Springs, CO, Creede, CO, Sandia, CO, Carlsbad, N.M. this also extends into a global system of tunnels and sub-cities.

Los Alamos National Laboratory—New Mexico. Genetics Research and bio-technology focused on mind control programming, genetics engineering as in cloning and DNA mapping known as "the exploration of the human genome."

Edwards Air Force Base—Mojave Desert, Southern California— underground base extends as far as 2 miles down.

Neu Schwabenland, Antarctica

Alsace-Lorraine Mountains area of France-Germany

Death Valley—Panamint Mountains region, California. Huge "Federation" base, Andro-Pleidian Nordic base (non-interventionists)

Camp Hero near Montauk Point—Long Island, NY. 8-Level Base that connects to the ITT center in New Jersey. The ITT center also has a connection/link to the Sub-Global Network.

Underwater bases off the coast of **Florida and Peru.**

Mt. Shasta—Telosian-Agharti Alliance (humanoid base). The central metroplex of Telos is said to consist of a multi leveled complex over 5 miles deep and at least 20 miles in circumference. Telos has subterranean connections via tube shuttles to at least 100 other subterranean cities below North America as well as to cites below South America such as the city of Posid below the Matto Grasso region of Brazil. The site of Alien and human meetings. Includes the infamous Telos City where it has been said that every president since Grover Cleveland has visited there.

Underground base in the **Plumas National Forest** in Northern California

Ehachapi, California—major center for joint alien/human activity

Kirtland Air Force Base—replicates UFOs, also an underground research facility where abductees have been and are taken (just as Dulce and others where their implanted, used and abused and sent back home). Located in the Manzano Mountain range south of Kirtland, AFB.

Blue Lake, New Mexico—base under the lake. UFOs seen both entering and and exiting the water.

Human UFO Fleet Bases:

Norton Air Force Base, California—UFO Garage. Known to store at least 3 of them.

Craft are stored in SE Dulce, Durango, CO, Taos, NM and the main fleet is stored at Los Alamos (under).

Los Alamos and the mountainous regions east and southeast of it in and around the Santa Fe National Forest are allegedly the major nest of Reptilian and Grey forces in North America, although there

are a number of large dens scattered throughout the underground networks between Dulce and Area 51.

Dulce is a major through point as a central infiltration zone for surface operatives, as well as an operational base for abduction-implantation-mutilation agendas and also a major convergence for sub-shuttle terminals, UFO ports, and so on.

1,700 paved miles of roads under Dulce and Northern New Mexico, towards Los Alamos is another 800 miles of tunnels.

The underground highways travel in the same directions as the shuttle system known as the Terradrive Shuttle or the Sub-Global System. One must be a very high ranking Mason, 33 or higher, Corporate-Intelligence Agent, or an Alien to gain access to the shuttle/tube system. Even though it's paid for from your tax dollars!

What follows is information on frequency weaponry, its history, its usage and its capabilities. This is the really dangerous stuff as far as man-kinds weaponry goes. Some of these weapons are space based weapons and some of these weapons are actually based on the moon itself. There is an ancient prophecy that states there will be a thousand years of peace when the Lion will sleep with the Lamb.

Once again I'm going to go out on a limb here to say that this is one way of saying that, things will be good for man if he can use technology for the greater good of everyone on the planet and not use it as a weapon for the destruction of everything to include the planet.—(Sabre)

Historical Background of Scalar EM Weapons by Lt. Col. T.E. Bearden (retd.), 1990 Copyright

Colonel Bearden is a nuclear engineer, wargames analyst, and military tactician with over 26 years experience in air defense systems, tactics and operations, technical intelligence, antiradiation missile countermeasures, nuclear weapons employment, computerized wargames and military systems requirements.

[This was a paper hastily whipped together, some years ago, to be able to send an abbreviated background paper to correspondents. The history prior to this paper runs from Hamilton and his quaternions to Maxwell and his quatemion theory, to Heaviside's vector curtailment of Maxwell's theory, to Nikola Tesla, to Whittaker, to Einstein and relativity, to Kaluza and Klein, and to the beginning of quantum mechanics.]

We begin our history in 1939 at *T.H. Moray's lab in Salt Lake City*. In that year, a Russian agent obtained detailed drawings of Moray's specialized amplifier which extracted energy from the powerful quantum mechanical fluctuations of vacuum. *Moray's radiant energy device* weighed 55 pounds and produced 50 kilowatts of power without conventional input. Numerous demonstrations are documented by engineers, scientists, and community leaders. After extensively testing Moray's device and obtaining the drawings by subterfuge, the Soviet agent destroyed the device. Moray, a truly great pioneer who was unjustifiably ignored in his time, had expended several hundred thousand dollars and exhausted his funds on the first unit. He was never financially able to rebuild it.

Thus in 1939 the Soviets obtained the secret, detailed drawings for a 29-stage electromagnetic device far ahead of its time. Moray had made the first germanium transistor, an amorphous pellet of multiple, finely powdered ingredients, sintered under heat and pressure to lock-in stress, and containing minute interfaces (very tiny built-in cracks) which acted as tiny scalar EM interferometers. Over the years Moray had painstakingly constructed 29 special tubes, each comprised of a blown quartz envelope containing several (usually three) of his transistor pellets. Only one in 300 of Moray's tubes would work, and thousands had been built to obtain the 29 good ones used in Moray's radiant energy amplifier. Each of Moray's pellets in the tube produced a vacuum state far from thermodynamic equilibrium, and the tri-assembly functioned as a macroscopic scalar interferometer and collector. Moray's 55-pound amplifier curved local spacetime and produced 50 kilowatts of usable load power from the curved vacuum source itself. Additional power could be taken from intermediate stages as well.

In short, Moray had produced a "Prigogine" transistor, still slightly more advanced than the transistors of today. [Ilya Prigogine received the Nobel Prize a few years ago for developing a new thermodynamics of systems far from thermodynamic equilibrium. Negative entropy is possible in such a system.]. Greatly stimulated by Nikola Tesla's repeated pronouncement that the ether was a sea of energy, Moray had painstakingly developed his special transistors, and arranged an assemblage of them to form a scalar interferometer and collector of the disintegrated "energy" (virtual particle flux) of vacuum, integrating it into real observable energy fed to a load. Some day *Moray*, together with Tesla, will be recognized as the first great scientific geniuses who succeeded in engineering scalar electromagnetics, electrogravitation, local general relativity, and Prigogine's new thermodynamics.

So in 1939 detailed drawings of Moray's unit were obtained and forwarded to Russia, along with details of experiments that the Soviet agent performed with the device in Moray's laboratory. Moray's lab still stands in Salt Lake City, Utah, operated by his son, John Moray, who has faithfully carried on his father's work.

However, in 1939 World War II was looming ominously on the horizon. Japan had started her expansive march in the Far East in 1931. In Germany, the Nazi Party and Adolf Hitler had risen to power in 1933. Italy had moved against Ethiopia in 1935-1936. From 1936 to 1939, Italy, Germany, and the Soviet Union were unofficially involved in the Spanish avil War, testing weapons for a wider new conflict in the future. The Sino-Japanese War in China (1937-1942) was proceeding. Germany had seized Austria and most of Czechoslovakia to the dismay of the West.

On September 1, 1939, Germany invaded Poland, and two days later Great Britain and France declared war. British and allied troops were forced off the continent, however, and miraculously evacuated at Dunkirk in May-June 1940. In June 1940 Italy, under Mussolini, entered the war and France surrendered shortly thereafter. In the latter half of 1940 a valiant Britain held firm against an onslaught of German bombers, and the restless Nazi juggernaut began to face toward Russia.

Thus in the cataclysmic events of 1939-1940, a Russia looking full
in the face of a Nazi threat and raging war in Europe had little time or
inclination available to pursue the bizarre, puzzling drawings and notes
obtained from Moray's lab. The pace of events in Europe continued
inexorably, occupying all the energy and attention of the Soviet
Union. In attempting to sidestep attack by Hitler, Russia had signed a
Nonaggression Pact with Germany in August 1939. Consolidating her
position, Russia had engaged in a stubborn, bloody war with valiant little
Finland from November 1939 to March 1940, suffering heavy casualties
and leading Hitler to conclude that Russia was militarily weak. In April
1940 Germany had occupied Norway and Denmark, and in May-June
1940 German armies had disastrously routed the British and French at
Dunkirk. As 1940 wore on, Hitler had moved into Romania, Bulgaria,
Yugoslavia, and Greece. A beleaguered Russia signed the Soviet-Japanese
Neutrality Pact of April 13, 1941 so that Japan would not pin Germany
in striking the Soviet Union. By early spring of 1941, German attack
on Russian forces was imminent, with daily overflights of Russian lines
and probing incidents.

On June 22, German divisions attacked in a mighty blitzkrieg,
encountering little resistance, and the war between Russia and Germany
had begun in earnest. As war raged on the Eastern Front and in Europe,
on December 7, Japan attacked Pearl Harbor, and the United States
entered the war against Japan and Germany.

The history of World War II is well-known and need not be repeated
here.

However, during the war the U.S. developed the atomic bomb in a
feverish and highly secret technical effort. During the Potsdam conference,
the U.S. exploded its first atomic bomb in the New Mexico desert. At the
conference, Stalin was informed of the bomb and its first successful test by
Truman, who had replaced the deceased Roosevelt as U.S. President. An
impassive Stalin showed no apparent concern, but inwardly he knew that
his secret plan to conquer Europe a year or two after the war—by which
time the West would have beaten its swords back into plowshares—now
had to be abandoned. A short time after, the U.S. dropped two bombs
on Japan by stunning surprise. Faced with total and certain annihilation,
the heretofore fanatical Japanese capitulated.

As one aftermath of the U.S. development and use of the atomic
bomb, Stalin—ever the total dictator—lashed his Soviet Academy of
Sciences furiously. Undoubtedly he pointed out that the destiny of

Communism had been frustrated by this great technical breakthrough made by the Americans (with help from the British and from political European refugees.) He informed his scientists in no uncertain terms that the next such breakthrough had better be Soviet. He planned to do exactly what the U.S. had done: Find a new area for a great technical breakthrough and superweapon, put the entire resources of the nation behind it, develop it in great secrecy, and thrust it upon the U.S. and its allies at the eleventh hour. The West would then be forced to capitulate, just as was Japan. We had already shown Stalin the successful scenario by which to win the next war.

Ironically, Stalin's "great technical breakthrough" was to come from Germany's radar scientific team, taken to the Soviet Union after the war. This team had drastically advanced the theory of radar cross section and radar absorbing material (RAM). They were on the verge of discovering phase conjugate time reversed radar waves—which would enable the great new superweapons Stalin sought.

Meanwhile, the Soviets plotted a delaying campaign until they were ready to strike the West. Furious action was required to catch up to the U.S. in atomic weaponry. Obviously it was only a matter of time until long range rockets, as evidenced by the German Vl and V2, would be mated to atomic warheads, thrusting the decisive role of artillery into the strategic arena, far beyond the wildest imagination of then-current thinking. Also, air defenses against the American and British bombers were desperately needed. Guided rockets capable of shooting down American bombers enroute to critical Soviet areas were am absolute necessity. Atomic warheads were needed against massed bomber fleets, as had been thrown against Germany during the war.

The Soviet state must launch a superhuman effort to rapidly achieve its desperately needed missiles and atomic weapons, for it was totally vulnerable until that had been achieved. Meanwhile, the old methods that had proven so effective when one cannot fight a stronger foe head-on—guerrilla warfare, insurgency, and terrorism—must now be massively employed on a global scale to focus the attention and physical exertion of the U.S. dragon elsewhere. Third party "wars of attrition" would threaten American interests around the globe, and U.S. forces would be siphoned off in debilitating struggles against guerrilla wars and in distant lands. These sapping wars would "bleed the dragon" and weaken it, while the Soviet Union prepared a special technical sword with which to administer the final coup de grace.

Stalin was an absolute dictator, brooked no opposition, and killed his opposition. Knowing what was good for them, the Soviet Academicians did not dare to debate or protest against his ultimatum to search every field of knowledge, no matter what. Instead, Academy scientists vigorously turned to a massive search for the new breakthrough area. Scientific literature from the West was hauled to Russia by the shipload. Thousands of Soviet PhDs and engineers were put to work in huge analysis institutes, sifting through the literature and digesting it—and carefully noting anomalies and areas which should be followed up. Nothing even remotely approaching such a technical digestion and analysis effort has ever been attempted in the West.*

From this search it seems clear that the Soviet discovered Whittaker's 1903 and 1904 papers also prescribing the engineering method.

Two short years after WWII, Soviet planning for Stalin's new thrust was finalized, and the cold war was launched around the world. A great dialog began among the leading Soviet military leaders to hammer out radically new strategy for utilization of new weapons of mass destruction. Feverish effort to develop and deploy the first defensive missiles was underway, as was a crash effort to develop long range ballistic missiles, large bombers and atomic weapons. At the time, German and Soviet radar scientists probably began to discover—and puzzle over and work out the theory and hardware for-time reversed EM waves and phase conjugate mirrors.

The Moray drawings (and related agent reports) must have been resurrected. Nikola Tesla's notes, taken from his hotel room at his death in 1943 and later turned over to Yugoslavia, must have passed into Soviet hands also. Soviet espionage was quite successful in easing the task of developing atomic weapons. Soviet rocketry, funded and staffed massively, forged ahead, particularly in big rockets and solid propellants. The great postwar arms race was on, so far as the Russians were concerned.

Several years after WWII, then, the Soviets should have been embarked on experimentation with a Moray prototype amplifier. They should have been deeply involved in time-reversed EM wave experimentation to rediscover what Tesla actually had done in wireless transmission of energy without loss. Also, the earlier Kaluza-Klein unified electromagnetics and gravitation theory(1921-1926) could not have eluded the Soviet scientists, who have consistently led the world in nonlinear mathematics.

Of necessity, the search for a new breakthrough area would have initiated intensive review of the foundations of physics and electromagnetics, in an effort to discover any "holes" that might exist. Thus the short debate at the turn of the century, that established Heaviside's limited version of Maxwell's EM theory—largely because of Western repugnance for Maxwell's use of Hamilton's difficult quaternions—would not have gone undetected. Careful review of that situation would immediately have revealed that

(1)   Maxwell's theory, which was worked out in quaternions rather than vectors, had actually not been completely captured by its translation into Heaviside's vectors by none other than Heaviside himself. [Heaviside was the cofounder of modern vector analysis and almost single-handedly produced what today is the recognized vector form of Maxwell's equations. He also nearly single-handedly transformed Maxwell's theory into its modem form.].

(2)   The effects of stresses in the vacuum medium had been eliminated by Heaviside, while Maxwell certainly had pointed out additional terms required to address this issue. [A Soviet team addressing this omission, would have inevitably connected it to Whittaker and Kaluza-Klein theories, revealing electrogravitation (scalar electromagnetics).].

(3)   A quaternion, which contains a scalar part in addition to a vector part, could maintain** its non-zero scalar part when the vector was zeroed, whereas in vector analysis, when the vector is zeroed, it is assumed that no residue exists. [This is also directly connected with the mistaken assumption that an abstract vector space itself can have no substructure.]

** *and even vary and oscillate, its internal EM stress.*

(4)   The direct connection could be made with Whittaker theory to produce scalar vacuum stress gravitation when the EM vector fields were zeroed.

(5)   The West had missed this unified theory because of its total aversion to quaternions, and its elevation of the curtailed Maxwell's theory to cult status.

(6)   Investigative experiments to explore bucking EM force fields which neatly zero sum would have almost immediately revealed

highly anomalous behavior of materials and circuits. [Such simple investigations of zero-vector EM force field summation do not seem to appear in the Western literature at all, so far as can yet be established.]***

*** *with the single exception of the Aharonov-Bohm effect, which has finally been proved after 27 years of controversy, and Hooper's work which was obscurely published.*

Something very much like that, together with quantum mechanical ideas, would have had to be applied to explain the operation of the Moray device. Note that **every one of the suppositions above was available to any thorough Soviet search**—and the Soviet scientists certainly made the search, more massively and thoroughly than has ever been done before or since. Since the evidence is overwhelming that the Soviet scientists developed electrogravitation and scalar EM weapons, it seems logical that their search succeeded along these lines or similarly.

About 1950-1952, the Soviets developed EM machines that could influence the brain and nervous system directly. This included the Lida machine, which can induce a catatonic state into a mammal such as a man, a cat, etc. U.S. scientists, obtaining one of these devices in the 1980s, reported that it utilized a 40 MHz carrier, and produced unusual waveforms (showing the multiple frequency content). Since the U.S. scientists do not possess scalar EM detectors, they have no measurements or knowledge of possible scalar components in the Lida's output signal. According to one U.S. scientist, the device was used by North Korean interrogators in brainwashing U.S. prisoners in North Korea during the Korean War, and was highly effective.

By the mid-50s, the Soviets should have been well along in prototype building, weaponizing, and development of larger devices and starting new weapons, based on what I have called **scalar electromagnetics**—unified EM/gravity field theory a la Whittaker or simply **electrogravitation** using negative energy and time-reversed EM waves.

In the winter of 1957-58, a monstrous nuclear accident occurred at a Soviet facility in the Ural Mountains near Kyshtym. [Inexplicably, for decades U.S. intelligence analysts feverishly resisted—and even suppressed—the evidence in their possession, adamantly insisting that such an event never occurred.]. Witnesses reported that **atomic wastes in the nearby storage site exploded.** Since scalar EM research and development would have been underway in the Soviet Union for several years, we can construct a scenario that will produce that nuclear event: Suppose a large nearby radar had been modified to radiate a scalar EM wave. Suppose it is also used to produce a standing scalar EM wave, after Whittaker.

As the radar continues to radiate, such a standing scalar EM wave is "charged up" to a higher and higher "potential", just like charging a capacitor. That is, in the standing wave, its "locked-in energy density of vacuum" is being steadily increased, simply by continuing to transmit with the radar in the scalar EM mode. However, what is actually being "charged up" or produced is a gravitational potential standing wave (and this G-potential is increasing.) Further, suppose the radar and its attached standing scalar wave (G-potential) are energized, and suddenly the radar transmitter suffers an electrical failure. The "scalar potential" represented by the charged-up vacuum energy (which can be enormous) in the standing wave would immediately discharge to earth in a sharp electrogravitational pulse (EGP). When a strong EGP pulse—which is absorbed by, and reradiated from, atomic nuclei rather than orbital electrons—struck stored radioactive nuclear wastes, the radioactive nuclei would all decay immediately, just as if internally exposed to an atomic bomb. A full nuclear explosion of the stored nuclear wastes—in the dirt and dirty—would occur, creating a terrible nuclear "accident."

That is exactly what caused the Kyshtym accident in 1957-58. *Failure of a large scalar EM transmitter produced a giant EGP "arc discharge" into the earth, striking the atomic wastes stored nearby.* The resulting explosion and radioactive fallout contaminated a very large area, which is still contaminated to this day. Hundreds—perhaps thousands—of casualties resulted.

So we can be confident that, in 1958, the Soviets were already experimenting with large scalar EM Whittaker beam weapons. We can also be confident that, after the Kyshtym disaster, the Soviets immediately realized the cause of the accident, and developed safety circuits to hold the standing wave from collapse in the event of electrical failure of the transmitter. Means to slowly "drain off" the accumulated EG potential safely into the earth—much like slowly discharging a capacitor—would also have been developed and incorporated into the beam transmitter.

Anomalous microwave radiation of the U.S. Embassy in Moscow began during the late 1950s, and has repeatedly occurred to this day. Twin beams have been noted in the radiation, suggesting interferometry. Scalar EM components on the carriers would not be detected by normal microwave instruments, although the weak carrier signals would. Further, the strength of the scalar EM wave's infolded components can be essentially independent of the strength of the carrier. Apparently a wide variety of physical effects were experienced by personnel in the Embassy, including several U.S. Ambassadors.

[Note that extensive tests of the electromagnetic transmission of cellular disease patterns between cell cultures have been reported by Soviet researchers, led by Kaznacheyev. *Kaznacheyev reported that the effects were due to near-UV photons emitted by the diseased cells and irradiating the test cells.* Such effects have been replicated at the University of Marburg, West Germany, using infrared emissions from diseased cells to irradiate test samples. At the University of Sydney, Australian researchers have accomplished the effect at a distance of over 100 feet, using what can only be called scalar EM radiation.].*

* Dr. Fritz Popp has discovered the virtual-state (i.e., scalar EM!) master control communication system for the cells in the human body.

The "Moscow radiation" has represented an intelligence probe for all these years: radiating a high-level U.S. target—the U.S. Ambassador to the Soviet Union—and inducing physical effects assured the involvement of high U.S. governmental agencies and official, including the CIA, DIA, NSA, the National Security Council, and the President. Eventually, the U.S. scientific community would also be asked to evaluate the signals.**

** *For a full explanation, see T.E. Bearden, Gravitobiology: A New Biophysics. Aug. 1989.*

The response of the U.S. Government and the counteractions taken (or not taken) at the U.S. Embassy in Moscow reveal whether or not the U.S. is aware of scalar electromagnetics and scalar interferometry, with **very high confidence.** Since apparently no U.S. scalar EM countermeasures have been taken at the Embassy, the Soviets have long been assured that the U.S. knows nothing of scalar EM, has not developed scalar EM weapons secretly, and possesses no defenses against the surprise use of scalar EM weapons against U.S. forces and installations.

In **January 1960**, Khrushchev announced the development of a new, fantastic weapon—one so powerful it could wipe out all life on earth if unrestrainedly used. **The New York Times** printed part of the story. Khrushchev, of course, was referring to the newly emerging scalar EM weapons. So in early 1960 the Soviets were in at least what we call the engineering development stage for large scalar EM beam weapons, which would be deployed when finished.

On **May 1, 1960** Soviet defensive radars—rigged as prototype scalar EM beam weapons—probably downed Francis Gary Powers's high-flying U-2 reconnaissance plane over the Soviet Union, precipitating a major diplomatic incident. At the time, no Soviet surface to-air missile could reach the high-flying U-2, but of course it was extensively tracked on Soviet defense radars. Powers reported that a flash occurred behind his plane **and persisted**—almost certainly the signature of time reversed wave real-time holography. Eisenhower first denied, then was forced to admit, the photo-reconnaissance nature of Powers' mission when Khrushchev revealed that the Soviets had captured Powers alive. Khrushchev then cancelled a major summit meeting with President Eisenhower. Powers was imprisoned, and released in February 1962 in exchange for Rudolf Abel, a convicted Soviet spy.

In the **fall of 1962**, with his new superweapons nearing deployment, the ebullient Khrushchev could no longer contain his eagerness. He "jumped the gun" and attempted to change the international balance of power at a single stroke. The Soviets started inserting MRBMs—weapons which would blanket the U.S. like a glove—into Cuba at a feverish pace. From information clandestinely fed by Soviet Col. Oleg Penkovsky, John F. Kennedy was aware that Khrushchev's bombers and ICBMs were in woeful shape.***

*** *Penkovsky knew nothing of the Soviet Weapons program in energetics, using time-reversed EM radar wave weapons.*

Not suspecting the Soviet superweapons, he knew that Khrushchev was bluffing—and proceeded to call the Soviet bluff, precipitating the Cuban Missile Crisis. Khrushchev was caught with his pants down. His superweapons were not yet ready, and the U.S. could have easily destroyed the Soviet Union. Reluctantly he backed down and removed the missiles, after first blustering enough to get an agreement from Kennedy not to attack Cuba.

In the Communist world, this was a serious loss of face by Khrushchev. He knew his days were numbered unless he did something dramatic to recoup stature.

Accordingly, as soon as his new superweapons were deployed and ready, Khrushchev did that "something dramatic." On **April 10, 1963** he destroyed the *U.S.S. Thresher* with one of his new weapons. Scalar EM (electrogravitational) beams, focused through the ocean to interfere on the Thresher under the surface, recreated spurious EM energy in the sub's electrical control circuits, jamming them so that the sub lost control, sank to crush depth, and imploded. Spurious electromagnetic "splatter" surrounding the immediate vicinity of the targeted area left a signature of intense EM interference with multiple systems and multiple frequencies of the **U.S.S. Skylark,** surface companion of the Thresher. This anomalous EM interference was so virulent that it required over 1-1/ 2 hours for the **Skylark** to transmit an emergency message back to headquarters that the sub had been lost.*** The death of the **Thresher** was Khrushchev's first blow.

*** Some electronic systems mysteriously malfunctioned, then later completely recovered spontaneously. This again is clearly a scalar EM weapon signature.

The next day, **April 11, 1963**, Khrushchev struck his second blow. Two deployed Soviet scalar EM weapons fired massive scalar EM pulses through the ocean. The two massive pulses met and interfered deep under the surface, 100 miles north of Puerto Rico. The resulting giant underwater EM explosion hurled a mushroom of water half a mile up into the air, and the anomalous explosion was observed (and later reported to the FBI and the Coast Guard) by the pilot and crew of a US. Jetliner passing nearby, enroute from Puerto Rico to Florida.

Thus we know the year—and indeed probably the very month—that Khrushchev's new "fantastic weapon" was deployed: April 1963.

In the 1960s, Curtis detected a previously unknown, anomalous, weak electromagnetic radiation pattern over the ocean. This is the type of pattern consistent with the introduction of scalar signals with a small' impure normal EM component or residue. Such cruder implementation would have been expected early-on in the Soviet program.

On **June 17, 1966** from the air near Teheran, Iran, *several airline pilots sighted a brilliant sphere of light,* "sitting on the horizon," so to speak, deep within the Soviet Union. The intensely glowing sphere expanded to enormous size, dimming as it expanded, always remaining "sitting on the horizon." The pilots observed the phenomenon for 4 to 5 minutes. A CIA report on the incident was released under the Freedom of Information Act.

In **1967-68**, anomalous holes appeared in clouds over the U.S., possibly associated with the beginning of early *Soviet weather engineering over the U.S.* That winter was particularly severe.

In the **late 1960's**, *Lisitsyn reported that the Soviets had broken the "genetic code" of the human brain.* He stated the code had 44 digits or less, and the brain employed 22 frequency bands across nearly the whole EM spectrum. However, only 11 of the frequency bands were independent. This work implies that, if 11 or more correct frequency channels* can be "phase-locked" into the human brain, then it should be possible to drastically influence the thoughts, vision, physical functioning, emotions, and conscious state of the individual, even from a great distance.

*To translate this to VHF or radar/radio frequencies, see Bearden, Gravitobiology, 1989.

It may be highly significant that

(1) up to 16 of the giant *Soviet woodpecker carriers* have been observed by Beck and others to carry a common, phase-locked 10-Hz modulation, and
(2) such a 10-Hz signal has been demonstrated by Beck, Rauscher, Bise, and others to be able to physically entrain or "phase-lock" the human brain, if stronger than the Schumann resonance of the Earth's magnetic field.

A human brain entrained by a common, phase-locked 10-Hz modulation on 16 carrier frequencies would effectively have 16 frequency channels phase-locked into it. The potential for using the *Woodpecker transmitters* to phase-lock an appreciable percentage of human brains in a targeted area, and then induce effects in the populace similar to—and even more drastic than—the effects induced in U.S. Embassy personnel in Moscow, should be strongly pointed out. Coupled with the Kaznacheyev work on EM transmission of cellular death and disease, using the *Woodpecker signals* to induce death and disease in the targeted populace may also be a distinct possibility. Modification of DNA/RNA—and viruses themselves—is also a possibility.

Many other Soviet scalar EM testing incidents are documented in Bearden, *Fer-de-Lance*. 1986, and in *AIDS: Biological Warfare*. 1988. Repeated presentations on the Soviet scalar EM weapons and their testing have been given at national symposia since 1978. The exact mechanisms utilized by these superweapons have also been given. A complete and consistent series of many anomalous events exists over the years, showing the continuing development, testing, and deployment of massive scalar EM weapons by the Soviet Union.

Actual use of the weapons to *create a giant interference grid over the U.S. and severely affect world weather* is documented in a videotape, Bearden, **Soviet Weather Engineering Over North America**, 1985. This videotape also includes several other major accidents, and also U.S. weather satellite photographs of anomalous exhausts from the Soviet Union's Bennett Island.** The exhausts are often in jets 150 miles long and nearly horizontal (about a degree and a half in elevation). They have been photographed coming from Bennett Island and the sea nearby, since 1974.

** Which actually belongs to the U.S.

Note that Bennett Island need not be the site of the actual scalar EM weapon itself. Energy extracted by a scalar EM howitzer (endothermic mode) from a distant region could be caught temporarily, then "relay fired" to Bennett Island by the same howitzer, switched into its exothermic mode. Thus the howitzer could act as an energy transmission relay between a distant endothermic target site and a separate distant exothermic exhaust site such as Bennett Island. Such exhausts have also been photographed corning from Novaya Zemlya.

On *January 23, 1974 a mysterious explosion over North Wales* rocked a 60-miles radius area. Associated anomalous light phenomena were also seen. Hundreds of anomalous booms, aerial rumbles, and aerial explosions, many accompanied by flashes or anomalous shaking of ground structures, buildings, windows, etc., have occurred over the U.S. and other Western nations in the 60s, 70s, and 80s. Many were simply Soviet tests.

In **June 1975,** Brezhnev called for a ban on weapons of mass destruction more terrifying than nuclear arms. He stated the need for an "insurmountable barrier" to the development of such weapons.

In July he repeated his strange proposal to a group of visiting U.S. Senators. Ponomarev, a Soviet national party secretary, again raised the same issue to a delegation of visiting U.S. congressmen in August. At the United Nations' thirtieth Session of the General Assembly on Sept. 23,1975, Foreign Minister Andrei A. Gromyko strongly raised the same issue, warning that science can produce "ominous" new weapons of mass destruction. He urged that all countries, led first by the major powers, should sign an agreement to ban the development of these unspecified new weapons. He even offered a draft, entitled "Prohibition of the Development and Manufacture of New Types of Weapons of Mass Annihilation and of New Systems of Such Weapons." The first article provided that the types of these new weapons would be "specified through negotiations on the subject." By its fixation on nuclear weapons and its ignorance of scalar EM, the West may have lost its only opportunity to prevent the spread of scalar EM weapons "more frightful than the mind of man has ever imagined," to use Brezhnev's characterization.

With obvious lack of knowledge of scalar EM weaponry by the West and failure to achieve a total ban, the Soviet hawks again were able to prevail over Brezhnev and the more conservative Party leaders. The Soviets decided upon a massive buildup of arms, with the intention of being prepared to dominate the world in a decade.

In **July 1976**, communications around the Earth were interrupted by the sudden emergence of powerful Soviet transmissions in the communications band, from 3-30MHz. The chirped signal produces a characteristic sound in a receiver similar to a woodpecker's beak hitting a wooden block. The transmitters were immediately dubbed "woodpeckers", and the signals "woodpecker signals." The Western intelligence community dubbed these giant transmitting systems "over-the-horizon radars." These powerful systems were brought on full deployment and activated after the Soviets decided to go ahead with a decade-long buildup to prepare to dominate the earth. That decision was reached after Brezhnev's 1975 failure to obtain world agreement banning development of scalar EM weapons. Activation of the giant Soviet *Woodpecker weapon systems* meant that the Soviet Union now would deploy massive scalar weapons on an unparalleled scale.

On **July 28, 1976**, a great earthquake destroyed Tangshan, China, killing some 600,000 persons. Just before the first tremor in the early morning, the sky lit up like daylight, with multi-hued lights seen up to

200 miles away. Electrical signals were also associated with the quake. Note that, in 1912, Nikola Tesla stated in an interview that it would be possible to split the planet, by combining vibrations with the correct resonance of the earth itself. Tesla stated, "Within a few weeks, I could set the earth's crust into such a state of vibrations that it would rise and fall hundreds of feet, throwing rivers out of their beds, wrecking buildings, and practically destroying civilization. The principle cannot fail . . ." Tesla once set off a growing local vibration and shaking of the entire neighborhood around his laboratory, using a 10-lb. device. Tesla later improved on his concepts, calling this area "telegeodynamics". In 1935 he said: "The rhythmical vibrations pass through the earth with almost no loss of energy . . . It becomes possible to convey mechanical effects to the greatest terrestrial distances and produce all kinds of unique effects . . . The invention could be used with destructive effect in war . . ."

Later in 1976 Soviet scientists increased the power of the giant transmitters to perhaps 40 megawatts each. The scalar EM power contained in the normal EM carriers is unknown. Transmitting pulses at about 7 Hz, it appears that enormous standing waves were inadvertently set up in-phase in the atmospheric duct around the earth. That is, the powerful carriers were circling the earth about 7 times per second, coming back around in phase with the previous loop. The scalar content also was in phase. The in-phase scalar component apparently produced the phenomenon of "kindling"; i.e., of "charging up" the loop with Whittaker (1903) gravity potential. This potential was perfectly in phase with the cavity resonance of the earth-ionosphere, and so the liquid-filled crustal features of the earth and the ionosphere coupled to it. In short, the potential became self sustaining, and increasing as long as the transmitters kept pouring in carrier power. The result was giant standing waves of totally unsuspected magnitude. The frightened Russians hurriedly turned off the transmitters, but were then faced with an undamped, sustained oscillation of the giant potential. Thoroughly frightened, the Soviets radiated intense bursts of power at the waves, trying to break up their coherence. According to one source, these waves lasted several months under that bombardment before gradually dissipating. Note that experiments by Dr. Robert Helliwell at the U.S. Antarctic research camp known as Siple Station has shown fairly conclusively that radio waves can be magnified up to 1,000 times or more in the ionosphere, according to reports in the open literature.

On **September 10, 1976,** the crew and passengers of British European Airway Flight 831, over Lithuania enroute from Moscow to

London, *observed an intensely glowing, stationary ball of light above the clouds underneath the plane.* When alerted by the airline pilot, Soviet authorities on the ground curtly informed him to pay no attention and, effectively, to exit the area. This Soviet reaction indicates that the incident may have been a Soviet test of an unknown type of device—in short, a scalar EM interferometer. If so, the interferometer intersection zone was deliberately placed near the aircraft to stimulate the crew and passengers. By the later reaction of the British government to the incident, the Soviets could ascertain whether or not Britain was cognizant of scalar electromagnetics.

On **May 18, 1977**, the Soviets signed an agreement with the U.S. and 29 other countries, promising not to attack each other by causing man-made storms, earthquakes, or tidal waves. The Soviets had already tested weather control against the U.S. in 1967 and had been steadily using it against America for almost a year.

On **April 2, 1978**, an anomalous straight, not jagged, beam of "lightning" came down from the sky at a 45 degree angle to the ground and struck Bell Island, Newfoundland, causing a loud explosion and damage to some houses, etc. Two cup-shaped holes about two feet deep and three feet wide marked the major impact. Wires leading to a shed and coop nearby were vaporized. Both structures suffered considerable damage, but no burning occurred except for a slightly scorched spruce tree. The anomalous bolt did not discharge into prominent metal contact points, such as a metal chimney running down to an iron stove. Instead, all the wiring on the property was blown out. A number of TV sets in Lance Cove, the surrounding community, also **exploded** at the time of the blast. Weather men confirmed that atmospheric conditions at the time were not conductive to lightning. The blast was heard 45 kilometers away in Cape Broyle. Apparently U.S. Vela satellites picked up the event. The incident was investigated promptly by two representatives from a U.S. weapons laboratory at Los Alamos, according to the news media.

On **Nov. 21, 1977**, cloud indications of a huge standing wave was observed off the Pacific coast of America, reaching from Alaska to Chile. Satellite photos show cloud banks over this stretch of the ocean, lying offshore, and reaching for the whole of the distance. These clouds grazed the land slightly at California. There a straight black line, as through drawn by a ruler, appeared in the cloud mass. It was an opening in the clouds, one mile wide and 200 miles long. There was no known explanation. This was a phenomenon without parallel in past records.

A series of anomalous high altitude booms occurred off the East coast of the U.S. around the end of the year in 1977-78, though some activity had started as early as July 1977. Flashes associated with some of the booms were observed. On Dec. 27, President Carter called for a full report. No reason could be found. These explosions represented adjustment of the interference grid that had been established over the U.S. by the *crossed Woodpecker beams*, and artillery-type high burst registration of scalar EM howitzers. Numerous other adjustments of the grid and registration of howitzers have occurred over the U.S.

The peculiar "nuclear flashes" seen by the Vela satellites in September 1979 and December 1980 could have been due to a testing of a scalar EM howitzer in the pulsed exothermic mode. In the mode, scalar EM pulses meet at a distance, where their interference produces a sharp electromagnetic explosion (hence the "flash", very similar to the initial EMP flash of a nuclear explosion. Even in the vacuum of space, such an explosive eruption of energy from within the local spacetime vacuum itself may be expected to lift matter from the Dirac sea, producing a plasma. Prompt absorption and re-radiation of energy from this sudden plasma may be expected to present nearly the same "double peak" profile as does a nuclear explosion. This was the profile presented by the flashes. Note that the second flash detected was apparently of an "explosion" primarily in the infrared, almost certainly ruling out a conventional nuclear event. It does not rule out, however, pulsed distant holography using pumped EM giant time-reversed wave transmitters.

From Afghanistan in **September 1979,** *British war cameraman Nick Downie observed gigantic, expanding spheres of light deep within the Soviet Union,* toward the direction of Saryshagan missile test range. *Saryshagan apparently contains at least one directed energy or particle beam installation* which could possibly function as a scalar interferometer/scalar EM howitzer. Downie observed multiple incidents in the direction of Saryshagan during the actual month (September 1979) that the first anomalous flash was detected by U.S. Vela satellites. On the other hand, there also exists tenuous evidence that these Vela events may have been associated with weaponry of another nation, not hostile to the U.S. There is evidence that this second nation also possesses scalar EM weapons, as do two additional nations besides the Soviet Union.

In latter 1980, a most anomalous drought was induced in the United States. On Feb. 2, 1981, the **Washington Post** commented: "For the past four months, a single weather pattern has gripped virtually the entire

United States, causing a coast-to-coast drought unique in the annals of weather recording . . . The weather system causing the drought is one of the most unusual national patterns ever recorded." It was also one of the most *artificial* ones ever recorded!

On **January 20-21, 1982** a swarm of more than 1,400 earth tremors occurred in that two-day period in north central Arkansas. Beginning on Jan. 12, activity had started in the swarm area, and three quakes registering above above 4.0 on the Richter scalar were recorded. The strongest was on Jan 20, and measured 4.5. There had been no previous recorded earth tremors in the area. Long dormant faults near the Ouachita Mountains were suspected as being responsible. Note that strong scalar waves, passing through the Ouachita fault zones, could have stimulated such activity, since a fault is a natural scalar interferometer and thus a scalar transmitter/receiver. Scalar reception would result in increased electricity in the rocks in the fault zone, in turn increasing the mechanical pressure in the rocks. Since the Soviets were heavily engaged in scalar EM weather engineering operations over North America in 1982, these tremors may have been side effects of those operations.

On **June 18, 1982**, pilots and crews of Japan Air Lines Flights 403 and 421 reported sighting a giant, expanding globe of light in the North Pacific, 700 kilometers east of Kushiro. This was another test of a scalar EM howitzer/interferometer producing a "giant globular shell" of energy at a great distance. When small, the intense shell produces a very high EGP and also a very high EMP inside the matter of any object penetrating the shell. The EMP will dud any and all electronic equipment; explode high explosives, fuels, and combustibles; and render any modern weapon harmless. The high EGP will detonate a nuclear warhead immediately in a "full-up" nuclear detonation. It will also instantly kill any living creature, including every cell, bacterium, virus, and organism in its body. It will also detonate any ordinary, non-radioactive material with a low-order nuclear detonation of all its nuclei. As the globular shell is made very large, "energy density" in its shell is reduced. However, any nuclear material or device will still suffer a low-order nuclear detonation from the EGP, and any biological system will still be instantly killed. The EMP will still dud any electromagnetic equipment presently made. As can be seen, the globe can be used to defend an entire sector of the sky against any kind of incoming threat—with 100% effectiveness. A hemispherical shield can be placed over one's own field army for terminal stage defense, and/or over the opposing force for initial phase defense, *against most everything!*

On **July 20, 1982**, Soviet official Lysenko of the Soviet Embassy in Washington, D.C. stated publicly that, should nuclear disarmament fail, the Soviets would quickly introduce new weapons more powerful than nuclear arms, and these weapons would not be verifiable. U.S. Army Col. John Alexander, together with this author, was in the audience and heard Lysenko's statement.

In latter **December, 1982**, odd, anomalous atmospheric booms occurred over Ohio. These represented tests and adjustments of the scalar EM interference grid over the U.S. and the probable registration of scalar EM howitzers used in conjunction with the grid. (See FDL for complete details of this grid weapon.).

Anomalous "laser blinding" of U.S. satellites over the Soviet Union has occurred on several occasions. On one occasion, a satellite was blinded for up to four hours. While such non-damaging blinding would be difficult for a ground-based laser to accomplish, it would be simple for a scalar EM interferometer—or a ground-based scalar EM laser—to accomplish, since the amount of energy deposited upon and within the satellite could be precisely controlled and even directly monitored. A possibly related anomalous temporary disabling of two or three power supplies has been demonstrated upon the British satellite Ariel 6 when passing over British Columbia or the Caspian Sea, if the sun is shining.

In **mid-January,** 1983, so-called "sonic booms" shook the air over Pennsylvania.

In **latter February, 1983** anomalous booms occurred over New York, and anomalous tremors shook the Mississippi Gulf Coast. Anomalous booms also shook Ohio.

On **April 9, 1984** a gigantic mushroom cloud—glowing, like a "halo"—rapidly emerged from above the ocean off the coast of Japan. The cloud grew to 60,000 feet or so within two minutes, and reached an enormous diameter estimated at 200 miles. Several 80eing 747 jet airliners were in the general vicinity; at least one of them was piloted by a former B-52 pilot who took evasive action, since the phenomenon resembled a giant nuclear explosion. However, there was no blinding flash of light, and no massive shock wave overlook the aircraft. Walker et al investigated the seismic and underwater acoustic instrumentation data surrounding the period of the incident, and essentially ruled out all **known natural** phenomena. They concluded that the incident was either an as yet unknown natural phenomenon or man-made explosion. In fact, it was very probably a test of a giant scalar EM howitzer, used in the endothermic (heat energy withdrawing) mode. Sudden energy withdrawal in a region above the ocean resulted in a sudden low pressure, sucking up a giant cloud of moisture. Inrushing air pushed the cloud upward into a giant mushroom, much like a giant expanding thundercloud anvil wells up when it forms, only faster. Walker and colleagues have again examined the reported position of the cloud, and placed it much nearer to the Soviet Union—between the Kurils and Sakhalin. Four stages were observed: (1) a towering, cumulus-like cloud rose out of the stratiform layer, (2) the cloud tower faded and was replaced with a small semicircular halo segment, (3) the halo expanded to a full circle, and (4) the halo expanded further and dissipated. The diameter of the halo at maximum size is now estimated to have been at least 380 miles, and the altitude of its center (at maximum size) is estimated to have been greater than 200 miles.

Note that the upper edge of the shell would have been at an altitude greater than 380 miles. This sort of "Tesla shield" is an antimissile and antiaircraft defense shield. Any object penetrating the shell receives both an electrogravitational pulse (EGP) and electromagnetic pulse (EMP) **arising inside it, from within its local spacetime.** The EMP will dud all electronics and explode all high explosive (HE) materials. EMI shielding is ineffectual, since the EM energy pulse arises **everywhere within** the vehicle, warhead, and circuitry from spacetime itself. The EGP will immediately fission radioactive material. If a strong EGP is experienced, a nuclear warhead or warheads will explode "full order", instantly. In fact, even the nuclei of ordinary, nonradioactive material struck by a strong EGP will fission in a low order nuclear explosion. In the "small shell" or "small globe" variant, a sufficiently strong EGP will be experienced by a penetrating object to cause full order detonation of nuclear warheads. In the "large shell" or "large globe" variant, a weaker EGP will be experienced, causing low order detonation or merely disruption of nuclear warheads.* As can be seen this defense weapon is effective against all types of warheads (nuclear, HE, etc.) and all types of penetrating vehicles (bombers, ballistic missiles, cruise missiles, RPVs, artillery shells, etc.). The 1968 statement in **Voyennaya Strategiya (Military** Strategy) by V.D. Sokolovskiy that 100 percent defense against missiles and aircraft is possible, and that this capability had been achieved by the Soviet Union, but not by the West, is true.

It will also explode fuel, propellants, and HE explosives, dud all electronics, kill all personnel, etc. at the same time.

Thus the incident apparently represents three tests in one: (1) the initial "cold explosion," or test of the howitzer in the endothermic mode, and (2) the immediate switching of the howitzer to the exothermic mode and creation of a small hemispherical shell of energy (Tesla shield). (3) The shield was then expanded into a globular shell, and expanded to giant size. The globular part of the exothermic test was very similar to the Soviet test observed from Teheran on June 17,1966. The glowing hemispherical shell was similar to several previous Soviet tests observe over the ocean and reported in the open literature. For example, on Mar. 24, 1977, the H.M.V. Kinpurnie Castle observed a large, moderately luminous hemisphere of light formed over the ocean, and the formation of two luminous patches or globes, one inside the hemisphere and one outside. The phenomena disappeared after 10 minutes. The gigantic incident off the coast of Japan on April 9, 1984 was indeed the testing of a giant Soviet scalar EM weapon.

On the night of **July 26, 1984,** the pilot and crew of a Boeing 747 (American carrier) flying from Tokyo to Fairbanks, near the Kuril Islands, noticed a slowly expanding hemisphere of white light off to their left above the horizon. The shell of light continued to expand over a 10 minute period until ahead of them and to the right. The crew braced for a shock wave which never arrived. Their weather radar saw nothing out of the ordinary. The shell of light had sharp edges and was semitransparent so that stars became visible through it. This was another Soviet test of the Tesla shield. Numerous sightings of this phenomenon have been made by airline pilots flying in and out of Japan.

In early **December 1984,** significant adjustments of the scalar EM interference grid occurred. In the vicinity of Los Angeles and San Diego, anomalous aerial phenomena such as explosions, rumblings, airquakes, and buildings and windows shaking without seismic disturbances occurred. A sharp and unexpected weather front appeared, racking the area with high winds on the night of Dec. 12, producing wide-spread damage and power outages. The newly-formed front, steered by the dynamic *Soviet woodpecker grid*, moved on to produce significant snowstorms in other areas;: even Tucson received a snow "dump" of 20 inches. Prior to the Dec. 12 anomalous weather front, this author accurately predicted an impending sudden drastic change in the weather in a brief interview over Radio station KABC, Los Angeles after the Dec. 8-9 anomalous aerial and ground structure disturbances. Complete explanation of Soviet weather engineering is given in Bearden, "Soviet Weather Engineering Over North America," 1985, 1-hr. videotape.

In **January 1985**, an incident of U.S. Navy-dispensed chaff drifting toward San Diego, California from off-coast was associated with significant failures of electrical systems and components in the city. Power was interrupted to as many as 60,000 homes. A specialized structure such as a piece of sophisticated chaff will reflect scalar waves of frequencies within its cut bandwidth. If scalar frequencies within the chaff bandwidth were present on the *Woodpecker carrier grid* above and surrounding San Diego, a myriad of reflecting bits of chaff in the moving chaff cloud would produce myriads of random, invisible "fireflies" of electromagnetic energy kindling at a distance, from randomized scalar interferometry, in a zone surrounding the cloud. As these "firefly" pulses of EM energy occurred inside components of the electrical system, interference would occur, leading to anomalous electrical failures. That is apparently exactly what happened.

In **February, 1985** a China Airlines Boeing 747 aircraft enroute to Los Angeles suffered anomalous engine flameout. The jetliner fell 32,000 feet, and the pilot finally managed to restart the engines and make an emergency landing in San Francisco. In the incident, different instruments apparently disagreed with **each other, and** disagreed with the observations of the pilot and crew. This is a direct indication of the test of one type of scalar EM weapon detailed in **Fer-de-Lance.** Only a "mild" test was conducted, as the Soviets apparently did not yet wish to actually destroy the aircraft.*

Scalar interferometry (endothermic mode) causes electrostatic cooling in the internal combustion gases of the jet turbines. This reduces the pressure, and the thrust. When cooled sufficiently, the engines flame out and fail. They cannot be restarted until the scalar charge has time to drain away.

We mention in passing that, **in 1972**, at a secret meeting of the leaders of the communist parties of Europe, Brezhnev named the year **1985** as the target year that the Soviets would be free to do as they will, anywhere on the globe. He stated that by that year the Soviet Union would control the oceans, the atmosphere, and 90% of the land area. We accent that he said **control,** not invade, conquer or occupy. [In 1960, Khrushchev also laid out a time schedule that focused on readiness, with technical superiority, in 1985.]

In latter **April, 1985**, Frank Golden discovered the rather sudden Soviet activation of 54 powerful scalar EM frequencies (27 pairs, each pair 12 kilohertz apart) transmitted into the earth and utilized to stimulate the earth into forced electrogravitational resonance on all 54 frequencies. These represented 27 giant "power taps" into the earth, each tap extracting enormous energy from the molten core of the earth itself, and turning it into ordinary electrical power. Each giant tap is capable of powering 4 to 6 of the largest scalar EM howitzers possessed by the Soviet Union.

In and around **May Day 1985**, the Soviets conducted a massive, "full up" strategy exercise of the scalar EM weapon systems and communications strategic exercise, monitored by Frank Golden (and for several hours, by the author.). For the May Day 1985 celebration, the 40th anniversary of the end of WWII, the Soviets conducted a full-up demonstration of the scalar EM superweapons armada for Gorbachev, the recently selected leader. Apparently over 100 giant scalar EM weapons were activated. Twenty-seven gigantic "power taps" were established by resonating the earth electrogravitationally on 54 frequencies (27 pairs where the two are separated from each other by 12 kHz.). By alternating the potentials and loads of each of the two paired transmitters, electrical energy—in enormous amounts—can be extracted from the earth itself, fed by the "giant cathode" that is the earth's molten core. Scalar EM command and control systems, including high data rate communications with underwater submarines, were also activated on a massive scale. The exercise went on for several days, as power taps were switched in and out, and command and control systems went up and down. **This exercise represented the achievement of Brezhnev's 1972 statement that by 1985 the Soviets would be prepared to do as they wish, anywhere in the world.**

August 28, 1985. A Titan 34-D missile launched from Vandenberg Air Force Base blew up after launch. The loss has officially been attributed to failure of a high-powered fuel pump, causing a massive oxidizer leak and a smaller fuel leak. While the Soviets may have used a small EM missile to disable the pump, no evidence is available at this time to support that hypothesis. Note, however, that the loss date is well after the complete exercise of all the Soviet strategic weapons in late April-early May. It is also within the time of the "anomalous booms" accompanying full-up testing of the Soviet prelaunch ABM/antibomber system against shuttle launches from Cape Canaveral. The Soviets may very well have decided to test the first actual "quickest reaction time possible" destruction of a real, launched US. missile in August, 1985, followed by the decision to test the first actual "quickest reaction time possible" destruction of a real aircraft (simulated bomber) in December, 1985.

Nov. 26, 1985. Launch of the shuttle Atlantis occurred in the evening at Cape Canaveral. A mysterious light was hanging in the sky; it was the marker beacon for registration of a Soviet prelaunch phase ABM/antibomber system. Just prior to launch, an EM missile strike occurred in the vicinity, as an offset test of an associated scalar interferometer in the exothermic mode. A photograph of the anomalous incident is highly suggestive of just such an EM missile. Twelve minutes after launch when the shuttle was safely down range and out of the way, a huge atmospheric, rumbling explosion occurred over the area, and was heard for hundreds of miles up and down the coast. This represented the test of the "multiple missile launch" kill mode of the weapon system. The marker beacon, seen by hundreds of persons, was photographed as it was slewed away by the distant Soviet operator. Many booms and rumbles have occurred over Florida for several years, as the Soviets adjusted and tested the various modes of the scalar EM grid weapon system. The shuttle launch on Nov. 26, 1985 was the third shuttle launch in which the giant explosion occurred over the area some minutes after launch. The Soviets were using the shuttle launches as a convenient simulation of a "missile target launch", against which to test their prelaunch ABM/antibomber system.

On **Dec. 12, 1985** the same Soviet weapon tested against the previous NASA shuttle launches and against various aircraft deliberately destroy an Arrow DC-8 taking off from Gander Air Force Base, Newfoundland. At lift-off, the aircraft—carrying over 258 U.S. marines and aircrew—lost power and sank into the ground tail-low, killing everyone on board.* It was a tragedy of enormous proportions, and especially to the families and friends of the brave servicemen and crewmembers who lost their lives. Three Canadian witnesses to the crash were interviewed over the Canadian Broadcast network television news on April 8, 1986 at 10:00 p.m. No flame or smoke issued from the plane before its descent and crash. However, the witnesses reported that the aircraft was mysteriously glowing with a yellow or orange halo. That is a signature of the use of a scalar EM howitzer in the "continuous EM emergence" mode, similar to the manner in which several F-111's were downed in Vietnam. DC-8's electrical systems were interfered with by EM energy and EM jamming noise created throughout each increment of spacetime occupied by the aircraft. A powerful charge was rapidly built up on the aircraft structures and skin. The "yellow glow" seen by the witnesses was a corona due to the skin of the aircraft acquiring a high electrical charge. The loss of the engines was probably due to the distant Soviet operator applying a localized endothermic (energy extraction) beam to the engines. In fact, one eyewitness actually saw the crossed, glowing beams form in the clouds and a ball or beam of light then go from that glow and strike the aircraft. Instant fire to the plastics inside the aircraft occurred, emitting deadly gases. Half the occupants died of cyanide inhalation before the aircraft struck the ground and exploded. The stricken airliner passed directly over one observer who heard its engines roaring in painful labor, rather than with the full-throated roar of normal power. The distant Russian operator/gunner apparently tracked the aircraft down the runway, using two modes against it: (1) energy interference, and (2) energy extraction.

An anomalous hole—characteristic of an electrogravitational (scalar EM) strike—was in a section of the fuselage ahead of the engines.

On **Jan. 1, 1986**, a startled Frank Golden detected—and physically verified with a special technique—that a metal softening scalar EM signal had been added onto the *Soviet Woodpecker signals*. Thus, at that time the Soviet Union was preparing to exercise a metal softening test at some future time, in some location over the U.S.**

** *Note this signal could also have been present for the Dec. 12 destruction of the Arrow DC-8 at Gander AFB, Newfoundland.*

On **Jan. 28, 1986** the Challenger disaster occurred. The Challenger was positively killed by the Soviet Union, using the scalar EM weapons through the *Woodpecker grid*. A host of indicators occurred.

1.   The anomalous cold weather in Florida was definitely engineered by the USSR,
2.   Specialized cloud patterns associated with Soviet grid engineering and weather engineering were observed and photographed in Los Angeles, California and Huntsville, Alabama, beginning several days before the incident, and particularly on the evening before the launch,
3.   The normal "pivot point" for turning the jetstream northeastward was moved south from Huntsville, Alabama to Birmingham,

Alabama by the Soviets; this was to force the jetstream much further south, and consequently move unusually cold air into the Florida panhandle, exposing the shuttle to undue cold stress,

4. About 4 hours before launch; all Soviet ships off shore suddenly left the area at speed; this was the first shuttle launch not "shadowed" by Soviet ships,

5. On the morning of the launch, higher frequencies were added to the interference grid to enable much-enhanced localization. The brains of small birds are very sensitive to these higher frequencies, due to their small diameter (wavelength) as a scalar EM receiver. If they remained in the area of localization, the birds would be in intense pain or killed. On the morning of the launch, national TV network news, announcers, noted that no **birds at all were flying in** the area, something which had never happened before.

6. At the time of the shuttle's destruction, a giant radial cloud pattern was actually in the general vicinity;

7. Three previous shuttle launches—the last on Nov. 26, 1985—had been used as direct test targets for tests of the Soviet launch phase ABM/antibomber defense system so the weapon was "zeroed in" on the launch site,

8. The *anomalous destruction of the Arrow DC-8* on Dec. 12, 1985 had already indicated a Soviet decision to elevate the testing to the actual destruction of targeted vehicles,

9. At the time of the launch, anomalous electromagnetic phenomena occurred in nearby restaurants,

10. A metal-softening signal was on the grid, as previously detected by Golden. This signal would be detected by the launch flame, and after ignition would result in a steady weakening of the metal in and around the booster flame. Note that the metal was already cold-stressed beyond what it had been tested to withstand,

11. Almost immediately after ignition, the booster seals vented, giving evidence that the cold stress and the metal-softening signal were weakening the system,

12. An anomalous 10 second or so burnthrough of the weakened booster occurred,

13. An anomalous "light"—possibly a Soviet EM missile—was observed to play on the rocket before it blew up,

14. An anomalous "flat plate" earthquake over about 11 states occurred within days after destruction of the shuttle. This type of quake is

strongly suspicious, and it was probably the test of the grid/howitzers in a ground wave interferometry mode,

15.   Substantial winds and air turbulence over the launch site increased the stress on the Challenger as it rose through this region. With a giant cloud radial in the area, one strongly suspects that the turbulence may have been deliberately created or augmented by the Soviet scalar EM grid,

16.   A few days later, from 1-4 February 1986, many birds inadvertently flew into the new, localized "pivot point" at Birmingham, Alabama, encountered the high frequency components, and dead birds fell from the sky in substantial numbers. Many different kinds of birds were involved.

17.   Most significant of all, General Daniel Graham has reported that, on the evening after the death of the Challenger, the Soviet KGB gave a party and celebrated the success of their perfect active measures against the Challenger! Note that all development, deployment, employment, and command and control of the Soviet scalar EM weapons are under the KGB. Finally, a U.S. classified investigation of the Challenger disaster was ordered by Congress, but its results have not been made public. Beyond any doubt the Soviets destroyed the Challenger, and killed the seven brave astronauts aboard the spaceship.

On **Apr. 18, 1985** a Titan 34-D missile, launched from Vandenberg Air Force Base, blew up 5 seconds after launch. Video cameras inadvertently caught the Soviet marker beacon, used with the scalar EM howitzer that destroyed the Titan, up and above the explosion, moving independently. Engineer Ron Cole personally examined the video frame by frame and asserted that the hovering ball of light was completely separate. This positive signature leaves no doubt that the Soviets destroyed that missile also. The previous Titan 34-D missile launch there in August, 1985 also blew up just after lift-off.

On **May 3, 1986** a NASA Delta rocket carrying a critically-needed weather satellite failed. The rocket suffered an anomalous "command-type" shutdown of its main engine during launch and began to veer off course, causing the range safety officer to destroy it. On a network video tape replay (several replays in succession), immediately after the loss of the rocket, this author observed that an anomalous light moved up from beneath the rocket and struck it, seconds before the destruction.

Subsequent attempts to locate that particular videotape have met with failure. Several other persons were watching the replay with me at the time, and saw the anomalous light strike the rocket. The light plus the "internal command surges" are strongly indicative of time-reversed EM pulses and interferometry from a pumped phase conjugate mirror adjunct operating with an over-the-horizon tracking mode.

Also in **May, 1986** a European Ariane missile was lost shortly after launch. There is at least one published news analyst report that the French Government concluded that the Soviets had interfered with the missile and caused its loss. No further evidence is available to this author regarding this incident, except one unevaluatable report from a wealthy financier that when he visited the Soviet Union, the French President discussed that issue with Gorbachev.

In **October, 1986**, the Iceland Summit conference was held between Reagan and Gorbachev. The Soviets offered a breathtaking "zero-option" proposal, intended to lead to dismantling of most major strategic nuclear weapons, which would greatly ease the way to much more drastic employment of Soviet EM weapons. In short, if nuclear disarmament could be engineered, the U.S.'s "dead-man fuzing" against extensive employment of Soviet weapons would be almost eliminated. In that case, with its deployed, massive scalar EM weapons unleashed, the Soviet Union immediately holds the winning hand, and could rapidly proceed to dominate and control most of the earth. The Soviet offer was conditioned on the severe limitation of SDI to the laboratory, however, and Reagan balked at this. The reason for adamant Soviet insistence that the SDI genie must not be tested in space is that, should the U.S. develop scalar EM weapons—such as high energy scalar lasers—and deploy them as SDI modifications, the power would be enormously increased. With one or two shots, such a laser could devastate a whole republic of the USSR.* If tested in space, prototype SDI launch and deployment vehicles would be available, even though the SDI system was not yet completed and deployed. In that case, the U.S. could possibly quickly launch several scalar EM weapons in retaliation for Soviet strikes against the U.S.—and from space even one weapon could destroy the Soviet Union. Thus, unless the SDI genie is chained up in the lab, a Soviet scalar EM initiative to dominate the world would involve unacceptable risk to the Soviet Union. At this writing, Gorbachev has mounted a great propaganda campaign to try to stimulate U.S. citizens, scientists, and political activists to demand nuclear disarmament, clearing out the U.S. dead-man fuzing.

Their EGPs, however would detonate nuclear warheads and nuclear facilities. The resulting fallout would devastate the earth. Such a use would be suicidal. But it would be a first-strike weapon, if a madman pulled the trigger.

During the **spring of 1986**, abnormally strong Soviet weather engineering occurred over the U. S., causing a drastic drought in the southeastern U.S. This drought was broken by a colleague who used an extremely powerful scalar EM device to redirect jetstreams. A most unusual and unique signature of the "blocking" against the Soviet scalar EM actions resulted: Two huge circulations developed in the atmosphere, clearly showed as two adjacent giant "holes" in the swirling cloud cover over the middle and eastern U.S. Between these two giant holes, the cloud circulations formed a stream of clouds, moving to the south, looking very: much like a giant vertical "bar" of a huge "Y-shaped" cloud flow.* Several national weathermen commented on this unusual pattern, which had not been observed before. The pattern continued, day after day, as the blocking continued.

* On the weather maps.

My colleague suffered a drastic illness and a serious operation, which interrupted the blocking operation. Immediately the jet streams changed back and serious drought returned to the southeast U.S., devastating a substantial portion of farm crops in the region. After some time, my colleague recovered sufficiently to resume blocking operations, breaking the drought and restoring the rains once again.

In **latter October and early November, 1986,** significant giant radial cloud patterns, associated with Soviet weather engineering using the scalar EM interference grid, were seen and photographed over California in the greater Los Angeles area. On Monday, Nov. 10, 1986 an anomalous, giant "fireball" was seen by hundreds of persons to move from west to east. It was seen over 4 states, and accompanied very sharp changes in the EM grid. An anomalous winter storm was in progress, with cold air spilling down from Canada. The storm penetrated very deeply southward, breaking cold records in a wide area of the mid-United States. At the same time, heat records were being broken in Florida, and generally throughout the southeast. Frank Golden verified by direct measurement that the electrogravitational field of the earth was agitated and most dynamic—in short, significant activity was being introduced by the Soviet scalar EM grid. (The activity was about half what it is when the Soviets are inducing an earthquake.).

On **Wednesday, Nov. 12, 1986**, the author photographed three giant radial cloud patterns—one a "twin"—in Huntsville, Alabama, associated with the anomalous storm, now labeled the "Siberian express" m many news reports.

On Thursday, **Nov. 3, 1986** at Huntsville, the author observed another twin giant radial, gently moving along from west to east. At dusk, another single giant radial was observed at Huntsville. Also, air control radars along the corridor from Los Angeles to Arizona suffered a mysterious failure or "power outage" not long after the "meteor" incident. Reports were received from Montreal, Canada that anomalous TV interference was experienced in that area about the same time the air control radars experienced anomalous failure. This may show that the agent causing the radar failures and TV interference extended across a wide area of North America. Note that the normal *intersection of the Woodpecker beams* covers such a broad area of North America in the interference zone of the beams. The so-called "meteor" (actually, a large, rapidly moving light was seen and noise was heard) of Nov. 10 may well have been a large scalar EM ball from the *Woodpecker grid weapon transmitters*, performing a simulated test of the multiple-vehicle kill mode, using a large "electromagnetic missile" created and moved by the associated scalar howitzers. Complete explanation of the operation of these systems is contained in Fer-de-Lance and other books and papers.

As **this background summary is cutoff in mid November 1986,** it does not include evidence of indications that exist for Soviet scalar EM enhancement or biological warfare against the West. The reader should be aware of the potential scalar BW threat, and the general lack of any effort to collate evidence in that area. [That area has been covered in the author's book, AIDS: **Biological War**fare, Tesla book Co., 1988. For complete coverage, see the author's *Gravitobiology: A New Physics,* 1989.]

For the present Soviet tactic, see the author's "Glasnost: 29th Move of a 30-Move Chess Game," *Raum & Zeit.* in publication, and "Political Manipulation of Unified Field Theory," *Raum & Zeit.* in publication.

ADDITIONAL NOTES AND REFERENCES

14. For details of the mysterious explosions of six major Soviet missile ammunition storage sites in seven months, see photograph, *Jane's Defence Weekly,* 2(3), July 28, 1984, p. 92; see also Mark Daly, "Goa and Goblet SAMs in Severomorsk explosion," *Jane's Defence Weekly,* 2(6), Aug. 18, 1984, p. 224; Derek Wood, "Soviets' northern fleet disabled . . . 'not viable' for six months," *Jane's Defence Weekly*, 2(1), July 14, 1984, p.3; Derek Wood, "Six explosions in the past seven months," *Jane's Defence Weekly,* 2(1), July 14, 1984, p. 3; John Moore, "The aftermath of Severomorsk," Jane's Defence Weekly, 2(6), Aug. 18, 1984, p.224.

Note from Webmaster: it doesn't stop in 1986.

What follows next is unclassified technical warfare information and information on its current uses for Psychological Warfare.
-Sabre

The State of Unclassified and Commercial Technology
Capable of Some Electronic Assault Effects
Eleanor White, P.Eng.
Last Update: January 26, 2008
http://www.raven1.net/uncom.htm

Note: The information content of this article has not been updated since December 1999. The information here applies regardless of the year it is read. Slight changes in wording will be made from time to time as activism to expose crimes done with this equipment may change the current terminology.

To help the reader appreciate the importance of this matter . . .

2012 A FAMILY BRIEF

"We need a program of psychosurgery and political control of our society. The purpose is physical control of the mind. Everyone who deviates from the given norm can be surgically mutilated.

"The individual may think that the most important reality is his own existence, but this is only his personal point of view. This lacks historical perspective.

"Man does not have the right to develop his own mind. This kind of liberal orientation has great appeal. We must electrically control the brain. Some day armies and generals will be controlled by electrical stimulation of the brain."

Dr. Jose Delgado (MKULTRA experimenter who
demonstrated a radio-controlled bull on CNN in 1985)
Director of Neuropsychiatry, Yale University Medical School
Congressional Record No. 26, Vol. 118, February 24, 1974

CONTENTS

This document is organized so that a narrative article appears at the top, followed by appendices.

- BLACK = NARRATIVE ARTICLE
- BLUE = APPENDICES

I. LIMITATIONS

The author acknowledges that this article falls short of a rigorous academic paper. This is explained by the fact that organized stalking and electronic harassment targets are kept in a sort of "barely alive" condition, with significant health problems, and either unable to work or just barely able to hold a job with limited earning potential.

Furthermore, since the perpetrators constantly work to prevent the public from knowing anything about organized stalking and electronic harassment, evidence is obtainable with great difficulty, and often the only evidence is of lower quality than would be accepted for a scientific treatise.

In short, everything in this article represents a struggle against immense odds. We ask readers to understand this.

Up to Contents

II. INTRODUCTION

Electronic assault and control technology had it's start in the 1950s, as an obscure branch of the CIA's MKULTRA project group. Just as organized crime is not stopped by hearings and court cases, neither did this originally obscure branch of MKULTRA activity, when the institutional/drug/child abuse phases were exposed by the U.S. Senate's Church-Inouye hearings in the late 1970s. No criminal proceedings followed, and only two civil law suits (Orlikow and Bonacci) have succeeded.

This assembly of unclassified and commercial literature is to show investigators and concerned citizens that in spite of the tightest possible information blackout imposed in the early 1970s, enough of the classified electronic harassment and control technology has leaked out to show that significant classified accomplishments are overwhelmingly likely, and in need of disclosure, here at the end of the 20th century.

It is hoped that government and media, who have shied away from this topic for decades, preferring the warm fuzzy feelings that "this can't be true", will read about the unclassified and commercial devices and understand the implications of continued turning the other way.

Up to Contents

III. E-HARASSMENT EFFECTS

Since government-backed electronic harassment and control is classified at the highest levels in all technologically capapble governments, the description of effects is taken from the personal experiences of the "2,000 or so" known organized stalking with electronic harassment targets. The experimentees without exception report that once the harassment begins, in virtually all cases it continues for life. It continues in every city, state, and country the target moves to. It continues in prisons and hospitals, even when the target is dying of cancer.

Many targets are young seniors, and some in their 70s and 80s.

Some have children and the children are often subjected to the same "testing" as their parent(s). Pets are not only tortured, but even killed, painfully and violently.

The effects pattern:

This article is about unclassified/commercial technologies which can produce some of the effects of the harassment equipment:

- All harassment consists of unique, carefully engineered-unprovable events to produce psychological stress in the victim. There are no events which do not fit that apparent purpose.
- In every series of stress event type, ONE introductory event of very high energy/effect is staged. The obvious purpose is to be certain the victim KNOWS this is external harassment, and not just "bad luck". From that time forward, the perpetrators appear to apply "Pavlovian training" so that they can get the victim to "jump" (or react in some way) to the same effect at a tiny fraction of the initial "introductory" event.

- This type of harassment started during the Cold War, and shows every characteristic of military and intelligence psychological warfare operations.
- This type of harassment points to CONTROL of the test subject. Endlessly repeated words generated inescapably within the skull are just one hypnosis-like experience.
- The total number and type of crimes which make up organized stalking and electronic harassment simply could not take place without cover being supplied by government. Bizarre court decisions in which the target is forced into involuntary mental health treatment and a criminal spouse (for example, a pedophile father discovered "doing" his own child) clearly show that organized stalking and electronic harassment are heavily protected by all levels of government and the medical establishment.

The effect types categorized:

Here is a list of most of the common effects. It is not exhaustive, but is intended to show the reader how the perpetrators' pallette of stress effects is broken down. Indent levels are used to show categories and sub-categories:

1.   Invasive At-a-Distance Body Effects (including mind)

    a.   Sleep deprivation and fatigue

        i.   Silent but instantaneous application of "electronic caffeine" signal, forces awake and keeps awake
        ii.   Loud noise from neighbours, often SYNCHRONIZED to attempts to fall asleep
        iii.   Precision-to-the-second "allowed sleep" and "forced awakening"; far too precise and repeated to be natural
        iv.   Daytime "fatigue attacks", can force the victim to sleep and/or weaken the muscles to the point of collapse

    b.   Audible Voice to Skull (V2S); sometimes non-voice sounds

        i.   Delivered remotely, at a distance
        ii.   Made to appear as emanating from thin air

        iii.   Voices or sound effects only the victim can hear

   c.   Inaudible Voice to Skull (Silent Sound)

        i.   Delivered by apparent at a distance radio signal; manifested by sudden urges to do something/go somewhere you would not otherwise want to; silent (ultrasonic) hypnosis is possible
        ii.   Programming hypnotic "triggers"—i.e. specific phrases or other cues which cause specific involuntary actions

   d.   Violent muscle triggering (flailing of limbs)

        i.   Leg or arm jerks to violently force awake and keep awake
        ii.   Whole body jerks, as if body had been hit by large jolt of electricity
        iii.   Violent shaking of body; seemingly as if on a vibrating surface but where surface is in reality not vibrating

   e.   Precision manipulation of body parts (slow, specific purpose)

        i.   Manipulation of hands, forced to synchronize with closed-eyes but FULLY AWAKE vision of previous day; very powerful and coercive, not a dream
        ii.   Slow bending almost 90 degrees BACKWARDS of one toe at a time or one finger at a time
        iii.   Direct at-a-distance control of breathing and vocal cords; including involuntary speech
        iv.   Spot blanking of memory, more than normal forgetfulness

   f.   Reading said-silently-to-self thoughts

        i.   Engineered skits where your thoughts are spoken to you by strangers on street or events requiring knowledge of what you were thinking
        ii.   Real time reading subvocalized words, as while the victim reads a book, and BROADCASTING those

words to nearby people who form an amazed audience around the victim

g. Direct application of pain to body parts

   i. Hot-needles-deep-in-flesh sensation
   ii. Electric shocks (no wires whatsoever applied)
   iii. Powerful and unquenchable itching, often applied precisely when victim attempts to do something of a delicate or messy nature
   iv. "Artificial fever", sudden, no illness present
   v. Sudden racing heartbeat, relaxed situation

h. Surveillance and tracking

   i. Rapping under your feet as you move about your apartment, on ceiling of apartment below
   ii. Thru wall radar used to monitor starting and stopping of your urination—water below turned on and off in sync with your urine stream

2. Invasive Physical Effects at a Distance, non-body

   a. Stoppage of power to appliances (temporary, breaker ON)
   b. Manipulation of appliance settings
   c. Temporary failures that "fix themselves"
   d. Flinging of objects, including non-metallic
   e. Precision manipulation of switches and controls
   f. Forced, obviously premature failure of appliance or parts

3. External Stress-Generating "Skits"

   a. Participation of strangers, neighbours, and in some cases close friends and family members in harassment

      i. Rudeness for no cause
      ii. Tradesmen always have "problems", block your car, etc.
      iii. Purchases delayed, spoiled, or lost at a high rate
      iv. Unusually loud music, noise, far beyond normal

b.   Break-ins/sabotage at home

    i.   Shredding of clothing
    ii. Destruction of furniture
    iii. Petty theft
    iv. Engineered failures of utilities

c.   Sabotage at work

    i.   Repetitive damage to furniture
    ii. Deletion/corruption of computer files
    iii. Planting viruses which could not have come from your computer usage pattern
    iv. Delivered goods delayed, spoiled, or lost at a high rate
    v.   Spreading of rumors, sabotage to your working reputation
    vi. Direct sabotage and theft of completed work; tradesmen often involved and showing obvious pleasure

---

Illustration of the bodily effects

Up to Contents

==

IV. MAJOR TECHNOLOGY CLASSES

These technology classes are for the UNclassified and commercial equipment which can emulate the "real" classified electronic harassment equipment.

Effect section 2, "Invasive Physical Effects at-a-Distance", clearly establishes the existence of remote precision manipulation of objects which is far beyond the capabilities of unclassified and commercial equipment at the time of writing.

REMOTE PHYSICAL MANIPULATION is not covered in this article, but the reader should know that both NASA and IEEE have noted successes in creating very small antigravity effects (which are not due to simple magnetism.)

TRANSMISSION METHODS FOR NEURO-EFFECTIVE SIGNALS:

- pulsed microwave (i.e. like radar signals)
- ultrasound and voice-FM (transmitted through the air)

While transmission of speech, dating from the early 1970s, was the first use of pulsed microwave, neuro-effective signals of as yet unknown type can now cause many other nerve groups to become remotely actuated. At time of writing, that technology appears to be classified.

PAVLOVIAN HYPNOTIC TRIGGERS:

A [Pavlovian] hypnotic trigger is a phrase or any other sensory cue which the victim is programmed to involuntarily act on in a certain way. The 50s-70s MKULTRA survivors can still be triggered from programming done decades ago. A name "manchurian candidate", from a novel by John Marks, is used to describe a person who carries Pavlovian triggers.

One of the main goals of the institutional/drug/child abuse phases of the CIA MKULTRA atrocities (1950's through 1970's) was to implant triggers using a "twilight state" (half-conscious) medication and tape recorded hypnosis. The ultimate goal was to have the acting out of Pavlovian triggers erased from the victim's memory.

Using a combination of Joseph Sharp voice to skull and Lowery Silent Sound technologies, these triggers can now be planted with the subject being unaware, or just barely aware of a very high tone in his or her hearing sense.

THROUGH-WALL SURVEILLANCE METHODS:

So-called "millimeter wave" scanning. This method uses the very top end of the microwave radio signal spectrum just below slightly into the infra-red frequency region. To view small objects or people clearly, the highest frequency that will penetrate non-conductive or poorly—conductive walls is used. Millimeter wave scanning radar can be used in two modes:

- passive (no signal radiated, uses background radiation already in the area to be scanned, totally UNdetectable)
- active (low power millimeter wave "flashlight" attached to the scanner just as a conventional light mounted on a camcorder), or, the use of archaeological ground penetrating radar

THOUGHT READING:

Thought reading can be classed as a "through wall surveillance" technology. Thought reading, in the unclassified/commercial realm, can be broken down as follows:

- magnetic skull-proximity or throat (vocal cord proximity) reading
- EEG (electroencephalogram) interpretation, particularly by way of inside the brain pickup wires

The ability to read and display what the eyes are seeing using implanted pickups is one form of technology almost the same as "thought reading".

NASA and DARPA have been reported in the mainstream press as working on technology which can pick up silently-said-to-self thoughts through pickup coils against the throat.

BRAIN ENTRAINMENT:

The reverse of biofeedback. Those low frequency electrical brain rhythms which are characteristics of various moods and states of sleep can not only be read out using biofeedback equipment or EEG machines, but using radio, sound, contact electrodes, or flashing lights, the moods and sleep states can be generated or at least encouraged using brain entrainment devices.

Brain entrainment signals cannot carry voice, which is a much higher frequency range. Brain entrainment can, however, be used to "set up" a target to make him/her more susceptible to hypnosis by inducing a relaxed and/or sleeping state. (This is the same state used by remote viewers.)

These major technology classes can produce some of the observed electronic harassment effects, FROM HIDING AND UNDETECTABLY, with the exception of remote physical manipulation.

IMPLANTATION is sometimes used to assist the above technologies but with current devices. Implants do not appear to be required, based on large numbers of people who have had no surgery, no unexplained wounds, and no missing time.

Diagram showing the overall method, based entirely on unclassified 1974 technology, of how SILENT hypnosis may be transmitted to a target without the target's being aware. This technique is probably the most insidious, because it allows months and years of programming and Pavlovian trigger-setting, while the target cannot resist because the target can't hear the programming.

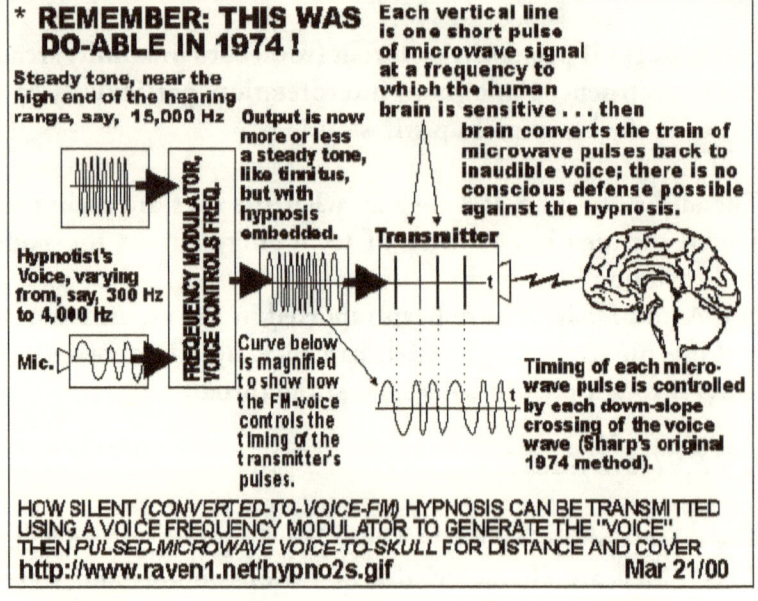

HOW SILENT *(CONVERTED-TO-VOICE-FM)* HYPNOSIS CAN BE TRANSMITTED USING A VOICE FREQUENCY MODULATOR TO GENERATE THE "VOICE", THEN *PULSED-MICROWAVE VOICE-TO-SKULL* FOR DISTANCE AND COVER
http://www.raven1.net/hypno2s.gif   Mar 21/00

Up to Contents

V. PULSED MICROWAVE

Pulsed microwave voice-to-skull (or other-sound-to-skull) transmission was discovered during World War II by radar technicians who found they could hear the buzz of the train of pulses being transmitted by radar equipment they were working on. This phenomenon has been studied extensively by Dr. Allan Frey, whose work has been published in a number of reference books.

What Dr. Frey found was that single pulses of microwave could be heard by some people as "pops" or "clicks", while a train of uniform pulses could be heard as a buzz, without benefit of any type of receiver.

Dr. Frey also found that a wide range of frequencies, as low as 125 MHz (well below microwave) worked for some combination of pulse power and pulse width. Detailed unclassified studies mapped out those frequencies and pulse characteristics which are optimum for generation of "microwave hearing".

Very significantly, when discussing electronic harassment, is the fact that the PEAK PULSE POWER required is modest—something like 0.3 watts per square centimeter of skull surface, and this power level is only applied for a very small percentage of each pulse's cycle time. 0.3 watts/sq cm is about what you get under a 250 watt heat lamp at a distance of one meter. It is not a lot of power.

When you take into account that the pulse train is OFF (no signal) for most of each cycle, the average power is so low as to be nearly undetectable.

Frequencies that act as voice-to-skull carriers are not single frequencies, as, for example TV or cell phone channels are. Each sensitive frequency is actually a range or "band" of frequencies. A technology used to reduce both interference and detection is called "spread spectrum". Spread spectrum signals have the carrier frequency "hop" around within a specified band.

Unless a receiver "knows" the hop schedule in advance, there is virtually no chance of receiving or detecting a coherent readable signal. Spectrum analyzers, used for detection, are receivers with a screen. A spread spectrum signal received on a spectrum analyzer appears as just more "static" or noise.

My organization was delighted to find the actual method of the first successful UNclassified voice to skull experiment in 1974, by Dr. Joseph C. Sharp, then at the Walter Reed Army Institute of Research.

Dr. Sharp's basic method is shown in Appendix PM6, below. A Frey-type audible pulse was transmitted every time the voice waveform passed down through the zero axis, a technique easily duplicated by ham radio operators who build their own equipment.

A pattern seems to be repeated where research which could be used for mind control starts working, the UNclassified researchers lose funding, and in some cases their notes have been confiscated, and no further information on that research track is heard in the unclassified press.

Pulsed microwave voice-to-skull research is one such track.

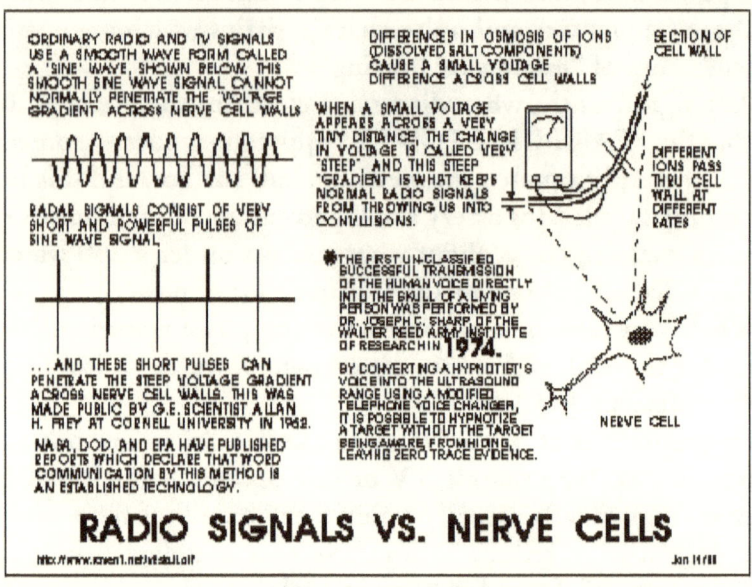

Illustration showing the principle behind pulsed microwave voice-to-skull

Appended articles:

PM1 http://www.raven1.net/lida.htm, photo and description of the old medical device, the Korean War vintage LIDA machine, a radio frequency BRAIN ENTRAINMENT device developed by Soviet Russia as a drugless tranquilizer. BRAIN ENTRAINMENT IS INCLUDED IN THE RADIO FREQUENCY SECTION BECAUSE THE MOST INSIDIOUS METHOD OF BRAIN ENTRAINMENT IS TO SILENTLY USE RADIO SIGNALS.

PM2 http://www.raven1.net/frey.htm, Human Auditory System Response To Modulated Electromagnetic Energy, Allan H. Frey, General Electric Advanced Electronics Center, Cornell University, Ithaca, New York

PM3 http://www.raven1.net/v2s-nasa.htm, NASA technical report abstract stating that speech-to-skull is feasible

PM4 http://www.raven1.net/v2s-kohn.htm, DOD/EPA small business initiative (SBIR) project to study the UNclassified use of voice-to-skull technology for military uses. (The recipient, Science and Engineering Associates, Albuquerque NM, would not provide me details on the telephone)

PM5 http://www.raven1.net/bioamp.htm, Excerpts, Proceedings of Joint Symposium on Interactions of Electromagnetic Waves with Biological Systems, 22nd General Assembly of the International Union of Radio Science, Aug 25-Sep 2, 1987, Tel Aviv, Israel SHOWS BIOLOGICAL AMPLIFICATION OF EM SIGNALS, pointing to relative ease with which neuro-electromagnetic signals can trigger effects

PM6 http://www.raven1.net/v2succes.htm, Excerpt, Dr. Don R. Justesen, neuropsychological researcher, describes Dr. Joseph C. Sharp's successful transmission of WORDS via a pulse-rate-modulated microwave transmitter of the Frey type.

PM7 http://www.raven1.net/russ.htm, FOIA article circulated among U.S. agencies describing the Russian TV program "Man and Law", which gives a glimpse into the Russian mind control efforts. (Dr. Igor Smirnov, a major player, was used as a consultant to the FBI at the Waco Branch Davidian standoff.)

Up to Contents

VI. ULTRASOUND AND VOICE-FM

Ultrasound is vibration of the air, a liquid, or a solid, above the upper limit of human hearing which is roughly 15,000 Hz in adults.

Voice-FM uses a tone at or near that upper limit, and the speaker's voice VARIES the frequency slightly. Either a "tinnitus-like sound" or nothing is heard by the target.

Ultrasound/voice-FM can be transmitted in these ways:

- directly through the air using "air type transducers"
- directly to the brain using a modulated microwave pulse train
- through the air by piggybacking an ultrasound message on top of commercial radio or television

The use of commercial radio or television requires that the input signal at the transmitter be relatively powerful, since radio and TV receivers are not designed to pass on ultrasound messages. However, the average radio and TV receiver does not simply stop ultrasound, rather, the ability to pass ultrasound messages "rolls off", i.e. decreases, as the frequency is increased.

Today's radios and TVs can carry enough ultrasound messaging to be "heard" by the human brain (though not the ear) to be effective in conveying hypnosis. This was proven by the U.S. military forces in the Gulf War.

Ultrasound's (and voice-FM's) main advantage in mind electronic harassment work is that it can carry VERBAL hypnosis, more potent than simple biorhythm entrainment.

The brain CAN "hear" and understand this "inaudible voice", while the ear and conscious hearing cannot. Once you can convey hypnotic suggestion which cannot be consciously heard, you have eliminated a major barrier to the subject's acceptance of the words being transmitted.

In previous decades, "subliminal advertising" using voice and images at normal frequencies were "time sliced" into an apparently normal radio or TV broadcast. This apparently did not work well, and now voice-FM "subliminal learning tapes" commercially available have superseded the old time slice method.

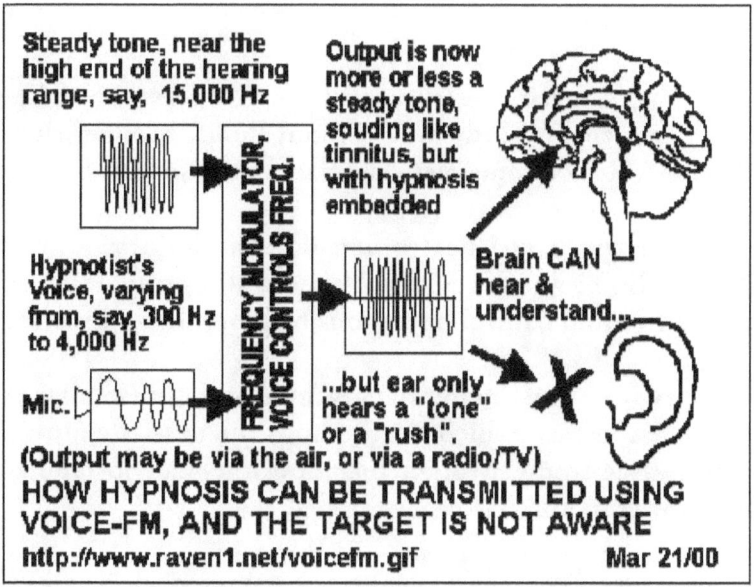

Steady tone, near the high end of the hearing range, say, 15,000 Hz

Output is now more or less a steady tone, souding like tinnitus, but with hypnosis embedded

Hypnotist's Voice, varying from, say, 300 Hz to 4,000 Hz

FREQUENCY MODULATOR, VOICE CONTROLS FREQ.

Brain CAN hear & understand...

Mic.

...but ear only hears a "tone" or a "rush".

(Output may be via the air, or via a radio/TV)

HOW HYPNOSIS CAN BE TRANSMITTED USING VOICE-FM, AND THE TARGET IS NOT AWARE

http://www.raven1.net/voicefm.gif   Mar 21/00

Illustration showing the operation of "silent sound" with the human hearing system, using near-ultrasound, FREQUENCY MODULATED voice

---

One method for projecting either audible voice or voice-FM over long distances, virtually undectable if line of sight, is the "acoustic heterodyne" or "HyperSonic Sound" system, patented by American Technologies Corporation, San Diego CA, *http://www.atcsd.com*

Illustration showing the principle of an ultrasound projection system capable of true ventriloquism at a distance, by American Technologies Corporation (licensor), Akai Japan (licensee)

Appended articles:

US1 http://www.raven1.net/silsoun2.htm, ITV Silent Sound report with comments by Judy Wall, Editor, Resonance, newsletter of MENSA's bioelectromagnetic special interest group

US2 http://www.raven1.net/commsolo.htm, an article by Judy Wall outlining instances of UNclassified, openly-admitted-to, electronic mind control operations by government agencies.

US3 http://www.raven1.net/armyparw.htm, an SBIR (small business initiative contract) which clearly shows intent to use ultrasound as an anti-personnel weapon, including one-man portability and with power to kill

US4 http://www.raven1.net/ssnz.htm, a commercial New Zealand company, Altered States Ltd., sells tapes which perform "suggestions" (i.e. hypnosis but not called such) using the Lowery patent voice-FM method, to hypnotize without the subject being aware. This is a key feature of neuro-electromagnetic involuntary experiments.

US5 http://www.raven1.net/acouspot.htm, a page originally from the MIT Media Lab's acoustic engineer, Joseph Pompeii. Describes a similar technique under commercial and military development (American Technologies Corp., San Diego) under the trade name "Hypersonic Sound". Shows that sound can be focussed to the extent of targetting just one person in a crowd, acoustically, using ultrasound.

Up to Contents

VII. THROUGH-WALL RADAR

When "millimeter wave" microwave signals are received, the waves are so small that they can display a two-dimensional outline of an object. Lower frequency radar can only show a "blip" which indicates an object's presence or motion, but not it's outline.

Incoming millimeter wave signals are channeled on to a plate with a two-dimensional array of elements sensitive to millimeter wave frequencies, in exactly the same way a camcorder focusses light on to array of sensitive light pickups. Each of the sensitive elements is scanned in a definite order, just as with a TV camera and screen, and a picture showing the outline of an object is formed.

If no signal is sent out by the scanner, it is called "passive" millimeter wave radar. If the subject is illuminated by a separate source of millimeter wave signals, it is an "active" scanner. Since passive systems can penetrate clothing and non-conductive walls UNDETECTABLY, it is obvious that with just a small millimeter wave "flashlight", non-conductive walls can be scanned through and still very little detectable signal is present.

Millimeter wave through-clothing, through-luggage is currently in use at airports, but can be a powerful tool for stalking groups wishing to harass a target at home.

Appended articles:

TWR1 http://www.raven1.net/lads.htm, LADS, Life Assessment Detector System, a product of VSE Corporation, can scan through more than a hundred feet of non-conductive or poorly-conductive material to detect a beating human heart

TWR2 http://www.raven1.net/nij_p44.htm, Prototype version of the "radar flashlight", which is a more portable version of the LADS system above. Can also be used to illuminate a subject for use with a Millivision thru-clothing/thru-nonconductive wall scanner

TWR3 http://www.raven1.net/millitec.htm, October 1995 blurb from Popular Mechanics, with photos showing hidden guns used for demo purposes (Millitech sold the rights to Millivision)

TWR4 http://www.raven1.net/psradar.htm, March 22 text taken from Patriot Scientific Corporation's web site, their ground-penetrating radar section. Patriot's GPR overcomes the limitation of the Millivision passive radar, i.e. inability to penetrate partially conductive walls.

Up to Contents

VIII. THOUGHT READING

"Thought reading", at least of words "silently said-to-self", appears to be not that difficult to do. NASA is experimenting with sensitive pickup coils on the throat of a pilot, for example.

Thought reading by way of pickup wires or coils on the head is an enhanced version of computer speech recognition, with EEG waves being substituted for sound waves.

The easiest "thought" reading is actually remote picking up of the electro-magnetic activity of the speech-control muscles in the vocal cords.

When we "say words to ourselves, silently", or, read a book, we can actually FEEL the slight sensations of those words in our vocal muscles—all that is absent is the passage of air. Coordinated speech signals are relatively strong and relatively consistent.

The other kind of "thought reading", i.e. "MINING" someone's brain for information from a distance is SPECULATIVE. We targetted individuals have no way to verify that is happening, however, we do know that we are "fed" hypnotic signals to force consistent "neutral" content (but of different character than prior to becoming test subjects,) DREAMS.

These forced, neutral content ("bland" content) dreams occur every single night and may represent the harassers' (or experimenters') efforts to have our experiences portray themselves in such dreams, in effect, MINING our experiences. Again, this is SPECULATION, but it seems very logical.

Appendix TR4, referenced below, confirms the ability of current unclassified technology to actually see what a living animal sees, electronically. It is therefore extremely likely that these forced dreams can be displayed on the experimenters' screens.

Finally, among the "couple of thousand" known organized stalking and electronic harassment, we sometimes have strangers either tell us what we are thinking, say they can pick up our broadcast thoughts, or tell us about events inside our homes at times when they could not have seen from the outside. BUGS and covert cameras are not used, and they have been searched for.

Appended articles:

TR1 http://www.raven1.net/thotuncl.htm, Commercially available thought-reading devices, both implant-style and non-implant

TR2 http://www.raven1.net/ratrobot.htm, Implanted rats can control devices with their thoughts

TR3 http://www.raven1.net/ebrain.htm, from the July 1973 issue of Popular Electronics, a system to read EEG signals (the stuff of which thought reading is made) at a distance by passing a radio signal through the human head and analyzing the passed-through signal. THIS IS NOT PROVEN TECHNOLOGY but it is one experimenter's suggestion.

TR4 http://www.raven1.net/elecvisn.htm, an article describing electronically reading a cat's brain waves and constructing a real-time image on screen from the EEG traces

TR5 http://www.raven1.net/m_switch.htm, the text from a site describing a mind-controlled "switch", which can not only turn appliances on or off, but also adjust controls like volume.

Up to Contents

IX. IMPLANTS

Electronic implants are actually one of the older forms of electronic mind control technology. Implants can either receive instructions via radio signals, passing them to the brain, or, can be interrogated via external radio signals to read brain activity at a distance.

Implants are currently used, however, many of our group have had no surgery, unexplained wounds, or missing time since becoming targets, and we suspect that both implant and non-implant technology may be in use.

Implants are significant for these reasons:

1. Their use, since World War II and continuing to the present day, associated with MKULTRA atrocities, is a crystal clear indication that a MOTIVE POOL of unethical researchers has existed through the late 1970s. The same people, none jailed, are still working, by and large. The reader can see that the existence of the same motive pool is overwhelmingly likely, given that no social changes have occurred which would prevent that.

2. The fact that to date (autumn 1999) no victim who has had implants removed has ever been able to get custody of the removed implant shows that research programmes using implants are still quite active and obviously quite important to someone. Implants, even though not in the hands of the victim, were photographed and are quite real:
    See *http://www.ritualabuse.net/MCF/*, the Mind Control Forum for details on involuntary experimentees' implantation and removal experiences.

3. The use of implants shows that, in the field of involuntary human experimentation, not every perpetrator group has access to the most sophisticated (implant-less) technology. Since implants for beneficial purposes are actively being promoted by NIH, it is obvious they will not disappear any time soon.

Appended articles:

IMP1 http://www.raven1.net/centneur.htm, an article showing that human implantation is being done and even encouraged by the U.S. NIH (National Institutes of Health). While this public information is for the public good, it is a small step to move from publicly known and VOLUNTARY implantation to CONCEALED implantation for INvoluntary and criminal purposes.

IMP2 http://www.raven1.net/italydoc.htm, a testimonial by an Italian psychiatrist who has been assisting involuntary experimentees;

this doctor began by assisting [Satanic or other] ritual abuse victims. Apparently involuntary brain implantation is alive and well in Italy, why not elsewhere?

IMP3 http://www.raven1.net/telectro.htm, a project abstract by AF, awarded to perform unclassified research and development of human implants which can read both physio- and PSYCHO-parameters.

IMP4 http://www.raven1.net/stimocvr.htm, an excerpt describing human implantation for purposes of two-way communication with the brain by way of implants and FM VHF radio. Blows away any doubts that human implantation has not been done, and even more, that the U.S. military are involved.

IMP5 http://www.raven1.net/sattrack.htm, describes an unclassified human implant satellite tracking system, ostensibly for benevolent use. (No method for avoiding unethical uses is described.) Applied Digital Solutions, Inc., Palm Beach, Florida.

Up to Contents

X. CONCLUSION

Conclusion? While the documentary evidence in this report, attached as appendices below, does not exactly "prove" we are being targeted by intelligence/defence contractors using classified electronic weapons, it certainly eliminates the argument that such devices are impossible, don't exist, or that government has "no interest" in them, or that the "were tried years ago but didn't work".

Add in the experiences of victims of the Tuskegee untreated syphilis experiments, the feeding of radioactive food to uninformed U.S. citizens, and the atrocities perpetrated under the institutional/drug/child abuse phases of the CIA's MKULTRA programmes, and you have more than enough grounds to petition for an independent, open investigation of electronic harassment capable technologies.

No doubt there were citizens of ancient Pompeii who argued that Vesuvius could not possibly erupt in their lifetimes.

Eleanor White

If any doubts as to the importance of this issue remain, please see below what the U.S. NSA (National Security Agency) says would be the result of releasing information on electronic mind control, which is one motive for developing weapons which can be used for electronic harassment:

Up to Contents

XI. APPENDICES

UP TO THIS PAGE, THIS REPORT HAS BEEN A NARRATIVE AUTHORED BY ELEANOR WHITE. THE APPENDICES ARE A COLLECTION OF THE BEST QUALITY FACTUAL MATERIAL FROM OFFICIAL SOURCES OUTSIDE THE ORGANIZED STALKING AND ELECTRONIC HARASSMENT GROUP. THIS MATERIAL MAY BE INDEPENDENTLY VERIFIED FROM REFERENCES PROVIDED.

Up to Contents

APPENDIX PM1 . . . THE LIDA MACHINE

Associated Press (Exact date not shown on copy but tests took place 1982/83) Loma Linda (Veterans Hospital research unit) San Bernardino County

An old medical, Russian-made device that transmits pulses of 40 MHz radio signal at pulse rates designed to match relaxed and sleeping states originally.

The machine, known as the LIDA, is on loan to the Jerry L. Pettis Memorial Veterans Hospital through a medical exchange program between the Soviet Union and the United States.

Hospital researchers have found in changes behaviour in animals.

"It looks as though instead of taking a valium when you want to relax yourself it would be possible to achieve a similar result, probably in a safer way, by the use of a radio field that will relax you" said Dr. Ross Adey, chief of research at the hospital. [Dr. Adey is now deceased.]

[Missing one line on the photocopy] . . . manual shows it being used on a human in a clinical setting, Adey said.

The manual says it is a "distant pulse treating apparatus" for psychological problems, including sleeplessness, hypertension and neurotic disturbances.

The device has not been approved for use with humans in this country, although the Russians have done so since at least 1960, Adey Said.

Low frequency radio waves simulate the brain's own electromagnetic current and produce a trance-like state.

Adey said he put a cat in a box and turned on the LIDA.

"Within a matter of two or three minutes it is sitting there very quietly . . . it stays almost as though it were transfixed" he said.

Tho hospital's experiment with the machine has been underway for three months and should be completed within a year, Adey said.

Eleanor White's comments (Dr. Byrd's statement follows):

1. Heavy "fatigue attacks" are a very common experience among involuntary neuro-electromagnetic experimentees. The LIDA device could, right out of the box, be used as a fatigue attack weapon, FROM HIDING, thru non- or semi-conductive walls.
2. If the LIDA machine is tuned for tranquilizing effect, then it might also be tuned for "force awake" and other effects too.

This device is an electronic harassment weapon, AS IS. A TV documentary stated the Russian medical establishment considers this 1950s device obsolete. (Wonder what has taken it's place?)

Below is a statement from Dr. Eldon Byrd, U.S. psychotronic researcher who funded Dr. Adey's work with the LIDA machine:

"The LIDA machine was made in the 1950's by the Soviets. The CIA purchased one through a Canadian front for Dr. Ross Adey, but didn't give him any funds to evaluate it.

"I provided those funds from my project in 1981, and he determined that the LIDA would put rabbits into a stupor at a distance and make cats go into REM.

"The Soviets included a picture with the device that showed an entire auditorium full of people asleep with the LIDA on the podium. The LIDA put out an electric field, a magnetic field, light, heat, and sound (of course light and heat are electromagnetic waves, but at a much higher frequency than the low frequencies of the electric and magnetic fields mentioned above).

"The purported purpose of the LIDA was for medical treatments; however, the North Koreans used it as a brain washing device during the Korean War. The big question is: what did they do with the technology? It could have been improved and/or made smaller. It is unlikely that they abandoned something that worked.

"Direct communication with Ross Adey: While he was testing the LIDA 4, an electrician was walking by and asked him where he got the "North Korean brain washing machine". Ross told him that is was a Russian medical device.

"The guy said he had been brain-washed by a device like that when he was in a POW camp. They placed the vertical plates alongside his head and read questions and answers to him. He said he felt like he was in a dream. Later when the Red Cross came and asked questions, he responded with what had been read to him while under the influence of the device. He said he seemed to have no control over the answers.

"The LIDA is PATENTED IN THE US. Why? They are not sold in the US—the only one I know that exists is the one that was at Loma Linda Medical Center where Adey used to work. Eldon"

Involuntary neuro-experimentation activist Cheryl Welsh, Davis CA, sent in this clipping from an article by Dr. Ross Adey but without complete bibliographic references:

"Soviet investigators have also developed a therapeutic device utilizing low frequency square wave modulation of a radiofrequency field. This instrument known as the Lida was developed by L. Rabichev and his colleagues in Soviet Armenia, and is designed for "the treatment of neuropsychic and somatic disorders, such as neuroses, psychoses, insomnia, hypertension, stammering, bronchia asthma, and asthenic and reactive disturbances".

It is covered by U.S. Patent # 3,773,049. In addition to the pulsed RF field, the device also delivers pulsed light, pulsed sound, and pulsed heat. Each stimulus train can be independently adjusted in intensity and frequency.

The radiofrequency field has a nominal carrier frequency of 40 MHz and a maximum output of approximately 40 Watts. The E-field is applied to the patient on the sides of the neck through two disc electrodes approximately 10 cm in diameter. The electrodes are located at a distance of 2-4 cm from the skin.

[Eleanor White's comment: The fact that Dr. Ross Adey mentioned an "audience" being put to sleep by the LIDA suggests that the "E-field" electrodes may not play an essential role. The radio signal appears to be the primary cause of the sleep/trance effect.]

Optimal repetition frequencies are said to lie in the range from 40 to 80 pulses per minute. Pulse duration is typically 0.2 sec. In an 8 year trial period, the instrument was tested on 740 patients, including adults and children. Postivive therapeutic effects were claimed in more . . ."

Up to Contents

APPENDIX PM2—FREY'S PAPER

Human Auditory System Response To Modulated Electromagnetic Energy

ALLAN H. FREY

General Electric Advanced Electronics Center
Cornell University Ithaca, New York

TRANSCRIPTION, Courtesy of MindNet Archives, Mike Coyle posted at http://www.ritualabuse.net/MCF/

Frey, Allan H., Human Auditory system response to modulated electromagnetic energy. J. Appl. Physiol. 17(4): 689-692. 1962.

(*) Asterisks indicate unreadable characters in the original copy.

NOTE: In 1962, frequencies were expressed as kiloCYCLES, megaCYCLES, etc., with abbreviations being kc, mc

—The intent of this paper is to bring a new phenomena to the attention of physiologists. Using extremely low average power densities of electromagnetic energy, the perception of sounds was induced in normal and deaf humans. The effect was induced several hundred feet from the antenna the instant the transmitter was turned on, and is a function of carrier frequency and modulation. Attempts were made to match the sounds induced by electromagnetic energy and acoustic energy.

The closest match occurred when the acoustic amplifier was driven by the rf transmitter's modulator. Peak power density is a critical factor and, with acoustic noise of approximately 80 db, a peak power density of approximately 275 mw / rf is needed to induce the perception at carrier frequencies 125 mc and 1,310 mc. The average power density can be at rf as low as 400 _u_w/cm2. The evidence for the various positive sites of the electromagnetic energy sensor are discussed and locations peripheral to the cochlea are ruled out.

Received for publication 29 September 1961.

A significant amount of research has been conducted with the effects of radio-frequency (rf) energy on organisms (electro-magnetic energy between 1 kc and ** Gc). Typically, this work has been concerned with determining damage resulting from body temperature increase. The average power densities used have been on the order of 0.1-t w/cm2 used over many minutes to several hours.

In contrast, using average power densities measured in microwatts per square centimeter, we have found that ****r effects which are transient, can be induced with rf energy. Further, these effects occur the instant the transmitter is turned on. With appropriate modulation, the perception of different sounds can be induced in physically deaf, as well as normal, in human subjects at a distance of inches up to thousands of feet from the transmitter. With somewhat different transmission parameters, you can induce the perception of severe buffeting of the head, without such apparent vestibular symptoms as dizziness or nausea. Changing transmitter parameters down, one can induce a "pins-and-needles" sensation.

Experimental work with these phenomena may yield information on auditory system functioning and, more generally, in the nervous system function. For example, this energy could possibly be used as a tool to explore nervous system coding, possibly using Neider and Neff's procedures (1), and for stimulating the nervous system without the damage caused by electrodes.

Since most of our data have been obtained of the "rf sound" and only the visual system has previously been shown to respond to electromagnetic energy, this paper will be concerned only with the auditory effects data. As a further restriction, only data from human subjects will be reported, since only this data can be discussed meaningfully at the present time. The long series of studies we performed to ascertain that we were dealing with a biological significant phenomena (rather than broadcasts from sources such as loose fillings in the teeth) are summarized in another paper (2), which also reports on the measuring instruments used in this work.

The intent of this paper is to bring this new phenomenon to the attention of physiologists. The data reported are intended to suggest numerous lines of experimentation and indicate necessary experimental controls.

Since we are dealing with a significant phenomenon, we decided to explore the effects of a wide range of transmitter parameters to build up the body of knowledge which would allow us to generate hypotheses and determine what experimental controls would be necessary. Thus, the numbers given are conservative; they should not be considered precise, since the transmitters were never located in ideal laboratory environments. Within the limits of our measurements, the orientation of the subject in the rf field was of little consequence.

Most of the transmitters used to date in the experimentation have been pulse modulated with no information placed on the signal. The rf sound has been described as being a buzz, clicking, hiss, or knocking, depending on several transmitter parameters, i.e., pulse width and pulse-repetition rate (PRF). The apparent source of these sounds is localized by the subjects as being within, or immediately behind the head. The sound always seems to come from within or immediately behind the head no matter how the subjects twists or rotates in the rf field.

Our early experimentation, preformed using transmitters with very short square pulses and high pulse-repetition rates, seemed to indicate that we were dealing with harmonics of the PRF. However, our later work has indicated that this is not the case; rather, the rf sound appears to be incidental modulation envelope on each pulse, as shown in Fig 1.

Some difficulty was experienced when the subjects tried to match the rf sound to ordinary audio. They reported that it was not possible to satisfactorily match the rf sound to a sine wave or to white noise. An audio amplifier was connected to a variable bypass filter and pulsed by the transmitter pulsing mechanism. The subjects, when allowed to control the filter, reported a fairly satisfactory match. The subjects were fairly well satisfied with all frequencies below 5-kc audio were eliminated and the high-frequency audio was extended as much as possible. There was, however, always a demand for more high-frequency components. Since our tweeter has a rather good high-frequency response, it is possible that we have shown an analogue of visual phenomenon in which people see farther into the ultraviolet range when the lenses is eliminated from the eye. In other words, this may be a demonstration that the mechanical transmission system of the ossicles cannot respond to as high a frequency as the rest of the auditory system. Since the rf bypasses the ossicle system and the audio given the subject for matching does not, this may explain the dissatisfaction of our subjects in the matching.

FIG. 1. Oscilloscope representation of transmitter output over time (pulse-modulated).

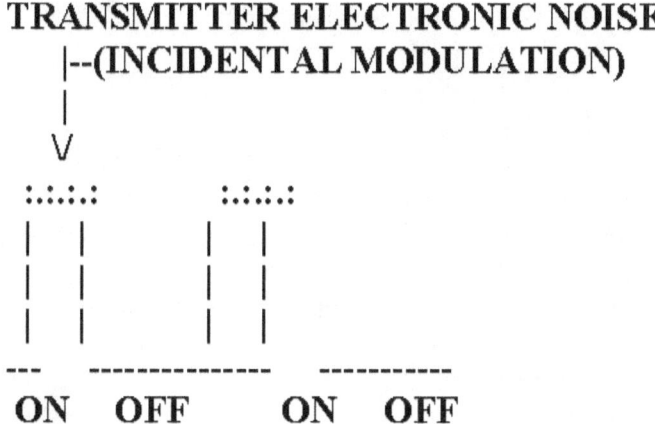

FIG. 2. Audiogram of deaf subject (otosclerosis) who had a "normal" rf sound threshold.

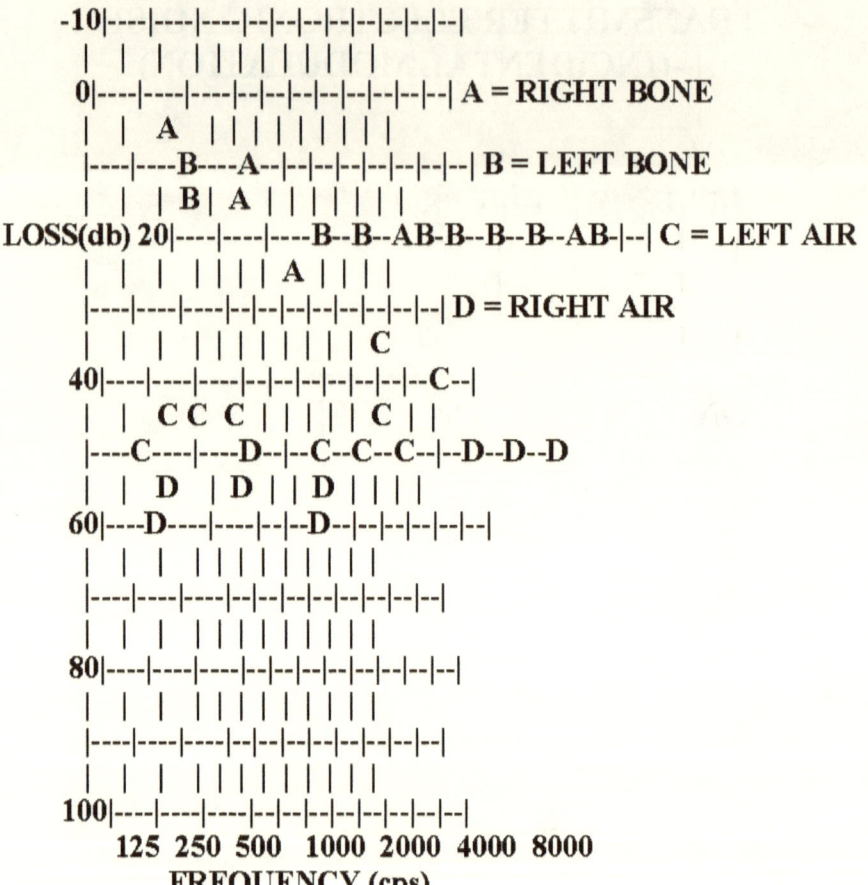

```
 -10|----|----|----|--|--|--|--|--|--|--|--|
    |   |   |  | | | | | | | | |
   0|----|----|----|--|--|--|--|--|--|--|--| A = RIGHT BONE
    |   |   A | | | | | | | | |
    |----|----B----A--|--|--|--|--|--|--|--| B = LEFT BONE
    |   |   | B| A | | | | | | |
LOSS(db) 20|----|----|----B--B--AB-B--B--B--AB-|--| C = LEFT AIR
    |   |   |   | | | | A | | | |
    |----|----|----|--|--|--|--|--|--|--|--| D = RIGHT AIR
    |   |   |   | | | | | | | | C
   40|----|----|----|--|--|--|--|--|--|--|--C--|
    |   |  C  C  C | | | | | C | |
    |----C----|----D--|--C--C--C--|--D--D--D
    |   |  D  | D | | D | | | |
   60|----D----|----|--|--D--|--|--|--|--|--|
    |   |   |   | | | | | | | | |
    |----|----|----|--|--|--|--|--|--|--|--|
    |   |   |   | | | | | | | | |
   80|----|----|----|--|--|--|--|--|--|--|--|
    |   |   |   | | | | | | | | |
    |----|----|----|--|--|--|--|--|--|--|--|
    |   |   |   | | | | | | | | |
  100|----|----|----|--|--|--|--|--|--|--|--|
        125 250 500  1000 2000 4000 8000
              FREQUENCY (cps)
```

TABLE 1. Transmitter parameters

Transmitter	Frequency, mc	Wavelength, cm	Pulse Width, _u_sec	Pulses Sec.	Duty Cy.
A	1,310	22.9	6	244	.0015
B	2,982	10.4	1	400	.0004
C	425	70.6	125	27	.0038
D	425	70.6	250	27	.007
E	425	70.6	500	27	.014
F	425	70.6	1000	27	.028
G	425	70.6	2000	27	.056
H	8,900	3.4	2.5	400	.001

FIG. 3. Attenuation of ambient sound with Flent antinoise stopples (collated from Zwislocki (3) and Von Gierke (4).

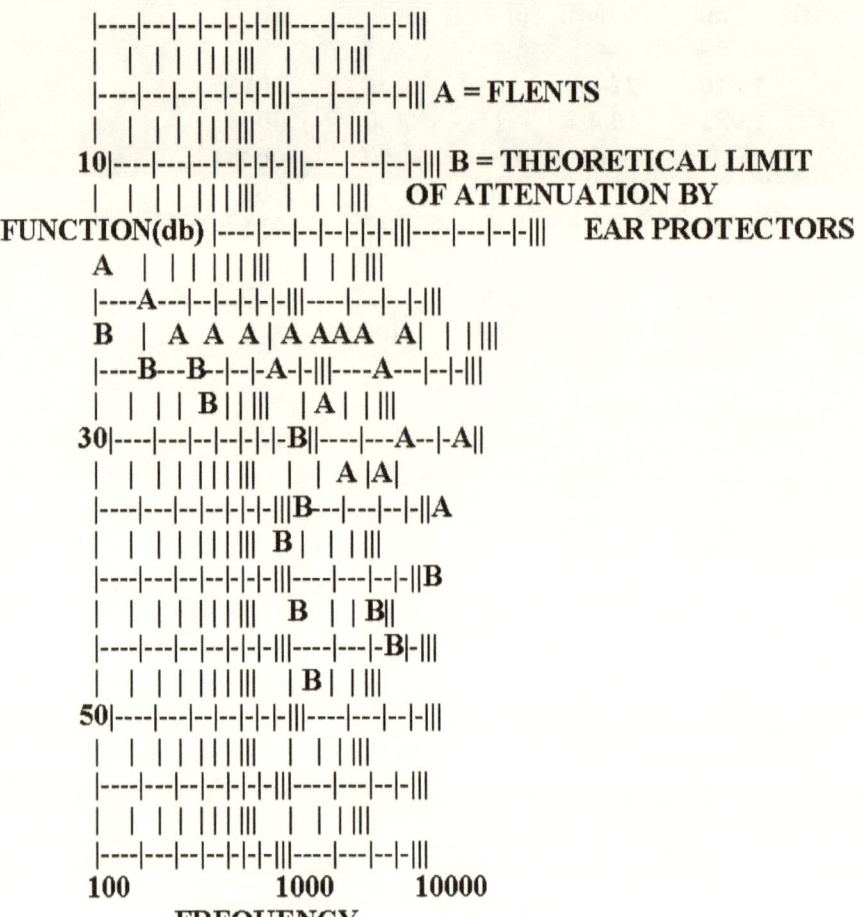

TABLE 2. Theshold for perception of rf sound (ambient noise level 70-90 db).

Transmitter	Frequency, mc	Duty Cy.	Avg Power mw, cm2	Peak Power Density, mw, cm2	Peak Electric Density v cm	Peak Magnetic Field Field turns, m	amp.
A	1,310	.0015	0.4	267	14	4	
B	2,982	.0004	2.1	5,250	63	17	
C	425	.0038	1.0	263	15	4	
D	425	.007	1.9	271	14	4	
E	425	.014	3.2	229	13	3	
F	425	.028	7.1	254	14	4	

FIG. 4. Threshold energy as a function of frequency of electromagnetic energy (ambient noise level 70-90 db).

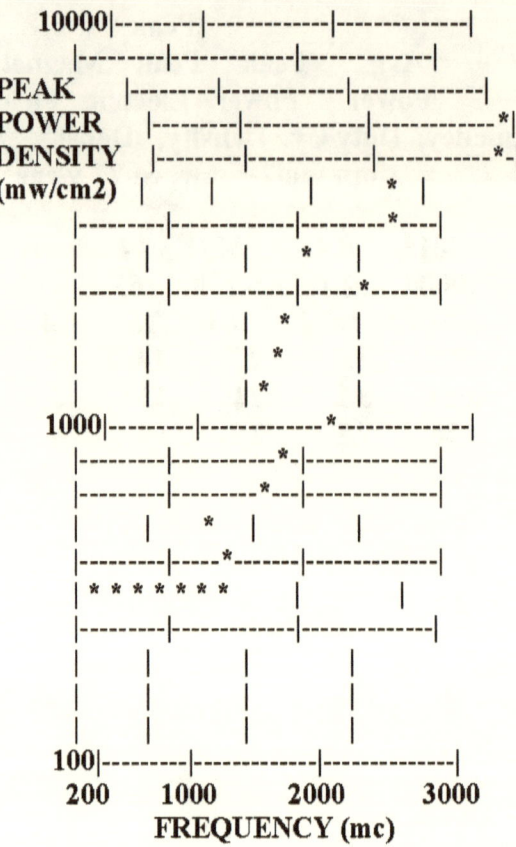

FIG. 5. Microwave power distribution in a forehead model neglecting resonance effects and considering only first reflections (from Nieset et al. (5), modified).

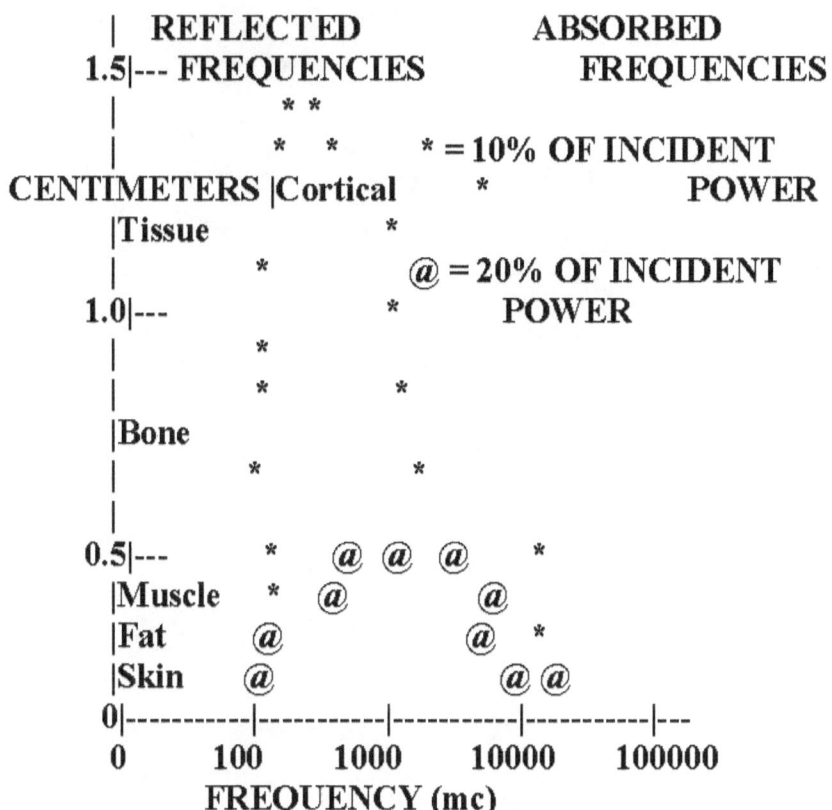

FIG. 6. Area most sensitive to electromagnetic energy (shaded portion).

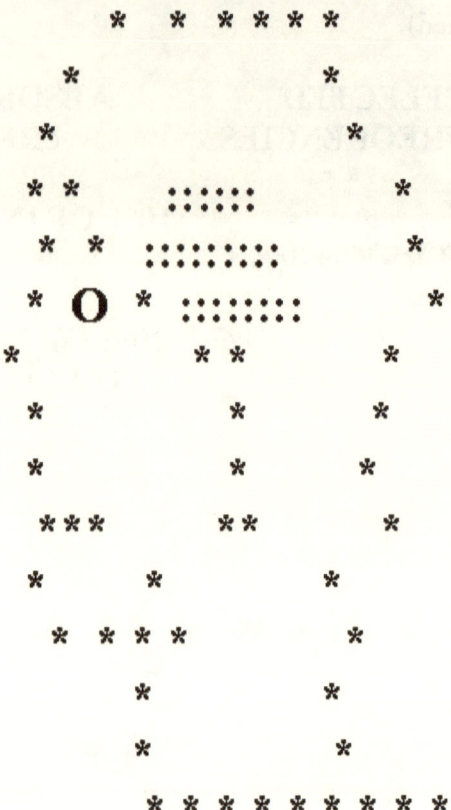

At one time in our experimentation with deaf subjects there seemed to be a clear relationship between the ability to hear audio above 5 kc and the ability to hear rf sounds. If a subject could hear above 5 kc, either by bone or air conduction, then he could hear the rf sounds. For example, the threshold of the subject whose audio gram appears in Fig. 2 was the same average power density as our normal subjects. Recently, however, we have found people with a notch around 5 kc who do not perceive the rf sounds generated by at least one of our transmitters.

THRESHOLDS

As shown in Table 1, we have used a fairly wide range of transmitter parameters. We are currently experimenting with transmitters that radiate energy at frequencies below 425 mc, and are using different types of modulation, e.g., pulse-repetition rates as low as 3 and 4/sec.

In the experimentation reported in this section, the ordinary noise level was 70-90 db (measured with a General Radio Co. model 1551-B sound level meter.) In order to minimize the rf energy used in the experimentation, subjects wore Flent antinoise ear stoppers whenever measurements were made. The ordinary noise attenuation of the Flents is indicated in Fig. 3. Although the rf sounds can be heard without the use of Flents, eventhough they have an ambient noise evel of 90 db, it appears that the ambient noise to some extent "masked" the rf sound.

Table 2 gives the thresholds for the perception of the sounds. It shows fairly clearly that the critical factor in the perception of the rf sound is the peak power density, rather than the average power density. The relatively high value for transmitter B was expected and will be discussed below. Transmitter G has been omitted from the table since the 20-mw/cm2 reading for it can be considered only approximate. The field-strength-measuring instruments used in that experiment did not read high enough to give an accurate reading. The energy from transmitter H was not perceived, even when the peak power density was as high as 25 w/cm2.

When the threshold energy is plotted as a function of the rf energy (Fig. 4), a curve is obtained which is suggestive of the curve of penetration of rf energy into the head. Figure 5 shows the calculated penetration, by frequency of rf energy, into the head. Our data indicate that the calculated penetration curve may well be accurate at the higher frequencies but the penetration at the lower frequencies may be greater than that calculated on this model.

As previously noted, the thresholds were obtained in a high ambient noise environment. This is an unusual situation as compared to obtaining thresholds of regular audio sound. One recent experimentation leads us to believe that, if the ambient noise level were not so high, these threshold fields strengths would be much lower. Since one purpose of this paper is to suggest experiments, it might be appropriate to theories as to what the rf sound threshold might be if we assumed that the subject is in an anechoic chamber.

It is also assumed that there is no transducer noise.

Given: As a threshold for the rf sound, a peak power density of 275 mw/cm2 determined in an ambient noise environment of 80 db. Earplugs attenuate the ambient noise 30 db.

If: 1 mw/cm2 is set equal to o db, then 275 mw/cm2 is equal to 24 db.

Then: We can reduce the rf energy 50 db to—26 db as we reduce the noise level energy from 50 db to o db. We found that—26 db rf energy is approximately 3 _u_w/cm2.

Thus: If an anechoic room, rf sound could theoretically be induced by a peak power density of 3 _u_w/cm2 measured in free space. Since only 10% of this energy is likely to penetrate the skull, the human auditory system and a table radio may be one order of magnitude apart in sensitivity to rf energy.

Up to Contents

RF DETECTOR IN AUDITORY SYSTEM

One possibility that seems to have been ruled out in our experimentation is that of a capacitor-type effect with the tympanic membrane and oval window acting as plates of a capacitor. It would seem possible that these membranes, acting as plates of a capacitor, could be set in motion by rf energy. There are, however, three points of evidence against this possibility. First, when one rotates a capacitor in an rf field, a rather marked change occurs in the capacitor as a function of its orientation in the field. When our subjects rotate or change the positions of their heads in the field, the loudness of the rf sound does not change appreciably. Second, the distance between these membranes is rather small, compared with the wavelengths used. As a third point, we found that one of our subjects who has otosclerosis heard the rf sound.

Another possible location for the detecting mechanism is in the cochlea. We have explored this possibility with nerve-deaf people, but the results are inconclusive due to factors such as tinnitus. We are currently exploring this possibility with animal preparations.

The third likely place for the detection mechanism is the brain. Burr and Mauro (6) presented evidence that indicates that there is an electrostatic field about neurons. Morrow and Sepiel (7) presented evidence that indicates the existence of a magnetic field about neurons. Becker (personal communication) has done some work indicating that there is longitudinal flow of charged carriers in neurons. Thus, it is reasonable to suspect that possibly the electromagnetic field could interact with neuron fields. As yet, evidence of this possibility is inconclusive. The strongest point against it is that we have not found visual effects although we have searched for them. On the other hand, we have obtained other nonauditory effects and have found that the sensitive area for detecting rf sounds is a region over the temporal lobe of the brain.

One can shield, with a 2-in.2 piece of fly screen, a portion of the stippled area shown in Fig. 6 and completely cut off the rf sound.

Another possibility should also be considered. There is no good reason to assume that there is only one detector site. On the contrary, the work of Jones et al. (8), in which they placed electrodes in the ear and electrically stimulated the subject, is sufficiently relevant to suggest the possibility of more than one detector site. Also, several sensations have been elicited with properly modulated electromagnetic energy. It is doubtful that all of these can be attributed to one detector.

As mentioned earlier, the purpose of this paper is to focus the attention of physiologists on an unusual area and stimulate additional work on which interpretations can be based.

Interpretations have been deliberately omitted from this paper since additional data are needed before a clear picture can emerge. It is hoped that the additional exploration will also result in an increase in our knowledge of nervous system functions.

REFERENCES:

Neider, P.C. and W.D. Neff. Science 133: 1010, 1961.

Frey, A.H. Aero Space Med. 32: 1140, 1961.

Zwislocki, J. Noise Control 4:42, 1958.

Von Gierke, H. Noise Control 2:37, 1956.

Nifset, R., Pinneo R. Baus J. Fleming, & R. McAfee. Ann. Rept. USAF

Rome Air Development Command, TR-61-65, 1961.

Burr, H., & J. Seipel, J. Wash Acad. Sci. 21: 455, 1949.

Morrow, R., & J. Seipel. J. Wash. Acad. Sci. 30: 1, 1969.

Jones, R.C., S.S. Stevens, & M.H. Laurie. J. Acoust. Sci. Am. 12: 281, 1940.

Up to Contents

APPENDIX PM3—NASA ARTICLE

TITLE: Effects of low power microwaves on the local cerebral blood flow of conscious rats

Original web link, folded for readability
http://techreports.larc.nasa.gov/ntrs/hget.cgi?recon?2044/3=/raid5/index/ star/80%2517043725%202044%20N19810004209recon1

Eleanor White's inserted note:

See the ITALIC text below. This abstract has implications FAR beyond "rats"

Document ID: 19810004209 N (81N12720) File Series: NASA
                     Technical Reports
Report Number: AD-A090426
Sales Agency & Price: CASI Hardcopy or Microfiche

Authors:
Oscar, K. J. (Army Mobility Equipment Command)

Published: Jun 01, 1980

Corporate Source:
Army Mobility Equipment Command (Fort Belvoir, VA, United States)
Pages: 10
Contract Number: None

NASA Subject Category: LIFE SCIENCES (GENERAL)

Abstract:

A decoy and deception concept presently being considered is to remotely create the perception of noise in the heads of personnel by exposing them to low power, pulsed microwaves. When people are illuminated with properly modulated low power microwaves the sensation is reported as a buzzing, clicking, or hissing which seems to originate (regardless of the person's position in the field) within or just behind the head. The phenomena occurs at average power densities as low as microwatts per square centimeter with carrier frequencies from 0.4 to 3.0 GHz. *By proper choice of pulse characteristics, intelligible speech may be created.* Before this technique may be extended and used for military applications, an understanding of the basic principles must be developed. Such an understanding is not only required to optimize the use of the concept for camouflage, decoy and deception operations but is required to properly assess safety factors of such microwave exposure.

Major Subject Terms:

AUDITORY PERCEPTION
BRAIN CIRCULATION
DECEPTION
MICROWAVES
PHYSIOLOGICAL EFFECTS
RADIATION EFFECTS

Minor Subject Terms:

BIOLOGICAL EFFECTS
HEMODYNAMICS
MILITARY TECHNOLOGY
RADIATION DOSAGE
SOUND LOCALIZATION
Language Note: English

NASA Access Help Desk

E-mail: help@sti.nasa.gov
Phone: 301-621-0390
FAX: 301-621-0134

Eleanor White's comments: I have received this report in it's entirety and I find that the article itself is about the title: Cerebral blood flow in rats. We involuntary experimentees are extremely lucky that the authors mentioned the success of microwave voice to skull transmission in this official document.

Below I have extracted some of the references to the full document which appear to be more promising and specific. This sub-list is guesswork since the references omit article titles, but any articles by Frey or Justesen are potentially useful.

1. Frey, A.H., Messenger, R. and Eichert, E., National Technical Information Service, Doc. No. AD747684 (1972)
5. Justesen, D.R., IEEE Spectrum 16, 67-68 (1979)
7. Frey, A.H., Feld, S.R., & Frey, B. Annals of N.Y. Academy of Science, 247, 433-439 (1975)
18. Frey, A.H., & Messenger, R., Science 181, 356-358 (1973)
25. Frey, A.H., & Feld, S.R., Journal of Comp. Physiology and Psychology, 89, 183-188 (1975)
27. King, N.W., Justesen, D.R., & Clarke, R.L., Science 172, 398-401 (1977)

Up to Contents

APPENDIX PM4—SEA/KOHN'S PROJECTS

Communicating Via the Microwave Auditory Effect

Web address:
http://es.epa.gov/ncerqa_abstracts/sbir/other/monana/kohn.html
Awarding Agency: Department of Defense
SBIR Contract Number: F41624-95-C-9007
Title: Communicating Via the Microwave Auditory Effect
Principal Investigator: Mr. Brian Kohn

Company Name:

Science & Engineering Assoc, Inc.
6100 Uptown Blvd NE
Albuquerque, NM 87110
Telephone Number: 505-884-2300
Business Representative:

Project Period:

Project Amount: $739,995
Research Category: Monitoring/Analytical

Description:

An innovative and revolutionary technology is described that offers a means of low-probability-of-intercept Radio frequency (RF) communications. *The feasibility of the concept has been established* using both a low intensity laboratory system and a high power RF transmitter. Numerous military applications exist in areas of search and rescue, security and special operations.

Supplemental Keywords: small business, SBIR,
See also: http://www.seabase.com
Last Updated: November 17, 1997

BRIAN KOHN'S PROJECT, EARLIER STUDY:

Program: SBIR
Agency: AF Field
Office: AL
TOPIC Number: AF93-026
Control Number: 93AL-185
Contract Number: F41624-93-C-9013
Phase: 1
Awarded In: 93
Award Amount: $37,806
Award Start Date: 17 MAY 93
Award Completion Date: 17 DEC 93
Proposal Title: Communicating Via the Microwave Auditory Effect
Principal Investigator Name:Brian Kohn
Principal Investigator Phone:505-884-2300
Firm SCIENCE & ENGINEERING ASSOC., INC.
SEA Plaza
6100 Uptown Blvd NE
SUITE 700
Albuquerque, NM 87110
Woman Owned: N
Minority Owned: N
Number of Employees: 95

Keywords:

MICROWAVE HEARING RF HEARING
BONE CONDUCTION THERMOELASTIC
COCHLEAR MICROPHONICS

Abstract:

In this research program, we plan to investigate a revolutionary new form of communication based on the microwave auditory effect. This proposed communication idea satisfies the requirements for an innovative, natural interface requiring no learning or training for efficient operation and effective communications. The purpose of the program proposed here is to extend the results of a recent feasibility study, performed for the Armstrong Laboratory/OEDR. *The study found that voice communications, via the microwave auditory effect, are highly feasible.* In Phase I of this SBIR, we propose to investigate the range of potential applications for this radically different form of voice communication and recommend hardware and systems concepts suitable for laboratory and brassboard demonstrations to be built under Phase II.

Up to Contents

APPENDIX PM5—BIOLOGICAL AMPLIFICATION

Electromagnetic Interaction With Biological Systems
edited by Dr. James C. Lin, University of Illinois
1989 Plenum Press, New York

Proceedings of the Joint Symposium on Interactions of Electromagnetic Waves with Biological Systems, held as part of the Twenty-Second General Assembly of the International Union of Radio Science, Aug 25-Sept 2, 1987, in Tel Aviv, Israel.

ISBN 0-306-43109-2
QP82.2.N64E44 1989
612.01448-dc19 88-38957
CIP

Eleanor White's comments on this posting:

This book focusses on NON-ionizing radiation, and contains detailed texts about NON-THERMAL effects. In other words, "right up our alley".

The main use of this book is to show that it is easy for electromagnetic signals to cause radio frequency hearing and other effects at LOW power levels. This in turn can be used to explain why DETECTION is so very difficult.

I don't understand the biological jargon, however, a few of the more plain-language paragraphs STUNNINGLY verify that with careful choice of signal frequency and modulation, not only can the body's cells detect the modulation envelope of an incoming radio signal (i.e. function as a "cellular crystal set") but even AMPLIFY these carefully formed signals. (Amplification of other effects, such as proneness to disease, is also covered in the book.)

"Detection", in terms of radio signal reception, means that some portion of the reciever "rectifies", that is, turns the AC of the incoming signal into varying DC. If this conversion is not done, voice to skull wouldn't work.

The book touches also on ways in which cells communicate, and shows that electromagnetic fields of relatively weak power levels can affect intercellular communication, which is, as I understand the subject, what the brain is "all about".

Bio-amplification is apparently why radio signals of very low average power ("MICROwatts" per NASA) can still produce audio effects, and no doubt plays a part in difficulties in detection.

When two more characteristics of voice to skull are factored in:

1. The carrier signal can be "hopped" continuously within the bioeffective bandwidth, known as "spread spectrum" transmission, and,
2. The voice modulation most effective for undetectable hypnosis is evidently a voice shifted just above normal hearing, but still audible to the brain,

. . . you have a recipe for incredibly difficult signals to detect.

This book is a mainstream publication, very well suited to use in our publicity and persuasion campaigns, and our dealings with authorities who claim radio signals don't affect living tissue except to heat it.

Finally, many thanks to Blanche Chavoustie for providing me photocopies of this book—a saintly work!

———————————

Page 110:

. . .

At that time [1953] excitatory mechanisms in nerver fibers and nerve cells were grouped under a common rubric of ionic equilibrium mechanisms. There was little interest in the possibility that functional organization of mebranes of cell bodies might involve threshhold sensitivities to both oscillating EM fields and to molecular stimuli *at energy levels substantially lower* than predicted by Hodgkin-Huxley models, and *substantially below typical threshholds in nerve fibers.*

Much recent research cited below has shown that imposed *weak* low frequency fields *(and radiofrequency fields amplitude—modulated at ELF frequencies)* that are *many orders of magnitude weaker in the pericellular fluid [fluid between adjacent cells] than the membrane potential gradient [voltage across the membrane] can modulate actions of hormone, anti-body neurotransmitter and cancer-promoter molecules at their cell surface receptor sites.*

From their electrical characteristics, these sensitivities appear to involve nonequilibrium and highly cooperative processes that mediate a major amplification of initial weak triggers associated with binding of these molecules at their specific cell surface receptor sites.

(Adey, 1983, 1986, 1987; Adey and Lawrence, 1984; Lawrence and Adey, 1982).

Page 122:

Cooperative Modification of Calcium Binding by RF Fields at Cell Surfaces with Amplification of Initial Signals

Initial stimuli associated with weak perpindicular EM fields and with binding of stimulating molecules at their membrane receptor sites elicit a HIGHLY COOPERATIVE modification of $Ca++$ binding to glycoproteins along the membrane surface.

As noted above, a longitudinal spread is consistent with the direction of extracellular current flow associated with physiological activity and imposed EM fields. This cooperative modification of surface $Ca++$ binding is an AMPLIFYING STAGE, with evidence from concurrent initial molecular binding events by imposed RF fields that there is a far greater increase in $Ca++$ efflux than is accounted for in the events of receptor-ligand binding (Bawin and Adey, 1976; Bawin et al, 1975; Liu-Liu and Adey, 1982).

Page 124:

...

Enzymes are protein molecules that function as catalysts, initiating and enhancing chemical reactions that would not otherwise occur at tissue temperatures. This ability resides in the pattern of electrical charges on the molecular surface. In the fashion of more familiar chemical catalysts, such as the hydrocarbon oxidation systems which function only at very high temperatures in automotive exhaust systems, a catalyst emerges unchanged from these reactions and is thus able to participate indefinitely in a specific reaction.

Activation of these enzymes and the reactions in which they participate involve energies millions of times greater than in the cell surface cell surface triggering events initiated by the EM fields, emphasizing the MEMBRANE AMPLIFICATION inherent in this trans-membrane signaling sequence.

Page 131:

...

Stimulus Amplification in Cooperative Systems

...

It is therefore clear that OBSERVED EM field interactions with cells and tissues based on oscillating ELF tissue gradients between 10 E-7 and 10 E-1 volts per centimeter would involve cooperativity MANY ORDERS OF MAGNITUDE GREATER than envisaged in the examples just cited.

In part this discrepancy appears to relate to far greater sensitivities to low-frequency EM fields [EW: ELF, that is, the "entrainment" frequencies] and to RF fields with low-frequency amplitude-modulation [EW: this includes radar hearing signals] than to imposed step functions or DC gradients [EW: common with contact electrodes, not of interest in mind control at a distance] used in many electrochemical experiments and models to test levels of cooperativity in biological systems. (Blank, 1972)

[EW: In plain language, both entrainment (ELF) fields and pulsating radar-like (RF) fields are a hell of a lot more influential on cells than some experimental work using DC and electrode methods.]

Page 95:

[EW: This section is not part of the demonstration that EM signals can be biologically amplified, as above. It's main interest is that a magnetophosphene "gun" was under consideration by the U.S. National Institute of Justice in 1993, along with a "fever" gun and a "convulsion" gun, both using microwave technology. As of 1999, nothing has been heard from NIJ on this development, however, page 95 here suggests that such a microwave weapon is feasible.]

Magnetophosphenes

An effect of time-varying magnetic fields on humans was first described by d'Arsonval (1896) [EW: Anyone doubt there has been some progress since 1896?] is the induction of a flickering illumination within the visual field field known as magnetophosphenes. This phenomenon occurs as an immediate response to stimulation by either pulsed or sinusoidal magnetic fields with frequencies less than 100 Hz, and the effect is completely reversible with no apparent influence on visual acuity. The maximum visual sensitivity to sinusoidal magnetic fields has been found at a frequency of 20 Hz in human subjects with normal vision.

[EW: Radio signals are a combination of electric and magnetic fields. To radiate a 20 Hz signal would require such huge antennas that it is impractical to do so. I'd recommend that if someone has the facilities and skills, I'd try some VHF (or microwave) pulsing at 20 Hz on an RF carrier at, say, the 2-meter (144-148 MHz) ham band with a duty cycle, say, of 20% pulse-ON time.]

Up to Contents

————————————

APPENDIX PM6—VOICE TO SKULL, 1974 SUCCESS

Microwaves and Behavior
Dr. Don R. Justesen
Laboratories of Experimental Neuropsychology
Veterans Administration Hospital
Kansas City, Missouri as published in
American Psychologist
Journal of the American Psychological Association
Volume 30, March 1975, Number 3

Eleanor's comments

   This LAYS TO REST ANY DOUBTS THAT VOICE TO SKULL TECHNOLOGY DOES NOT EXIST OR IS "IN THE FUTURE"! PERIOD!
   This article describes in precise terms how Dr. Joseph C. Sharp and staff transmitted the WORDS for the digits 1 to 10 using a modulated version of an Allan Frey type pulsed microwave transmitter. A detailed description of Frey transmitters can be viewed at:

Appendix PM2

The relevant text is below.

————————————

Page 396:

. . .

   The demonstration of sonic transduction of microwave energy by materials lacking in water LESSENS the likelihood that a thermohydraulic principle is operating in human perception of the energy. Nonetheless, some form of thermoacoustic transduction probably underlies perception. If so, it is clear that simple heating is NOT a sufficient basis for the Frey effect; the requirement for pulsing of radiations appears to implicate a thermodynamic principle.

Frey and Messenger (1973) and Guy, Chou, Lin, and Christensen (1975) confirmed that a microwave pulse with a slow rise time is INeffective in producing an auditory response; only if the rise time is SHORT, resulting in effect in a square wave with respect to the leading edge of the envelope of radiated radio-frequency energy, does the auditory response occur.

[Eleanor's comment: This is why we don't "hear" ordinary radio and TV signals.]

Thus the rate of change (the first derivative) of the waveform of the pulse is a CRITICAL factor in perception. Given a thermodynamic interpretation, it would follow that information can be encoded in the energy and "communicated" to the "listener".

Communication has in fact been demonstrated. A. Guy (Note 1), a skilled telegrapher, arranged for his father, a retired railroad telegrapher, to operate a key, each closure and opening of which resulted in a pulse of microwave energy. By directing the radiations at his own head, complex messages via the Continental Morse Code were readily received by Guy.

Sharp and Grove (note 2) found that appropriate modulation of microwave energy can result in "wireless" and "receiverless" communication of SPEECH. The recorded by voice on tape each of the single-syllable words for digits between 1 and 10. The electrical sine-wave analogs of each word were then processed so that each time a sine wave crossed ZERO REFERENCE IN THE NEGATIVE DIRECTION, a brief pulse of microwave energy was triggered.

[Eleanor's comment: This is, in effect a form of what is called pulse-RATE modulation.]

By radiating themselves with these "voice modulated" microwaves, Sharp and Grove were READILY able to hear, identify, and distinguish among the 9 words. [Typo?] The sounds heard were not unlike those emitted by persons with artificial larynxes.

Communication of more complex words and of sentences was not attempted because the averaged densities of energy required to transmit longer messages would approach the current 10 milliwatts per square centimeter limit of safe exposure.

The capability of communicating directly with a human being by "receiverless radio" has obvious potentialities both within and without the clinic. But the hotly debated and unresolved question of how much microwave radiation a human being can safely be exposed to will probably forestall applications within the near future.

. . .

Up to Contents

_____

APPENDIX PM7—U.S. GOVT DOCUMENT RE: RUSSIAN MIND CONTROL

A Warning to the World from Russian Psychotronic Researchers

Unclassified FOIA document, courtesy Cheryl Welsh, which is a bulletin circulated among U.S. government agencies reviewing an October 6, 1995 Russian TV news segment titled "Man and Law" on the topic of mind control technology. The images below complement another video produced by German ZDF TV and shown on December 22, 1998. The 1998 video's transcribed sound track, in English is posted at:

http://www.raven1.net/russvid.htm

_____

 -FAMILY BRIEFING CONTINUED-

_____

Now that you have read how frequency is used as a weapon, you might now understand how radiating the atmosphere around you is increasing the possibility of inducing toxicity in the form of cancer from within all it encounters.

Here's a question for you, why is the US switching from analog AM/ FM, UHF/VHF radio signals to an all digital cable system?

Is it so they can eliminate these signals locally, making it easier for them to detect the utilization of frequency weapons being used on the people of this country?

Or are they going to utilize HAARP on the people of this country to induce the zombie like trance state needed to transform the immediate populace into slaves to fulfill the agenda of the returning ET Reptilian race?

Or could we be wrong about the Government and its elitist factions? Might they actually be working to prepare a defense against this invasion? Possibly using these frequency weapons to stop the manifestation of these creatures into this dimensional reality by creating a frequency shield, or heliosphere around our entire country or even planet?

More interesting yet is, might they know that the 2012 alignment will open real "Star-Gates" all over this planet on every sacred site where there is a temple associated with a planet? And what if they know that these planetary temples will open micro-worm holes that will allow people to travel to other planets in the flesh just like traveling in a car like we do now to the nearest city?

I'm betting this is the case, and not only that but I'm also betting that not only can you travel from planet to planet but you can actually travel from solar system to solar system and even from galaxy to galaxy. Perhaps that's why they are trying to chip everyone, so they can track or even control what it is you can access and where it is you are allowed to even go.

Open your mind to the possibility. Could we be on the presipis of escaping this entire system of human suppression and monitary slavery?

One can only speculate at this point.—(Sabre)

What follows is information from Dr. Dan Burisch's web site *www. eaglesdisobey.net* and touches on what it is exactly that he is up to. I will provide a short brief for you at the end of this particular section, as for what it is that he is doing, it must be explained to be fully appreciated.— (Sabre)

What follows has been sourced from eaglesdisobey.net

Welcome Message:

Welcome to Eagles Disobey. This is the flagship website, for information and research of Dr. Dan Burisch and Marcia McDowell, Ph.D.

Here is the best place on the internet to find out information about Dan Burisch, because this site (originally started under the dot com extension, back in 1998) is the longest near continuous repository of information about Dan Burisch; his past and present research, as well as the history of his involvement with Majestic-12, Area-51, and extraterrestrial visitor, housed in a lab located deep beneath the desert floor at the Groom Lake facility (S-4).

Burisch's Ganesh Particle Acknowledged By Caltech

From BJWolf
BJWolf007@rogers.com
1-14-3

It's official . . . coming in at #8 of the most popular search strings in Caltech's Electronic Thesis and Dissertation (ETD) archive is . . . the Ganesh Particle—discovered by Dr. Dan Burisch and outlined in the Lotus Protocol.

Yes, doctoral candidates and post doctoral researchers seem to be expressing quite an interest in the "Ganesh particle"—the existence of which had not even been hinted until Dan Burisch wrote the Lotus Protocol and then followed up with his experiments at Frenchman Mountain. Dr. Burisch even coined the name "Ganesh particle"! Now, the esteemed academicians at Caltech seem to be particularly (excuse the pun) interested in what the professional critics and skeptics have tried to ridicule for so long. Look at the stats for yourself. You will find the top ten search strings here:

http://www.its.caltech.edu/~gscnews/nov02_main.html

If you go about half way down the page and click on the ETD archive, you will be taken to this:

Yep, there it is—coming in in spot #8. The particle that nobody had ever heard of before Dan Burisch proved it was here.

From the moment that Dan theorized its existence, the Ganesh Particle generated a great deal of interest from our 'black project' government and military organizations. As experimentation continued, and the particle and its mechanics were isolated Dan was put under increasing pressure which he railed against, eventually providing information and details about the project to Bill Hamilton and myself in a short series of meetings, just prior to the announcement of his death last October. What most people don't know is that Dan confided in me that the Ganesh particle had some unusual genetic properties which is why the "black project" officials were so interested in it—properties only seen in the alien DNA he worked with up at S4 (Area-51).

Dan had written on the white board (behind them) many of the properties of the Ganesh particle, and how the crossbridge structuring worked. (Bill Hamilton—left. Dr. Dan Burisch—right)

Dan had become well aquainted with the DNA from tissue samples he was ordered to collect fom "Captive".

rense.com
Drs. Burisch & McDowell
To Present 'Lotus'
At Caltech
From Marcia McDowell
3-4-8

Hello, Jeff . . .

Remember this publication which first hit your website?

http://www.rense.com/general33/searcasdh.htm

Burisch's Ganesh Particle Acknowledged By Caltech

From BJWolf
BJWolf007@rogers.com
1-14-3

Well, the few who yelled that I made a mistake should now rethink things. On March 1, 2008, in Laughlin, Nevada at the meeting of the International UFO Congress, Dan made the announcement about Lotus we have been hinting at for awhile.

In 2001, the Looking Glass viewed the Lotus Project being found in archives in 2091, by a graduate student and his professor from Caltech. This was told to the public as early as 2003 by Dan and I! (Dan presented this information while he was still a member of Majestic.) Many have been wondering how that could possibly happen? They called us "names."

Well, this is how it can happen. This is how it happens when people are telling the truth.

We are honored to announce that on March 10, 2008, at 4:15pm, Dan and I have been invited and are scheduled to give an initial presentation on Lotus at the California Institute of Technology (Caltech), in the Nobel Laureate-graced Beckman Institute Auditorium.

Our faculty representative, who invited us (he would certify that we requested nothing) on January 22, 2007, after initial contact with us in December 2006, had no idea of the Looking Glass prognostication until today. In fact, he only arrived at Caltech in the Fall of 2004—almost 2 years after you first published the article, listed above.

The public is invited and it is free.

The general public abstract and schedule may be accessed at:

http://www.acm.caltech.edu/colloq.shtml
http://www.acm.caltech.edu/colloquia/07-08/burisch_d.html
http://today.caltech.edu/eas/listing.adp? template=cce&sponsor_
    id=1201&sponsor_id=361&sponsor_id=142&sponsor_id=
    1743&sponsor_id=334&sponsor_id=323&sponsor_id=1603&ran
    ge=term&term=Win ter&year=2007-2008

10 March, 2008
ACM/BE COLLOQUIUM SPECIAL LOCATION—
Beckman Institute Auditorium

Dan Burisch and Marcia McDowell
Unacknowledged Special Access Projects, Retired

A Peculiar Silicate-Associated Phenomenon

I am also including some images for you. (The slides were the ones presented in Laughlin. You will see your website's name proudly presented. The 2002 conversation occurred, but it was your website's publication which was first made public.)

Congratulations, Jeff! Your website helped to add proof of the existence and efficiency of the Looking Glass. This is very important for everyone, as Dan and I have told them that the Looking Glass has revealed that we are not facing the catastrophe that a few want to try to sell to the world. We are on the precipice of a new renaissance of Light and Love, should we spiritually join together in, and Pray for, think about, meditate on (however your philosophy dictates) "Unity for Humanity." We are now in the time of passage, so it is very important that people know this information. Such a spiritual path has been also suggested by Al Fast Thunder, Elder of the Lakota, with whom we sat in Sacred Circle, last year. You may see Dan wearing the sacred beads of the Lakota, which were presented to him by them, in one of the photos from Laughlin. The people in that photo are (L to R): Dan, Rob Simone, myself, and Paola Harris.

Additionally, for your readers, our newest book, Emanation of the Solfeggio, with a forward by Dr. Leonard Horowitz, will be published soon by Dandelion Books! (Information on that can be tracked on our website.)

Dan and I have always respected you for presenting the truth. We hope to speak with you, again.

Marci
(aka, former pseudo—BJ Wolf)

Marcia Ann McDowell, Ph.D.
President, Eagles Disobey, Inc.
http://www.eaglesdisobey.net

-FAMILY BRIEFING CONTINUED-

Now I know there is really no information about what it is that Dan is doing other than the hint that was given about it showing some characteristics of the alien DNA that he himself and the other project officials were probably privy to at area 51 during their experiments on "Captive". But after watching Dan Burisch describe the research on project Lotus this is what I have come to understand.

What Dan has found is that there is a crystalline thread that runs throughout the entirety of the human body. This thread is responsible for channeling the electrical impulses from the human brain to the respective part of the body that the brain is signaling.

So what Dan is doing is synthetically growing this crystalline structure in a lab and sending an electromagnetic charge through it. To his great surprise when he did this the crystal began to grow flesh around it. Amazed by what he was witnessing and well aware of the fact that he did not know what it was that was being created, he aborted the experiment. He did this because he feared that he might be calling the abomination of GOD into existence, and rightly so.

Now what they realized they were doing was calling forth life from right out of the etheric realm, from nothingness they were bringing forth life. Anyone who knows anything about gene manipulation and biological engineering knows that this is a huge leap in the history of man-kind and science.

What he figured out is that there is a buffer in between this realm of 3 dimensional reality and the realm of the etheric plane. This buffer zone is like a little restaurant waiter that takes orders and allows out of the etheric kitchen only exactly that which has been mathematically ordered. In other words he had to mathematically figure out how much crystalline to use and how large or small of an electromagnetic charge to send through it, to figure out what it was that he was calling forth into this dimensional reality as a living life force. If he used a lot of crystalline and a large charge he might be asking the little waiter to molecularly build him the DNA strand of a zebra. So the little waiter would only allow certain particles from the etheric realm to move past him in order to construct the proper molecules in the appropriate mathematical order that would create the DNA strand of that which was being called forward into a 3 dimensional reality or existance, in this case a Zebra, which would manifest itself in the flesh around this electromagnetically charged crystalline strand.

Once again, this is just the way I understand it and as I said, this is a huge leap forward for Man. We are now on the cusp of being able to bring forth life from the etheric realm. No more embryonic testing would be needed and gene manipulation could be a thing of the past. Biological engineering would take on new meaning and this would lead to the understanding of a new realm unlocking the door to an entirely new science. Now, if I understood him correctly, and if this is the case, then we are on the way to truly playing GOD. I think that this is a great advancment. An exciting direction towards the future development of a new science. What a leap!—(Sabre)

 -THE GREATEST INFORMATION-

Almost 70 years ago our government made contact with extraterrestrials who warned man-kind of an impending disaster that would almost assuredly wipe out 2/3rds of the worlds population.

From that point in time the Military Industrial Complex has been working on digging a hole deep enough to provide man-kind with a safe spot that would enable a very educated portion of the population to survive so that others could one day learn from them and perpetuate their crafts. Doctors, Surgeons, Scientists, Astrologists, and more, all of the people at the top of their respective fields are to be moved to safe locations underground in the event the need arise as has been seen and fore told.

However the Military's industrial complex did not put all of their eggs into one basket. They launched several programs into multiple fields allowing for a multitude of discoveries on many different fronts.

The greatest of these scientific understandings was this, we can create our own reality and we do have control over our future. But it is up to each one of us to do the right thing in the face of any crossroads when good confronts evil. This is what must be done by us to create harmony here on this plane of dimensional realities with the understanding that everything we do here affects the planet and even the alternate dimensional realities around us.

With that being said I am happy to deliver the greatest news that has been delivered to man-kind during this tumultuous time by MAJ Dan Burisch, PhD and that is that we have survived the predicted cataclysmic events as seen through the varying projects associated with this time line. This fantastic news follows.—(Sabre)

This is an excerpt from Project Camelot

- We have just received the following urgent notification from *Marci McDowell*, on behalf of *Dan Burisch*. It is important to note that there is no immediate danger at this time. However, we are posting this in the spirit of peace and unity for all peoples of the planet. This is a wake-up call for all. As it appears that we are not on the catastrophic timeline however, this does not lessen the urgency for all to come together in peace during this time.

Marcia Ann McDowell, Ph.D.
President, Eagles Disobey, Inc.
Co-Founder, "Eagles Disobey"

Further information taken from the Eagles Disobey website:

PURSUANT TO TAU IX-6 PROTOCOLS FOR THE H-1-E MAJ: EAGLES DISOBEY WILL STAND IN ABEYANCE FOR 72 HOURS FOR THIS PUBLIC NOTIFICATION. ONCE THE 72 HOURS EXPIRES, EAGLES DISOBEY WILL RESUME NORMAL OPERATIONS.

DR. BURISCH WILL LIKELY BE ENTERING A 1-2 WEEK RETREAT, TO MAKE A VERY SERIOUS DECISION. PER PREPLANNED DECISIONS, DR. BURISCH HAS CALLED FOR A "DEVICE" TO BE BROUGHT TO HIM IMMEDIATELY, AND FOR THE ENTIRE MAP DATABASE FROM THE FORMER MAJESTIC. (THIS DOES NOT MEAN THAT HE WILL REVEAL IT. IT MEANS HE HAS IT AT HIS DISPOSAL SHOULD CONDITIONS WARRANT.) DON'T ANYONE EVEN ASK: DR. BURISCH'S LITTLE ONES ARE REMAINING WHERE THEY ARE—IN LAS VEGAS, NEVADA. WE DO NOT BELIEVE THAT THIS WILL RESULT IN T2. DR. BURISCH WILL HAVE CONTACT ONLY WITH DR. McDOWELL, HIS PRIVATE RELIGIOUS ASSOCIATE (A RABBI), AND DR. GOLDSTEIN OUR PSYCH. PER OUR PREPLANNED PROCEDURES, IN THE EVENT OF THIS HAPPENING, DR. GOLDSTEIN WILL ACT AS THE OFFICIAL PUBLIC REPRESENTATIVE, THE VOICE FOR EAGLES DISOBEY. THE PUBLIC WILL HEAR MORE BY MARCH 28, 2009. SHOULD OTHER EVENTS OCCUR, EAGLES DISOBEY WILL IMMEDIATELY REVERT TO AN ALERT/EMERGENCY MODE AND ALL IMPORTANT INFORMATION WILL BE IMMEDIATELY TRANSMITTED TO THE PUBLIC.

On March 13, 2009, Dan found the date of the J-Rod's and Orions' T2 catastrophe event, buried in the words of Chi'el'ah (the J-Rod he met), and in their "Shadow Language." A careful study was conducted, and the date was found to be the likely valid date of the T2 event. Millions, over 10 million worldwide, responded to the call by Eagles Disobey to focus upon Unity for Humanity, during the time of the "Cycles Cross." On March 28, 2009, at 12:00 noon Universal time, a meeting was held with the Echelon Representatives of the adjourned Majestic, and of the "New Group" which took over in Majestic's wake.

Most of the original crew, who were with us during the critical beginning "Unity" phases, in 2006, were also present. At that meeting, we announced, several minutes AFTER the date/time, that we had passed the date recorded in the J-Rod's and Orions' history, that we had survived the Cycles Cross without disaster!

As we approached the Cycles Cross, and WE THE PEOPLE (worldwide) focused, the Sun became ever more quiet.

Eagles Disobey maintains the records of the time and will steward those records to ensure that the future sees that we stood up and faced the specter of fear, and TOLD it to go away! When we did, "IT" RAN AWAY FROM US!

Now, April 1, 2009, this in:

Deep Solar Mininum   04.01.2009

April 1, 2009: *The Sunspot cycle is behaving a little like the stock market. Just when you think it has hit the bottom, it goes even lower.*

2008 was bear. There were sunspots observed on 266 of the year's 366 days (73%0. To find a year with more blank suns, you have to go all the way back in 1913, which had 311 spotless days: Prompted by this numbers, some observers suggested that the solar cycle had hit bottom in 2008.

Maybe not. Supported counts for 2009 have dropped even lower. As of March 31st, there were no sunspots on 78 of the year's 90 days (87%).

It adds up to one inescapable conclusions: "were experiencing a very deep solar minimum," says solar physicist Dean Pesnell of the Goddard Space Flight Center.

Click Here to see the entire article: *http://science.nasa.gov/headlines/y2009/01apr_deepsolarminimum.htm?list951772*

This presentation by NASA reads-

"Pesnell believes sunspot counts will pick up again soon, "possibly by the end of the year," to be followed by a solar maximum of below-average intensity in 2012 or 2013."

What a big difference from the earlier reports forecasting solar doom, no?

As presented much earlier by Eagles Disobey—WE ALL HAVE THE POWER TO DECIDE!

Eagles Disobey gives its heartfelt thanks and gratitude to all who focused on Unity during this precious time in human and life history. We are all victorious because we stood on the right side of the "great dividing line of history" and now embark on an unknown path, which the Eagles (Marci and Dan) feel will inevitably lead us to the stars and beyond!

-END NOTES-

WE HAVE PASSED THE CATASTROPHE TIME LINE DATE BOBBY!!

I don't know if you understand how beautiful that is but, HOLY SMOKES! What a fantastic ride. This whole UFO, ET thing has been going on since WWI and we have been going insane building underground bases for the survival of our species and we have averted it just by doing some things differently, and of course the tip off was put to good use. The Military is still going through with its plans and that's what they do, it's a good thing. They plan on continuing until approximately 2030, I believe at which point, they should open those underground bases up for tours.

Long story short, is that we have found out that we can actively affect our reality. Just like the American Indians did with the rain dance. They would dance with all of their thoughts tuned in to making it rain and they would dance in a circle while chanting, building energy in the area to the point of critical mass in which once obtained the thoughts would actually manifest themselves into this dimensional reality bringing forth that which was being called forward, . . . rain. Anyone can do it. Astralwalker was correct.

We can actually influence and affect our reality around us and that is what the biggest secret of the elite government is! It's not the fact that there are ETs, and it's not the technology that they brought with them, it's the fact that WE HAVE THE ABILITY TO MANIPULATE THE REALITY AROUND US, That's what the big secret is Bob. They are studying this at the University level all over the country right now, as it has already been done in the Black Budget Projects years ago.

According to one Rear Admiral for the Department of the NAVY, who stated upon his death bed that the biggest secret was not that there were aliens visiting us in UFO's, and it was not the technology that they had brought with them, it was the fact that man could create his own reality and that was the biggest secret that the Clandestine Government had to keep from the people. Had we found this out we could just walk away from the whole game that society has instilled as a control measure over us. Everyone knows that society has been structured for us and that we have NO CONTROL over its tightening grip. Just as those who have created this sociological structure know, that if we were to walk away from it, it would end all control over us. And it just does not get any simpler than that.

To those who are enlightened enough to realize that man is a creature of FREE CHOICE in this universe and that the glory of all that is good is written as the name of GOD across his heart. If you don't want to do something, don't. YOU DON'T HAVE TO, there is no-one here who can make you do something you don't want to or that you know in your heart to be wrong. There is no reason for war, NONE WHAT SO EVER, and anyone who tells you differently is selling something. Point them out to others and rebuke them, they are not here to let your spirit, or the spirit of God within you, do it, or GODS will, they are here to imprison you in the system that they are trying to sell, force upon you, or trick you into adopting.

The infamous Report from Iron Mountain which examines the possibility for peace accompanied with the desire to ensure the masses of man-kind are controlled into bending to the socially structured will of a guardian figure is still a working document at the highest levels.

Introduction of a fake threat from outer space in the form of an alien is introduced as a reason to even keep a military service enacted upon the planet. Possibilities of control have also been examined through the implementation of slavery. It is hypothesized that without an ARMY to back up the social system in power, no one has to conform to its over regulative, intrusive violations of invasions of privacy, freedom of speech, free thought, and the free right to procreate as intended by GOD.

Depopulation of the planet through the use of food as a biological and/or chemical weapon is even discussed as a consideration for population control. It is my understanding as well as the understanding of all who are following this madness, that this project could still be under way. For instance S.A.R.S. was thought to have been a biologically engineered weapon created to attack the Asian population. Another more recent possibility of the release of a bio-weapon is thought by many to have been the Swine Flu outbreak of May 2009 that led from Mexico up in to the state of California.

It reminds me of Hollywood's attempt to remake old movies into new ones in an attempt to make more money.

The mad men who have ordered these viruses to be reengineered and then released upon the general population are guilty of crimes against all humanity. And those who make declaration of these orders should be questioned, arrested, tried and made to suffer the consequences of being subject to enduring the taste of the virus that they themselves were ready to unleash upon the rest of us administered to their own family members as they are made to watch. If you are ready to unleash a global biological or chemical pestilence upon humanity, then you should be prepared for you and your family to be the first to suffer these atrocities, and I wouldn't wish that upon my worst enemy, it's just not Godly.

I emplore everyone, "Read the Report from Iron Mountain!"

In conclusion I would like to state, that this briefing is a compilation of technical reports, cited web-site information, and declassified government information, accompanied by my own personal expressed opinions, comments, and conclusions.

It is not my intent to misrepresent or belittle any of the cited works contained within this briefing nor is it my intent to undermine the credibility of all, or any of those responsible for the development and release of such works. To the contrary, I would like to applaud their every effort.

Although I do share their enthusiasm and even their opinions, once again it is not my intention under any circumstances to misrepresent, misquote them, or draw any conclusions on their behalf. Nor do I want to mislead the reader in anyway into thinking that the above aforementioned wonderful people share the same opinions as I do.

I know this is a lot of unconventional information to swallow, or to even wrap your head around, but the facts speak for themselves. This document was meant as a briefing that was to be delivered to my family concerning the possibilities of upcoming global catastrophies that, in some instances, may or may not be avoidable. That being said, my message to you is GET PREPARED AND STAY PREPARED.

The fact of the matter is that, yes we have made contact with Extraterrestrials, this has officially been disclosed, (*www.disclosureproject. com*) Yes there are special groups of people working with them. Yes we have back engineered their crafts and now have many of our own.

This is known because well over 400 individuals have come forward as "Whistle-Blowers" and given their testimonies to Congress under oath. Some of these video interviews can be seen courtesy of Project Camelot at *www.projectcamelot.org* These are people like Generals, Ministers of defence, and Polititians. Black Budget Operators with 180+ IQ's who hold university PhD's in multiple area's such as Astronomy and Planetary Science, Astro-Physics, Nuclear Engineering, Avionics, Molecular Engineering, Micro-Biology, and Bio-Chemistry just to name a few, and these are not low brow university degrees.

The people who have come forward to divulge their information didn't even ask for anything when they did it. They didn't ask for money, they didn't ask for fame and they surely didn't do it to make their parents proud of them. They did it because they knew it was right, and they did it because they knew they needed to. Some of the insiders that have come forward have become targets of opportunistic slander attacks and labeled as lunatics, or fanatics who are mentally unstable and in some cases, it has destroyed their careers. And despite knowing this could happen they have still come forward and done it, losing their families and in some instances even their lives.

If we are to believe that these people, who hold multiple University PhD's, and are recognized as academic leaders in their professional fields are crazy or delusional, then why did the US Government single them out and hire them, grant them clearance levels to access top secret project information and pay them unheard of salaries to perform work on these projects? Are we to truly believe that this caliber of individuals, who are respected among their peers, would give all of this up for nothing, with the prospect of becoming a target with a death sentence? To what end? For what purpose? Think about it. Would you do something like that unless you were **absolutely certain it was the truth?**

It is up to each one of us, as we have a human responsibility, to continue to ask questions regarding government work and its involvement with scientific research. The government is charged with doing work on our behalf. Do not give up. Continue to ask questions. Do the research for yourself. We can find the answers. Follow the sources that have been provided for you throughout the works contained within this briefing document. The truth is out there, and it's astounding.

These have been the most fantastic discoveries in our history, made by the brightest individuals on the planet, and as far reaching as they may seem, have led to man-kind finding its roots in the cradle of life, allowing him to step forward in time to answer his greatest and most sought after questions to have ever been asked. Where did we come from? How did we get here? What is our purpose? ARE WE ALONE?

To wit, I would answer . . . What do you think?!
-(Sabre)

-End-

www.ingramcontent.com/pod-product-compliance
Lightning Source LLC
Chambersburg PA
CBHW031813170526
45157CB00001B/45